Y0-AKS-853

Data Analysis for the Chemical Sciences

Data Analysis for the Chemical Sciences

A Guide to Statistical Techniques

Richard C. Graham
Professor of Chemistry

Richard C. Graham, Professor of Chemistry
Department of Chemistry
Towson State University
Towson, Maryland 21204

Formerly at
Department of Chemistry U.S. Military Academy West Point, New York

Library of Congress Cataloging-in-Publication Data

Pending

Printed in the United States of America
ISBN 1-56081-048-3 VCH Publishers, Inc.
ISBN 3-527-28114-2 VCH Verlagsgesellschaft

Printing History
10 9 8 7 6 5 4 3 2 1

Published jointly by

VCH Publishers, Inc.
220 East 23rd Street
New York, NY 10010-4606

VCH Verlagsgesellschaft mbH
P.O. Box 101161
D-6940 Weinheim
Federal Republic of Germany

VCH Publishers (UK) Ltd.
8 Wellington Court
Cambridge CB1 1 HZ
United Kingdom

To Hazel, for patience, love and devotion

To Gillian, Kimberley, Curtis, Nathaniel, Janet, Keith, Amber and Emilee who sacrificed time so Dad could finish this book

Preface

The chemist is often faced with the necessity of deciding whether the desired results were obtained from an experiment, a need not limited to just analytical chemists. How to make such a decision is the realm of the statistician, and chemists should understand some of the basic statistical tools that will help in making such decisions. This text has as a purpose the presentation of statistical and data analysis tools that can be used by all chemists. A chemist needs tools to determine suitability of synthesis methods, quality of analytical results, relationships between data, structure in a data set, spatial similarities in data sets, temporal trends of data sets, and patterns in data sets.

The author has long been associated with analysis of large data sets ranging from subsurface migration of chemical contaminants to atmospheric deposition of chemicals and patterns resulting from the deposition. In these analyses, use has been made of spatial and temporal variation techniques.

A general theme of this textbook is to follow a data point resulting from the analysis of a sample to characterization of relationships between many different samples. Acquisition of analytical data points and the statistics that can be used to characterize similar data points fits into a general class of statistics called univariate statistics. The discussion will include ways to calculate the central tendency of the data points (the true value), the dispersion of the data points (how close are multiple analyses of the same sample), the frequency distribution of the data points, and methods to classify the frequency distribution of the data points.

To obtain a set of data points for a sample, the chemist must be able to transform instrumental responses to real world amounts of analyte in a sample. Prior to calibration, the data may need to be smoothed to minimize the randomness of the data. Techniques such as Savitzky-Golay and recursive and nonrecursive digital filters will be discussed.

To obtain accurate data, the chemist must rely on calibration of the instruments. Different calibration techniques including familiar regression techniques will be portrayed. Although most calibration techniques have traditionally treated single analytes (i.e., the analyst obtains a single set of pairs of responses and concentrations of a single analyte), modern instruments have the capability of simultaneously providing responses to multiple analytes. Rarely are the responses independent, with the instruments exhibiting interaction of response between analytes. Techniques to correct for interactions will be presented.

There are also situations in which a single variable will be dependent on many other variables. Multivariate regression techniques can be used to obtain calibration and other information for such dependency. The concepts of accuracy, precision, and detection limits will also be discussed. Although the classical determiner of goodness of fit of data to a linear regression ($y = mx + b$) has been the correlation coefficient, other diagnostics such as analysis of the residuals from the regression will also be presented.

After many sets of analytical data have been obtained, the relationship between the data sets must be ascertained. Multivariate techniques (principal component analysis, and cluster analysis) and analysis of variance (both parametric and non-parametric) techniques will be presented to determine similarities/differences in sets of data that have multiple variables. The multivariate techniques may also be used to determine the structure within a single set of data when the data set is composed of multiple variables.

Of course before any samples are taken, appropriate experiments must be designed from which samples are obtained that meet the goals of the sampling program.

The goal of this book is to present the life cycle of a set of data from initial design of the data acquisition program to production of the final report of similarities/differences between data sets. Because of the plethora of statistical packages that are available, I will not attempt to show how solutions were obtained using any particular package. Most of the results in examples will have been obtained using the STATGRAPHICS statistical package, although the steps needed to gain a solution to an analysis will not be presented.

It is anticipated that this will not be an end of introducing statistical analysis to chemists but that it will awaken a need in the chemist to use the tools of statistics to aid in future experimentation. As the reader uses these techniques and encounters new uses for statistics, the appreciation for the usefulness of the need for statistical analysis will grow. The need will also arise to learn new techniques and principles. As you read this book and have the need arise for new techniques, please convey them to the author for inclusion in future editions of the text. Comments are welcome and examples to illustrate the principles are also welcome and solicited.

R. C. Graham
Towson, Maryland
January 1993

Contents

List of Figures

List of Tables

1
CHEMICAL DATA

The Need for Statistical Interpretation

A Short History of Statistics

Yule and Kendall [1] in 1953 quoted Hooper [2] who said that "Statistics" deals with "the science that teaches us what is the political arrangement of all the modern States of the known world." The early uses of statistics were then to tell the heads of states what the political arrangement was. By the early nineteenth century this representation of the character of a STATE was evolving into the use of numerical methods. The evolution continued when statistics came to use summary numbers to describe and often to compare sets of data values. It was soon recognized that comparisons could be made quantitatively only if tables of critical values were calculated. The calculation and tabulation of the necessary values for the tables were long and extremely tedious. The tables that were calculated were of necessity much abbreviated and discrete*. Also, the test statistics were calculated using not full data sets but only portions of the data sets. Much of the theory of sampling arose out of the desire to fully describe a data set with only a portion. The test statistics were often calculated as the summation of parameters.

The invention of digital and analog computers helped ease the tediousness of the calculation of tabular values. However, the power of the computer was available to only a few individuals and not to the "average" researcher. Thus, that "average" researcher continued to rely on printed tables for critical values against which to compare test statistics. Statistical calculations on mainframe computers have led to many numerical (as opposed to analytical) solutions to statistical problems. But those computers were not readily available to the

---

* Only having tabulated values for a few given critical values as opposed to continuous— having the ability to calculate a critical value for any desired level of significance.

ordinary researcher. The statistical calculations were still very expensive and very tedious to perform. Today that evolution is continuing with the proliferation of extremely powerful desktop computers to which many scientists now have access.

The availability of the personal computer has changed and will continue to change the way researchers view the need to apply statistical interpretation to their data. The evolution today is toward not using the tabulated values of critical comparisons but to calculate actual significance levels and *P* values. This leads to more precise and accurate estimates of the probability of obtaining a particular set of data.

As this growth of availability and power of desktop computers continue, more and more researchers will put this power to use at a cheaper price. The availability of powerful computer hardware is taxing programmers to make software that uses the hardware. Very powerful statistical software is available in the marketplace*. This software is bringing the ability to perform complex and intricate statistical calculations to the professional researcher. The need then exists for researchers to understand the statistical tests that can be applied to the sets of data they generate. As we examine statistical tests that chemists should use, that understanding should come.

Statistics as Applied to Chemistry

The general application of statistics to chemical problems has been called ***CHEMOMETRICS***. The field of chemometrics has been growing in the recent past and will continue to see a growth in the future. This growth will no doubt be fired by the availability of the personal computer and the software to perform the sophisticated statistical tests that should be performed.

The modern chemist has a multitude of instruments from which to acquire "data." Each type of instrument that can be found in the contemporary laboratory will probably include some type of computerized control. The chemist, with this volume of data available, must be able to deal with such amounts of data with rigid yet flexible tools; RIGID in the sense that the

---

*A listing of some of the available statistical packages is given in Appendix A.

mathematical tools conform to accepted standards of data analysis; and FLEXIBLE in the sense that a variety of mathematical tools are available for the data analysis.

The chemist and other disciplines* that use chemical data must have tools available to judge the validity of the data that comes from an analytical laboratory. Part of the goal of this text will be to present tools to judge the validity of analytical data from instrumental (and other) methods of analysis. Guidelines for the development of an analytical method will be presented. The guidelines will include methods of calculation of detection limits, calibration of instruments, calculation of the calibration curve, quality control on the analytical method, measures of precision and accuracy, measures of the confidence limits on the analytical results, and applicability and types of alternate methods of calibration.†

For analytical chemists to properly present the results of analysis of samples that are given to them, they must be assured that the samples actually represent the medium being sampled. Thus the sampler must also be concerned about the need to design a set of experiments that will properly represent the material being sampled. This is applicable to all areas of chemistry not just analytical chemistry. For example, the organic chemist when presenting a sample of a new compound to the analytical chemist to determine the hydrogen, nitrogen, and oxygen content of the material must have assurance that the sample being forwarded is representative. The organic chemist must also be able to have the tools to ascertain if the reported CHO contents of a sample agree with the theoretical; in other words, how close to the theoretical is close enough?

Of equal interest here is the need to properly determine the size of the sample that will be required. This size of sample is not necessarily limited to the mass of the sample, but is also related to the particle size and particle size distribution of the material (if it is a powder), and related to the number of samples to be submitted to ensure that the analytical results are of the required significance level. The environmental, the forensic, and the drug chemist must have the tools to design a set of experiments that will ascertain the effect of different types of treatments. One of the sections of the book will deal with the

---

* For example, environmental engineering, chemical engineering, and disposers of hazardous waste.

† Other than the traditional $y = mx + b$ type of calibration formulation.

design of experiments covering such classical designs as the Latin Square, the factorial, and the randomized block; in addition some of the newer techniques will also be covered.

Once analytical chemists have done their job, others will likely use these results to find similarities (or differences) between samples, sampling locations, different methods of synthesis of a compound, temporal and spatial dependence of the sampling design, and so on. Analytical results by themselves are of little consequence as isolated pieces of data. They must be assimilated and massaged to yield meaningful conclusions. Statistical methods to determine similarities (or differences) between data sets will be presented with appropriate data sets as examples. The data analysis will not be limited to any particular method of analysis such as ANOVA, Student's *t*-test, and other parametric or nonparametric tests. Many different tests will be presented. It will be the goal of the book to introduce ways to select the proper test through an understanding of the underlying assumptions of each test.

Development of an Analytical System

An analytical system should be considered to be much more than just the analytical instrumental technique from which a set of instrumental responses or concentrations is obtained. The ***ANALYTICAL SYSTEM*** must be considered as all operations and functions which are accomplished to produce a data set. Some of the operations that should be considered as the ***ANALYTICAL SYSTEM*** are

Design of the sampling program
Sampling
Sample transport
Subsampling at laboratory
Analysis of sample
Recording of analysis data
Quality control/quality assurance
Preliminary data reduction
Peak finding
Using calibration charts
Data compilation
Data comparison/decisions
Pattern recognition

Each of these topics will be discussed in this text with a progression from univariate to multivariate statistics.

Every operation has a variance associated with it and the total variance of the analytical system is a combination of the variances of the individual operations. As such the importance of understanding each step of the entire system must be gained. Each step in the system must be characterized with the appropriate parameters of the distribution that classifies the system.

To further underscore the utility and necessity of using and understanding statistical tools in chemistry, let us take as an example the development of an *analytical technique*. In this sense, *analytical technique* is only that portion of the analytical system that concerns itself with the chemical (traditional wet chemical or instrumental) operations necessary to calculate a concentration from a set of instrumental responses. The discussion will be generic and could be applied to a wet chemical or an instrumental technique. We will call this developing an *analytical method*.

Developing an analytical method

The need to develop quality analytical methods (not necessarily synonymous with the analytical system) is underscored by the remarks of John Taylor formerly of the National Institute for Standards technology at a seminar on Quality Assurance of Chemical Measurements: "A laboratory is expected to be able to specify the quality of its data in quantitative terms. This requires the existence of degree of quality assurance".[3]

Guidelines are available that detail the documentation necessary for an analytical system to be acceptable to various governing agencies. For example the U.S. Army's Toxic and Hazardous Materials Agency* (THAMA) has published an extensive manual for contractors working on their projects.[4] The requirements for a method to be used in analytical programs of THAMA are that acceptable univariate regression curves must be established for the analyte, and accuracy and precision and detection limit must also be established. The testing of the acceptability of a method must be performed across several days and not on a single day. They also require that all anticipated levels of the analyte be

---

* Aberdeen Proving Ground (Edgewood Area), Maryland 21010.

tested on each of those days.

Similarly, the American Society for Testing and Materials (ASTM) uses a peer review and voting procedure to determine the suitability of an analytical method. They require a minimum of the following for a method to be considered as a "standard" ASTM method[5]: documentation, performance measures (accuracy, precision), and detection limit. They also require round-robin interlaboratory testing to ensure that the methods that are developed are free of idiosyncracies of an individual laboratory and that the method will be transportable to other laboratories that desire to use the method. The American Public Health Association (APHA) publishes a set of ***Standard Methods for the Examination of Water and Waste Water.***[6]

Let us examine a few items that will help in the development of an analytical method and that will also serve as an introduction to topics that will be developed in greater detail in various chapters of the book.

Identification of the problem

The analytical chemist is often beleaguered with requests that sound something like "What's in this sample?"; "Tell me everything that is in this product by four o'clock this afternoon!"; or, "How bad is the contamination in this mixture?" These requests will come from supervisors, engineers, and others who are not as well versed in analytical chemistry as you. An analytical chemist presented with this type of a request must immediately begin to be a detective. Moreover, an instrumental analytical chemist is considered to be a detective of the highest order. Initially, the history of the sample must be determined. This history includes such things as its source, the reaction conditions, if any, what the sample is meant to represent and whether it is a representative part, the quantity of the total material available for analysis, what determination is desired, and what level of determination is required. This list can be extended greatly, but should indicate to you many of the problems that must be addressed at the outset of an analytical quest.

Before any further actions are attempted in this development of a laboratory analytical technique, many questions must be asked:

1. Is the sample inorganic or organic?

2. Are the constituents to be identified as to presence (qualitative) or presence and quantity (quantitative)?
3. To what level are the results to be obtained?
4. Is this to be a routine method or is it a single occurrence?
5. What are the quality control/quality assurance methods which must be applied to the method?
6. What are the possibilities to extend this analytical technique to other laboratories?
7. What is the desired (required) level of confidence that must be attached to the result?

Certainly other considerations must be taken into account after the initial problem is defined. The relative merits of traditional wet chemical methods and the newer instrumental methods must be considered. The level to which the analysis is to be determined will dictate which type of method must be used. In general, if the amount to be determined is greater than 10-50 mg, wet chemical methods are adequate; however, if the amount to be determined is less than 50 μg, one must rely on instrumental methods. The gray area in the range between 50 μg and 50 mg, where either type of method might suffice, is where the expertise and experience of a skilled analytical chemist becomes important.

Searching the Literature

The first step in developing any analytical technique must be a search of the available literature. Often this step is overlooked because of lack of knowledge of how to conduct a proper and complete literature search. The need to search the literature is paramount in developing a new method. A wealth of methods have already been developed at least to some partial measure. The literature search will serve not only as a resource for the development of the entire method, but it can possibly point out several new avenues and techniques that previously may not have been considered. An example of the literature search that proved fruitful was the need to analyze soil samples for residues of alkyl phosphonic acid esters. The methods that had been used were generally derivatization with *N*-nitrosoguanidine or other agents followed by injection on a gas chromatograph. Prior to derivatization, the soil sample had been extracted using a mildly basic solution since the acid was not fully esterified. Derivatization involved adding the agent to the aqueous solution and extracting with an alkane, typically hexane. The procedure was fraught with difficulties including poor reproducibility, poor extraction efficiencies from the soil and the aqueous extract, and poor recoverability of acids added to standard soil samples. Near this time, Small et al.[7] introduced the technique of ion chromatography, which has now become almost a household word in an analytical chemistry laboratory. The technique was tried with much success by Sarver et al.[8,9]

Many books exist on searching the chemical literature[10,11,12] and specialized literature such as environmental sciences[13]. Although computerized search routines are available (LOCKHEED, BRS, CHEM ABSTRACTS, and DIALOG), they are becoming available for general use through PC based software such as STN Express[14] which prepares the search off line and then automatically transmits the search request to the data base. These computerized search mechanisms are extremely useful when highly specialized literature searches are required from a large number of data bases. Additionally, even though these computerized data bases exist, they should be relied on for only a portion of and not the entire literature search.

Literature sources fall into four general categories: primary, secondary, tertiary, and nondocumentary sources. Of these, primary sources (the initial reports of a discovery, an experiment, a new instrument, etc.) are the most valuable. However, to find an original article by just perusing journals is rather time-consuming and can be frustrating. Included in primary sources are patent applications, journals, theses, dissertations, conference proceedings and research reports.

Generally, the original article is found by consulting a secondary source. These secondary sources include ***Chemical Abstracts***, ***The Science Citation Index, Advances in Chemistry***, ***Chemical Reviews***, and any other journals that have as their primary purpose the review of prior work as opposed to publishing original work. Other secondary sources include such reference books as ***Timmermans' Handbook***[15], ***The Handbook of Chemistry and Physics*** and other handbooks by the Chemical Rubber Publishing Company.*

---

* Chemical Rubber Publishing Company, Boca Raton, Florida.

The tertiary sources are the least well defined. These sources usually contain little subject matter themselves, but help to guide you into primary or secondary sources. Tertiary sources include trade journals, guides to the literature, book lists, locations lists, and lists of ongoing research. These sources can sometimes guide you to a new area by acquainting you with experts in a field or describing literature appropriate for examination.

The nondocumentary sources are the hardest to define in that these sources are normally developed by years of experience and association with colleagues. These sources include memberships in professional and learned societies, and contacts with professionals in industry, academia, and research organizations. Meetings "behind the scenes" at conferences, symposia, etc. These sources are invaluable aids for research direction and topics.

The literature search should yield several methods that may be applicable for a given determination. Usually the situation is that a method cannot be found for the exact determination needed to be performed; however, a method may have been reported for a compound similar to the one you are investigating. In these situations, the method becomes a candidate for evaluation. For example, suppose you are interested in determining the degradation products of DDT induced by photochemical means in hexane. The only methods that you can find are for the determination of DDT in soil and water. Can the literature method be adapted for your use? Special note should be made for any exotic or special equipment or chemicals. An example is the method, earlier indicated, for the determination of phosphonate residues in soils and water. This method required the use of *N*-nitrosoguanidine, which is extremely carcinogenic and explosive and as a consequence is no longer available. Although a method may appear to be very good, the availability of the chemicals or equipment may make the method worthless for your purposes. Hopefully, methods should be ones which can use existing equipment in the laboratory. Unless a method has been tried and proven, few supervisors will gamble large sums of money for purchase of new instruments.

Standard Analytical Reference Materials

The success or failure of **ANY** analytical method is dependent upon the STANDARD ANALYTICAL REFERENCE MATERIAL (SARM) that is available. Early in this century, a need was recognized in this country for the establishment of a repository for reference standards. The early beginnings of the National Bureau of Standards dealt with the measurement of length, time, and volume. The National Bureau of Standards has since been renamed the National Institute of Standards Technology (NIST), recognizing that the organization is much more than a repository of standards, but is also a leading agency in the production of standards for a wide variety of disciplines. They work closely with nongovernment associations such as the American National Standards Institute (ANSI), the American Society for Testing and Materials (ASTM), and the International Association of Electronic and Electrical Engineers (IEEE). Since those early beginnings, over 1000 different SARMs have been proposed and certified by the NIST.

All analytical methods are dependent on the standard analytical reference materials for the construction of calibration charts for the proper use of quality control, for interlaboratory comparisons, and for intermethod comparisons. The major reason for the failure of two or more methods or laboratory procedures is the lack of a common standard. An example of the wide disparity and interlaboratory variability is presented by Pippenger et al.[16] for the analysis of anticonvulsant drugs. Recognition of this wide variability led to the testing of new SARMs in the anticonvulsant drug area resulting in a new SARM soon to be released by the NIST.

The properties of SARMs are important criteria for the acceptance of a material as a SARM. Literally years of testing and analysis by several different methods is required before a SARM is accepted. Some of the important criteria that must be satisfied are as follows:

1. **STABLE:** A SARM must be stable under a given set of storage conditions. The value of the parameter must not change with time. If it does, the SARM (or possibly the method) becomes suspect. Testing for SARM stability may be approached in several methods. The most obvious method is to seal samples and retest for the constituent over a long period of time. This is obviously a long and expensive process. A second method is to prepare a sample and store it under aggravated storage conditions for a relatively short period. These aggravated storage conditions might include high temperatures, intense ultraviolet light, corrosive atmosphere or combinations of all of these conditions.
2. **WELL CHARACTERIZED:** The composition of the material must not

only be stable, but it must be well known and somewhat easy to procure. Ideally, the composition of the material will be determined by several different methods. Simple variations of the same method, e.g., changing gas chromatographic columns, is normally not sufficient to qualify as a new method. It must be recognized that purity is neither a sufficient nor necessary condition to guarantee acceptance as a SARM.

3. **HOMOGENEOUS**: The composition of the SARM must remain constant throughout the entire lot of the SARM. This compositional homogeneity must remain over a variety of subsampling techniques.
4. **TRACEABLE TO THE NIST**: Since the NIST is the standard approving agency in the United States, all secondary or other reference standards used must be traceable to a set of NIST standards.

The purity of a material does not necessarily constitute sufficient justification for the acceptance of a SARM. Essentially, the ultimate use of the SARM will dictate the form in which it must be prepared. If the goal of the method is to determine the purity of an organic compound by using differential scanning calorimetry, an acceptable SARM would be indium, for which the heating curve and phase transitions are known precisely. However, if the goal of the method is to determine the trace levels of a pesticide in leaves, the SARM would be leaves that had been analyzed for the pesticide by a variety of methods over a long period of time.

Sampling

Sampling a material can often present unique problems if strict probability formulas are rigidly adhered to. The goal of sampling and subsampling is to obtain a portion of the material from which, by analysis and statistical inference, the chemical composition of the entire bulk of the material will be obtained. The number obtained must be representative of the "true value" for the bulk. An example of the problems generated by strict probabilistic sampling formulas is obtained by application of the sample size formulas given by Harris et al.[17] and by Benedetti-Pichler.[18] These formulas will be discussed in more detail in later chapters.

$$N = p(1-p)\left[\frac{100}{\sigma}\right]^2 \quad (1)$$

and

$$N = \left[\frac{P_1 P_2}{P_{ave}}\right]\left[\frac{d_1 d_2}{d_{ave}}\right]^2\left[\frac{100}{\sigma^2}\right]^2 \quad (2)$$

Where

n is the number of particles required for a representative sample
p is the fraction of contaminated particles
d_1 and d_2 are the respective densities of the two particles
d_{ave} is the bulk density of the sample
P_1 and P_2 are the estimated levels of contamination in two types of particles
P_{ave} is the level of contamination in the entire sample
σ_s is the relative standard deviation of P_{ave} caused by a sampling error

Consider, for example, a soil sample. By making some basic assumptions about the particle size, the level of contamination, the desired level of precision, and the densities of the particles, one can obtain values for the required number of particles to have a representative sample. If the desired precision is 10%, the density of particles and bulk equal, a level of contamination at 1 ppm in particles of both types and the fraction of each is 0.50, one would require an analytical sample volume equal to six dump trucks of soil. As you can see, this would be ridiculous to analyze and then difficult to dispose of the residue. Alternative approaches to resolve this dilemma will be presented later.

Accuracy and Precision

Accuracy indicates how closely the "true value" is predicted by the analysis. This implies that a "true value" is known, which is rarely the case with original research. If a comparison is being made, a SARM may be considered a "true value," which underscores the importance of selecting a proper SARM.

Precision refers to the reproducibility of a method. The value obtained may not be accurate, but if the variability is low, one can compensate by stating that a systematic error exists. This compensation may be in the form of an adjustment of a calibrating knob on an instrument or by the addition or subtraction of the effect from the obtained value.

An analogy to accuracy and precision is rifle marksmanship (Figure 1-1). If you hit the bullseye, you are accurate. If your shot group is together, but low and to the right, your precision is good, but accuracy is poor. You are shooting well, but need to adjust your zero. The mathematical formulation and definition of these quantities may vary. Several will be presented in later chapters.

Minimum Detectable Level

In this day of sophisticated instrumentation, determining if a compound is present (qualitative) or how much of a compound is present (quantitative) is no longer sufficient. The question now being asked is "What is the minimum detectable level of a compound?" Near this minimum detectable level (MDL), two types of errors (commonly referred to as errors of the first and second kind) become important. Errors of the first kind (α error) are those where a hypothesis is true but is rejected as false; and errors of the second kind (β error) are those where a hypothesis is false but is accepted as true.

The Congress in recent years has mandated the protection of the environment through a variety of laws. Nearly all of these laws invoke requirements to have emission standards set by a specific government agency. For any institution, be it a federal or state government, an industry, or a university, to be in compliance with the law, certification of the standards must be made. Very often an emission standard is at or near the MDL and, therefore, it is imperative for an analytical chemist who is analyzing samples near the MDL to know exactly what the MDL is for a given compound for the method being used.

A variety of definitions have been proposed for determination of the MDL. Many of these definitions are similar and convey somewhat the same interpretation. However, there are some that are not based on strict statistical criteria and should be avoided. The chapter on detection limits and minimum detectable levels will discuss some of the various definitions and their relative merits.

One simple type of MDL that is calculated is that used in atomic

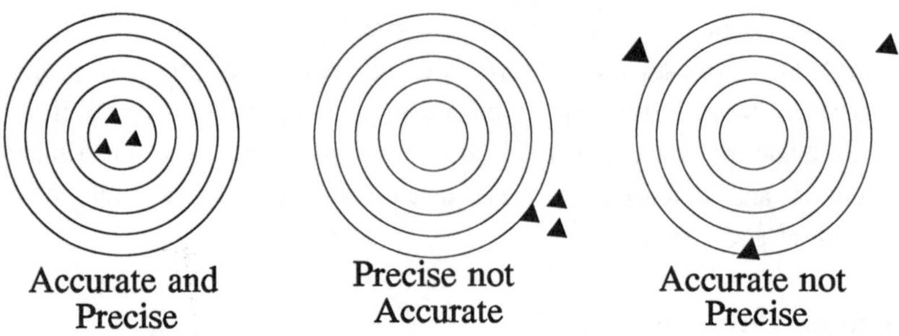

Figure 1-1. The relation between accuracy and precision.

absorption spectrophotometry, which requires analysis of only a single sample and a single blank. Several more sophisticated techniques are available that address this topic from the viewpoint of variance about a calibration curve. One example is given by Hubaux and Vos[19] and another is given by Schwarz.[20]

Interferences

Interferences are a major nemesis of analytical chemists trying to develop a valid analytical method. Once a method is ready for an application in the lab, the chemist must obtain and test natural samples. After testing the sample, the chemist must then consider tests for interferences. The initial test should be the spiking of the natural sample for determination of an increased response. The amount of the spike should be approximately equal to that which was found in the initial determination. Spiking should then be repeated three times with each spiking equal to approximately twice the amount of the previous sample. This method is known as the method of standard additions. Since this method does not always guarantee the absence of interferences, consideration should be made to analysis of selected samples for the analyte by an alternate method or by the Generalized Standard Addition Method. This is especially useful in such nonspecific analyses as chromatographic methods. The necessary statistical calculations will be presented in the chapter on methods of calculating concentration.

Ruggedness Testing

"All too often the weak link in the chain for developing new standard test methods is the paucity of the intralaboratory experiments...before they are submitted to interlaboratory study".[21] Prior to a method being adopted as a standard method, extensive intra- and inter- laboratory testing must be accomplished. Intralaboratory testing ideally consists of a factorial design for the testing in which many factors that may affect the method are varied. Among these factors are temperature, the analyst, time between samplings and the analysis, storage conditions of the sample, variations of the method from one day to the next, concentration effects, media effects, and other local environmental factors. Realistically, you may not be able to test all possible effects and must be satisfied with whatever testing the funding and time constraints will allow. In such circumstances, a reduced factorial design can be used. A method for such a reduced set has been proposed by several individuals including Youden and Steiner[22] after the technique of Plackett and Burman.[23] How do we analyze the data obtained from such experiments? We will learn that the most widely used statistical technique for such data sets is analysis of variance.

Documentation

Documentation is perhaps the weakest link in the development of a new analytical method. After a method has been developed, it must be presented in a manner that can be easily interpreted and used by another analyst. Details such as amount of dry reagents, volumes of liquids, times for heating or cooling, temperatures, and other common parameters must be included. Without sufficient details, months spent to develop a new method may have been wasted. As a minimum, a documentation format must address such items as applicability, MDL, precision, accuracy, equipment required, reagents, and analytical procedures.

Conclusion

As can be seen from the foregoing discussion, even as relatively uncomplicated a thing as developing an analytical method (after all of the chemical problems are resolved) requires an extensive amount of statistical interpretation of the obtained data. Documentation of the method requires that many different statistical operations be involved. As will be seen the statistical operations will range from simple descriptive statistics such as a mean or a standard deviation to much more complex such as an analysis of variance or separation of means. At the conclusion of using this text, the readers should be able to make use of various statistical techniques to draw conclusions about their data.

References

1. Yule, G.U., and Kendall, M.G. ***An Introduction to the Theory of Statistics***, Griffin, London, 1953.

2. Hooper, G ***The Elements of Erudition***, 1770, publisher unknown.

3. Taylor, J. Seminar on ***Quality Assurance of Chemical Measurements***, National Bureau of Standards, Gaithersberg, Maryland, 1983.

4. U.S. Army Toxic and Hazardous Materials Agency, Manual of Quality Control and Methods Documentation, 1986.

5. American Society Testing Materials, Philadelphia, Pennsylvania, 1989.

6. American Public Health Association, *Standard Methods for the Examination of Waste Waters and Waters*, 15th ed., 1980.

7. Small, H., Stevens, T.S., and Bauman, W.C. ***Anal.Chem.***, 1975, **47(11)**, 1801.

8. Schiff, L.J., Pleva, S.G., and Sarver, E.W. *Analysis of phosphonic acids by ion chromatography* in **Ion Chromatographic Analysis of Pollutants**, Mulik, J.D. and Sawicki, E., eds., Ann Arbor Science, Ann Arbor, Michigan, 1979, pp. 329-344.

9. Bossle, P.C., Martin, J.J., and Sarver, E.W. *High Performance Liquid Chromatographic Analysis of Alkyl Methylphosphonic Acids by Derivatization*, Technical Report ARCSL-TR-83002, Aberdeen Proving Ground, Maryland, 1983.

10. Mellon, M.G. ***Chemical Publications***, 4th ed., McGraw-Hill, New York, 1965.

11. Soule, B.A. ***Library Guide for the Chemist***, McGraw-Hill , New York, 1938.

12. Bottle, R.T. *The Use of Chemical Literature*, 2nd ed., Archon Books, Hamden, CT, 1969.

13. Morris, J.M., and Elkins, E.A. ***Searching the Literature of the Environmental Sciences —A Strategy***. State University of New York, College of Environmental Science and Forestry, Syracuse, NY, 1976.

14. STN Software for Searching the Chemical Abstracts.

15. Timmermans, J. ***Handbook of Physical Constants for Organic Compounds***, Vol. I and II. Elsevier, New York, 1965.

16. Pippenger, R. et al. ***Arch. Neurol.***. 1976, **33**, 351.

17. Harris, W.E., and Kratochvil, B. ***Anal. Chem.***, 1974, **46**, 313.

18. Benedetti-Pichler, A.A. ***Physical Methods of Chemical Analysis***, Berl, W.G., ed., Vol. 3. Academic Press, New York, 1956, p. 186.

19. Hubaux, A., and Vos, G. ***Anal. Chem.***, 1970 **42(8)**, 849.

20. Schwarz, L.M. ***Anal. Chem.***, 1979, **51(6)**, 723.

21. Wernimont,G., ***ASTM Standardization News***, December 11, 1978.

22. Youden, W.J.., and Steiner, E.H.. ***Statistical Methods for the Chemist.*** Association of Official Analytical Chemists.

23. Plackett, R.L. and Burman, J.P. ***Biometrika***, **1946**, **33**, 305.

2

PROBABILITY and STATISTICS

Introduction

Before we talk about statistics per se, the relation between probability and statistics must be understood. In the realm of chemical measurements, all results are not possible. In reality the potential results are quite limited. For example, it is not possible to have a sample whose concentration is less than zero for a given analyte. Therefore it can be said that the lower bound on concentration in any sample for any analyte is zero. Similarly, there is an upper bound of 100% for the concentration in any sample for any given analyte. We can thus ascertain that the probability of analyzing a sample and obtaining a result less than zero is exactly zero. Likewise, the probability of obtaining a result greater than 100% is also zero*. However, all concentration values between zero and 100% are possible. But are the possible concentration values equally likely? Are the values of the possible concentrations described by some other probability? In all likelihood, the probabilities are not equal. In general we will not know the probability of finding any given concentration when a sample is analyzed.

We have ali dealt with probabilities in the past and have some feel for some probabilities that we might see every day: the probability of obtaining a head when we toss a coin is exactly one-half (½), the probability of obtaining a 6 when throwing a single die is exactly one-sixth (1/6), and the probability of drawing a king from a deck of cards exactly 4/5 or one-thirteenth.

We must be able to calculate the probability for a given event to occur. The set of outcomes for all possible experiments that can occur for a given type of event (such as tossing a coin or throwing a die) is called the sample space. If an experiment is conducted, the probability that one of the outcomes will be

---

* We will see later because of the way that concentrations are calculated that occasionally the sample concentration may appear to be greater than 100% or less than zero.

observed is equal to one. This is similar to saying that one of the outcomes of the event will be observed. This is not meant to infer that any given event has a probability of one, but that one of the events in the sample space will occur. The probability that a given outcome will occur is much different and is dependent on the number of times that a given event occurs. The probability that a given outcome will be observed in a sample space must be greater than or equal to zero, however, it must also be less than or equal to one.

$$0 \leq P(A) \leq 1 \tag{1}$$

If the relative number of times that an outcome will be observed, [$n(A)$], is known and also the total number of event outcomes, n, in the sample space, then the probability of the outcome may be calculated from

$$P(A) = \frac{n(A)}{n} \tag{2}$$

An immediate adjunct of this very fundamental equation is that the sum of all the probabilities of a sample space must equal one:

$$\sum_{k=1}^{n} P(A_k) = 1 \tag{3}$$

Set Theory

It is not the intent of this text to give a full treatment of ways of calculating probabilities or of set theory, however, there are some important relations that must be presented since the concepts are crucial to later topics that will be presented. Understanding the concepts of independence of two events is the key to understanding the explanations involved in the calculation of confidence and significance levels. The concepts of intersections and unions are important to use when calculating the probability of two or more interconnected events. These are presented here in a concise manner for later reference and use.

Intersections and Unions

Consider a diagram in which a circle represents the probability that the outcome of an event will be observed such as shown in Figure 2-1A. The probability that the outcome will not be observed is given by the area outside the circle and within the rectangle. If a second event is also to be observed, the probability that the two events will occur simultaneously is influenced by whether the simultaneous occurrence is of a nature that both (AND) may occur or that either (OR) may occur. If it is required that both occur, then the probability that both will occur is the probability of the intersection of the two events and is shown in Figure 2-1B. In Figure 2-1B, the area that overlaps both outcomes is the probability that both events will occur. This probability will always be equal to or less than either probability

$$P(A \cap B) \leq P(A)$$
$$P(A \cap B) \leq P(B) \tag{4}$$

However, if at least one of the outcomes of the two events is to be observed then the probability is the UNION of the two events and is shown in Figure 2-1C. The shaded area within both circles represents the probability that either the outcome of either will be observed; Note that the area of the $A \cap B$ must be subtracted from the total area of the two circles since the intersection has effectively been included in both the probability of A and the probability of B. One of the probabilities must be subtracted.

$$P(A \cup B) = P(A) + P(B) - P(A \cap B) \tag{5}$$

Two events are defined to be mutually exclusive if the probability of intersection is zero and is shown in Figure 2-1D. Note that there is no common intersection between the outcomes of the two events.

$$P(A \cap B) = 0 \tag{6}$$

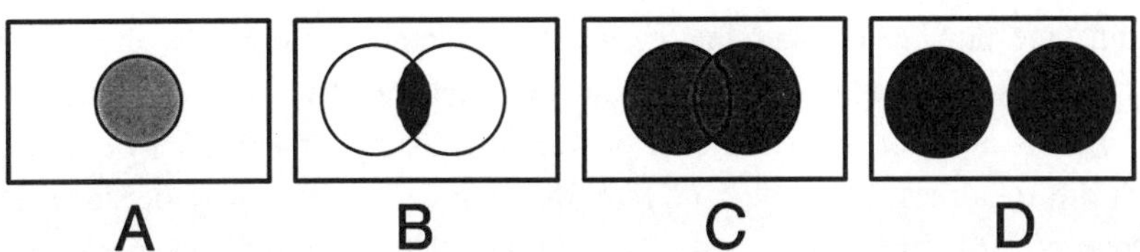

Figure 2-1. Probability of events. **A**. Single event. **B**. Events *A* and *B*. **C**. Event *A* or event *B*. **D**. *A* and *B* are independent events.

Conditional Probabilities

The intersection of two events implies that both events have occurred. This leads to several equations about conditional probabilities. The first of the conditional probability equations is that the probability of an event *A* occurring given that event *B* has already occurred is equal to the probability of the intersection divided by the probability of the event *B*.

$$P(A \mid B) = \frac{P(A \cap B)}{P(B)} \tag{7}$$

Two events are termed to be independent if the conditional probability of *A* given that event *B* has occurred is equal to the probability of event *A*:

$$P(A \mid B) = P(A) \tag{8}$$

Which also implies that the conditional probability of *B* given that event *A* has occurred is equal to the probability of *B*:

$$P(B \mid A) = P(B) \tag{9}$$

The result of this property is quite important in data analysis in chemistry since the independence of sampling is always assumed. Quite simply it states that a sample that is obtained for chemical analysis is independent of any other sample obtained for chemical analysis. This condition of independence leads to another quite important way of calculating the probability of two or more independent events:

$$P(A_1,A_2,A_3,\ldots,A_n) = P(A_1) \cdot P(A_2) \cdot \ldots \cdot P(A_n) = \prod_{k=1}^{n} P(A_k) \tag{10}$$

The equation for independence is a direct result of the product rule of finding the number of ways that events may occur. For example, the idea of throwing two dice and finding the number of independent pairs of numbers is 36. This is obtained as the number of ways that a single number may occur on each die (6) times the number of ways that a single number may occur on the second die (6). The product rule simply states that if more than event occurs, the number of ways that all of the events may occur is the product of the number of ways for each event:

$$\text{Number of ways} = n_1 \cdot n_2 \cdot ... \cdot n_k = \prod_{j=1}^{k} n_j \tag{11}$$

Combinations and Permutations

Two formulas that are quite important to calculate the number of ways that several events may occur are the formulas for **COMBINATIONS** and **PERMUTATIONS**. Combinations are the number of ways that a set of objects can be combined without regard to position. For example, if we are concerned only about the number of ways that *n* objects (call the name of the objects, *X*) can be taken *k* at a time, then the combination of X_1-X_2-X_3 is the same as the combination X_2-X_1-X_3 and the combination X_3-X_2-X_1. However, each of the three associations given above is a distinct association if order is considered. As can easily be discerned, the number of combinations is significantly less than the number of permutations.

In a given molecule of a protein, the order of the amino acid subunits is important since the order will determine the type of protein. Consider the number of ways that the four amino acid molecules may be put together. Since there are 26 amino acids, any one of the 26 may be placed in the first position leaving 25 for the second position, 24 for the third position, and 23 for the fourth position.

$$\text{Number of ways 4 of 26} = 26 \cdot 25 \cdot 24 \cdot 23 \tag{12}$$

This should be recognized as the beginning of a factorial series. In fact, it is the ratio of two factorials, 26! and 22!. To generalize, the equation becomes

$$P_{k,n} = \frac{n!}{(n-k)!} \tag{13}$$

Which is read the number of permutations of *n* objects taken *k* at a time is n factorial divided by *(n-k)* factorial. This equation implies that order is important. If order, however, is unimportant, then the result must be divided by *k*! (the number of ways of taking all of the *k* objects at a time). This gives the equation for the number of combinations of *n* item taken *k* at a time:

$$C_{k,n} = \binom{n}{k} = \frac{n!}{(n-k)! \cdot k!} \tag{14}$$

Relation between Probability and Statistics

How are probability and statistics related? In probability we know all there is to know about a given group of data elements since we know values for all the data elements. We can count the number of elements that have similar properties and can group those elements together in classes. We gave two examples above of a group of outcomes of a random event; throwing a die and tossing a coin. Since we can count the number of outcomes (2 possible for the coin, and 6 possible outcomes for the die), as long as the coin and die are fair, if we only throw the die once we can only obtain as an outcome 1, 2, 3, 4, 5, or 6 and each of these outcomes is equally likely or probable. A priori, we cannot determine what the outcome will be, but we can say something about what we expect to occur. We cannot obtain 1 and 2, 1 and 3, or any other given combination. Similarly with the coin, we can obtain only one of two outcomes, heads or tails. We cannot obtain both a head and a tail on a single coin toss.

The probability changes quite drastically when we start considering multiple events as discussed above. For example, what is the probability of obtaining a head and a tail in two coin tosses? of obtaining four 1s, five 5s and three 4s in 12 tosses of a single die? The reader should be able to calculate this probability using the above equations.

The foregoing was a discussion about probability predicting what the outcome of a given throw of a die or a toss of a coin would produce. But how did we know what the ratio of outcomes would be? Experience of having thrown dice many times or tossing a coin many times provides the answer. In a very large number of experiments, the relative numbers of obtaining a 1, 2, 3, 4, 5, or 6 when tossing a die are equal or the relative numbers of obtaining a head or a tail are equal. But if we were to throw a die only a small number of times or toss a coin a small number of times, we could predict what the relative numbers of each outcome will be from the results of such experiments. This is the purview of statistics.

Populations and Samples

Populations

Let us examine the concept of populations and samples. A population is taken as the entire set of data values that can be assigned. For example, the weight of each beaker in the chemistry stockroom; the height of all students at a university; the yield of all possible syntheses of 18-crown-6 (Figure 2-2). Populations may be assigned to one of several categories such as numerical, dichotomous, hypothetical, or conceptual populations. These categories may be reduced to two types: absolute and conceptual (hypothetical). An absolute population indicates that there are a finite number of elements in the group and each element of the group may be described. An absolute population also implies that we know all there is about the population; the number of elements in the population and the values that each element has. Absolute populations are limited to events that have already occurred.

On the other hand, the conceptual or hypothetical population results from a theoretical set of elements. The number of elements in the hypothetical population may be infinite or bounded. The conceptual population would refer to either the populations that are indeterminate or that may occur in the future.

In the examples above, the set of data comprised of the weight of each beaker in the chemistry stockroom would be an absolute population; the number of beakers in the chemistry may be counted and each weighed. Similarly, the set of heights of all students at any given university would comprise an absolute population. It may be difficult to ascertain all the information about a given population; for example, it would be difficult to obtain the names, birth dates, and birthplaces of every individual born in the United States in the year 1900, yet that would be an absolute population. However, the set of data that describes the yield of all possible syntheses of 18-crown-6 would be a conceptual or

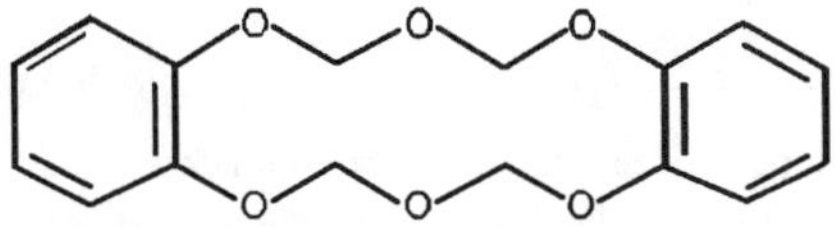

Figure 2-2. 18-crown-6, a macrocylic polyether.

hypothetical population since it is impossible to determine all of the yields for the synthesis of that particular compound. However, the height of students who would attend a university in the future is a conceptual population because we do not know who the students are. In chemistry, we deal almost exclusively with conceptual populations.

Samples

From the early beginnings of statistics, the concepts that were important were the populations.* The statisticians were interested in obtaining values that would characterize, describe, and summarize the whole of the population. From the knowledge about the whole, they were then interested in predicting what the composition of a smaller subset of the population would be. This small subset was called a sample. This prediction of the composition of the sample is termed probability. But just as important is looking at a small portion of the population and inferring what the entire population would look like. The numbers that are used to infer characteristics about the population are *statistics*.

Example

The outcomes of the coin toss or the throw of a die are examples of samples. The entire set of all possible coin tosses would be deemed the population.

Rarely does a chemist exhaustively determine all the possible values for a given material. This would be counterproductive and also very costly, not to mention destructive of the material trying to be characterized. But the chemist is more interested in the ability to take a small portion of the material that has been produced, a small portion of the environmental area to be characterized, etc. toward the end that the few measurements will be able to be used to characterize the whole from the measurement of a few. In chemistry, however,

* From the quote by Hooper in Chapter 1, it should be inferred that statistics had its origin in *populations* of countries. The term has never been changed and now applies to the concept of an entire set of data.

we generally do not know the characteristics of the population and must rely on the characteristics of the sample to allow us to say something about the population.

The relation between statistics and probability, shown schematically in Figure 2-3, shows an inverse and cyclical relation. Statistics and probability are the links between the population and the sample. Probability allows us to predict the composition of a sample based on the characteristics of the population. Statistics allow us to infer the properties of the population from the characteristics of the sample(s). Probability tells us what the sample might look like; statistics tells us what the population looks like. Statistics also uses probability to tell us what confidence we might expect in describing the population by the characteristics of a sample.

Rarely will the sample that has been taken from a population look exactly like the population. However, if enough samples are taken, the group of the samples will begin to look like the population. In general, we do not exhaustively determine all the values of a population, but instead rely on obtaining a sample that is representative of the population. The size of a sample (i.e., the number of members of the population we choose to analyze or measure), is always less than or equal to the number of members of the population. N will be used to denote the number of elements in a population and n to denote the number of elements in a sample, thus $n \leq N$. The early uses of statistics were to describe or summarize the population and have been given the name of **descriptive or summary statistics**. The values of the properties of the

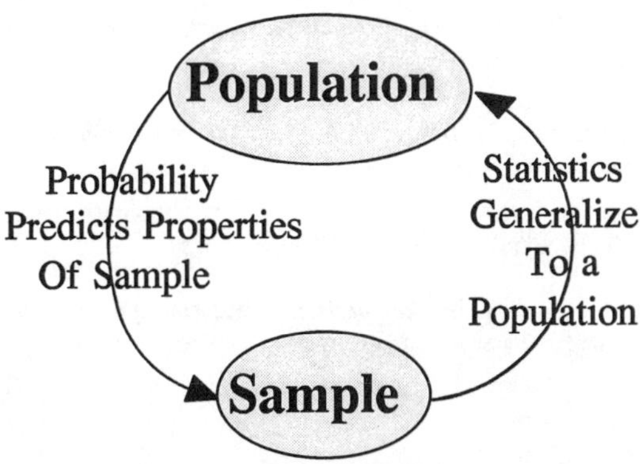

Figure 2-3. Relation between samples and populations AND probability and statistics.

elements of the sample are then used to ***ESTIMATE*** the parameters of the population. Each sample that is drawn from the population must be random which means "that each and every distinct sample of size *n* has exactly the same probability of being selected."[1]

In terms of chemical sampling, any sample that is taken from the environment, in the laboratory, in the medical clinic should have equal probability of being selected. In practice, chemical sampling, especially in the environment, is somewhat biased. For example, when the National Trends Network for monitoring acid rain was established in 1981–1982,[2] large regions of the country were excluded because they violated one or more of the siting criteria that had been established for the network.[3,4] A general purpose of a monitoring system is to determine if changes in the constituency of a process have occurred over time. It is of necessity that the monitoring then be conducted "down wind" or "down stream" of the monitored process. The process being monitored may be a manufacturing process where the quality of the finished product is being monitored, or it may be to determine if a factory is emitting pollutants in excess of the permitted amounts. There is little reason to monitor upstream of the process, except to determine a baseline in some cases such as pollution monitoring. The need to monitor a process downstream introduces a bias into the sampling scheme.

References

1. Winkler, R.L. and Hays W.L. ***Statistics—Probability, Inference and Decision***, 2nd ed., Holt Rinehart Winston, New York, 1975, p.273.

2. Wilson, J.W., and Robertson, J.K. ***Design of the National Trends Network***. U.S. Geological Survey Special Publication, 1983.

3. Graham, R.C. ***Water, Air and Soil Pollution***, 1990, **52**,

4. Graham, R.C., and Robertson, J.K. ***Intercomparison of Wet Atmospheric Deposition Samplers***. U.S. Geological Survey, Water Resources Investigation Report 90-4042, 1990.

3

EXPLORATORY DATA ANALYSIS

General

In the realm of chemistry, *DATA* may be obtained from a variety of sources. But just what constitutes *DATA*? How are *DATA* obtained? How are *DATA* used? How are *DATA* characterized? The chemist must wrestle with all of these questions and understand the answers to the questions. We will use *DATA* as meaning any set of numbers that represents a chemical phenomenon. Examples of sets of *DATA* are a baseline of a chromatogram, the peaks of a chromatogram, the pH values obtained from collecting and analyzing samples of atmospheric precipitation, the percentage yields of organic synthesis experiments, the values of the enthalpy of a reaction, the temperatures obtained in a calorimetric experiment, or any other like phenomenon. No field of chemistry is exempt from having data associated with it. All branches of chemistry have *DATA*; phenomenon that can, should, and in most cases **MUST** be quantified and characterized.

Random Variables

We considered in the notion of sampling that all samples should be selected at random with equal probability of obtaining any sample. Each time a sample is taken several numbers may be obtained on the analysis of the sample. Each one of these numbers may be linked with the sample space (the totality of all samples) with a particular rule of association. The rule that associates the sample space with the observed values is termed the RANDOM VARIABLE. A random variable may take on several values. The numbers (concentration, weight, absorbance, EMF) obtained from analysis of the sample are termed values of the RANDOM VARIABLE since the numbers obtained from analysis usually have varying values depending upon the source of the sample and also the time at which the sample is taken. These random variables represent and characterize the sample. Samples are generally not characterized by a single variable, but by many variables. Further these variables are termed RANDOM because the obtained value is dependent on which of the possible experimental outcomes result from the experimentation. Each of the possible outcomes of the experimental process is generally associated with a particular sample space and also has a particular rule that associates the possible outcomes with the sample space.

A RANDOM VARIABLE is a rule of association between an experimental sample space and the set of possible outcomes of the experiment. Generally, the rule that associates experimental outcomes with the sample space yields a single number. The value of the random variable will depend on the experimental conditions, including the temperature, time, barometric pressure, or any other of a host of controllable parameters. By knowing the relation between all of the experimental parameters, we should be able to predict the value of the RANDOM VARIABLE. Unfortunately, rarely do we know all of the conditions (let alone the interrelationship between those conditions) to allow us to make quantitative predictions. In general, we will make experimental measurements that will be influenced by the environmental conditions. The random fluctuations in the environmental conditions will lead to random fluctuations in the value of the measured RANDOM VARIABLE. Our goal will be to quantify the magnitude of those random fluctuations and to be able to use the magnitude of the fluctuations to determine the relation between the experimental conditions and the outcomes of the RANDOM VARIABLE.

Example

The rolling of dice is a process that produces RANDOM VARIABLES. If one blue die and one yellow die are thrown, 36 possible pairs of numbers result. These 36 pairs may be treated differently. For example, we can define a random variable, X=even sums of the two dice. This random variable, X, may take on the values:

$$x = 2, 4, 6, 8, 10, 12 \tag{1}$$

Another random variable which could be defined is Y=products of the two die whose square roots are integers. This random variable, Y, may take the values:

$$y = 1, 4, 9, 16, 25, 36 \qquad (2)$$

RANDOM VARIABLES may be either discrete or continuous. The two previous examples provided illustrations of discrete random variables. The major difference between a discrete random variable and a continuous random variable is whether the values of the random variable are being counted (discrete) or whether the values of the random variable are being measured (continuous). It is often thought that if the random variable can assume an infinite number of values that the random variable is continuous. This is not always the situation since even some discrete sets of values of random variables may have an infinite number of outcomes. An example of a random variable that can be considered as being discrete and continuous is the yield of a chemical synthesis. Any value from zero to 100% for the yield is possible indicating a continuum of values possible in the interval. The percent yield will depend on fine variations in the reaction conditions such as temperature, length of reflux, or concentration of reactants. As the conditions change, so does the yield for the reaction. However, since the yield is usually tabulated, the set of yields is discrete.

This at first might seem impossible to have an infinite number of yields of a particular synthesis. The key here is that it is POSSIBLE but not very PROBABLE. As we examined earlier, the probability of an event is the number of events that have that same outcome divided by the total number of outcomes:

$$P(A) = \frac{n(A)}{n} \qquad (3)$$

Since the number of syntheses with an infinite yield is zero, the probability of a synthesis having an infinite yield is also zero. Other examples of discrete random variables are the number of atoms in a molecule, the number of beakers in a storeroom, the densities of a series of organic compounds (such as the alkanes), the number of gas tanks in an automobile. In most of these examples, there is a difference between could be POSSIBLE and PROBABLE. Theoretically an infinite number of atoms in a molecule is possible, but it is not very probable. Examples of continuous random variables in chemistry would include the concentration of sulfate in rain, the depth of a lake, the temperature of a reaction, the enthalpy of a reaction, the density of a single organic compound, or the optical rotation of an optically active organic molecule.

Example

Consider a sampling program to determine the extent of contamination in an underground aquifer. The drilling program produces samples of soil and water. Each sample is analyzed. The substances for which the samples are analyzed (pH, specific conductance, sodium ion, dioxin) are the RANDOM VARIABLES. Each of the analytes may take on an infinite number of values. A continuous random variable may be bounded (e.g., $0 \leq pH \leq 14$) or may be unbounded ($0 \leq [Na^+] \leq \infty$). Not all of the values will be observed with equal probability. The distribution of the values of the random variable will allow us to distinguish between sets of samples as coming from different populations.

At this point, a distinction ought to be made about random samples and values of the random variable. Random samples indicate "that each and every distinct sample of size ***n*** has exactly the same probability of being selected".[1] But this does not imply that every value of a random variable has an equal probability of being selected. The probability of selecting a sample that has a particular value of a RANDOM VARIABLE for a randomly selected sample will depend on the relative number of all possible values that have that particular value. Therefore, although a sample may be randomly selected, the values of the random variable are not equally likely to be selected.

The set of probabilities of obtaining the values of a random variable is called the PROBABILITY DISTRIBUTION FUNCTION of the random variable. If the random variable is discrete, there will be one probability associated with each of the values of the random variable. If the random variable is continuous, there will be a continuous function that will describe the probability of obtaining a single value. The number of ways of obtaining a value will be denoted by $f(x)$; the cumulative number of ways will be denoted by $F(X)$. We can also calculate the probability of obtaining a given value and this will be denoted by $P(X)$. Of importance also is the need to find the probability of finding a number less than or equal to the value; this is denoted by $P(x \leq X)$. We will defer the discussion of the continuous probability distribution function until Chapter 5.

Example

Consider again the two dice that have previously been discussed. There are 36 possible combinations of the two dice that may be evaluated in a multitude of different ways. First consider the random variable, X = sum of the two dice. The values of the random variable are {2, 4, 5, 6, 7, 8, 9, 10, 11, 12}. Note that there are only 11 values, but there are 36 sums indicating that some of the values occur more than once. In fact the number of occurrences of each of the values is given in Table 3-1.

The probability distribution may often be expressed as a functional relationship expressing the probabilities more conveniently. In this example, the mathematical relation between the value of the random variable and the probability of obtaining that value is given:

$$P(x) = \begin{cases} \frac{1}{36} \cdot (x-1) & x=2,3,4,5,6 \\ \frac{1}{36} \cdot (13-x) & x=7,8,9,10,11,12 \end{cases} \tag{4}$$

Graphic Presentation of Data

Table 3-1. Number of Occurrences, Cumulative Occurrences, Probability of Number of Occurrences, and Cumulative Probability.

x	2	3	4	5	6	7	8	9	10	11	12
$f(x)$	1	2	3	4	5	6	5	4	3	2	1
$F(x)$	1	3	6	10	15	21	26	30	33	35	36
$p(x)$	1/36	2/36	3/36	4/36	5/36	6/36	5/36	4/36	3/36	2/36	1/36
$P(\leq x)$	1/36	3/36	6/36	10/36	15/36	21/36	26/36	30/36	33/36	35/36	36/36

One of the first things that ought to be done after a set of data is obtained is to examine the data graphically to see if there are trends or patterns in the data. Examining a set of data graphically is an art since there are many techniques to use. In general, there are a few ways that data should be graphically portrayed. In this section only the graphic representation of a set of UNIVARIATE data, i.e., numbers representing only one variable, will be discussed. Let us present a few concepts of examining data to obtain the most information from that set.

The goals of data analysis are to *describe* data, *summarize* data, *explore characteristics* of data sets, *discover relationships* in the data, and *make decisions* based on the data. From our earlier discussions it should be evident that these goals of data analysis may be related to the type of statistics that are being used. *DESCRIBE* and *SUMMARIZE* almost assuredly relate to the statistical field of *DESCRIPTIVE STATISTICS*. Exploring characteristics in the data relate to the field of *INFERENTIAL STATISTICS*. To discover relationships is largely inferential statistics but with a multivariate flavor. *DECISION MAKING*, on the other hand, is a whole realm of study that uses the statistics that are provided from the descriptive and inferential studies. Anyone using statistics must be sure that the statistics are sound and able to *truthfully* portray the original data and not present misleading conclusions.

As data are acquired, the chemist often desires to identify trends, patterns, and periodic functionality in the data. As an example, the chromatogram is of little use in its raw state. But it can provide some summary values such as location of peak (retention time), width of peak, area of peak, height of peak, or a value comparing peak shape to ideal peak shape. These values may then be used to obtain concentration (peak height or peak area), condition of column (from peak shape), and identification of material giving rise to the peak (retention time). These can all be considered as summarizing the chromatogram.

Although the goals of data analysis are relatively simple, the accomplishment of the goals is not always straightforward and easy. Tabular data are difficult for the mind to assimilate and recognize patterns or relationships. Tukey,* several years ago, introduced the concept of graphing data

*. John Tukey, statistician, pioneered many different fields of statistics, including time series analysis and exploratory data analysis.

before any statistical analysis is performed on the data. He termed these procedures ***Exploratory Data Analysis.*** but in actuality they are very powerful graphic procedures. The human eye can quickly recognize patterns and trends in numbers when presented in graphic form. The procedures are designed as a means to bridge the gap between descriptive and inferential statistics since the procedures incorporate elements of both. As these procedures are implemented, the chemist can be guided to the statistical operations to perform and also be guided to further experimentation that may be required.

Guidelines for Graphing

Before we actually discuss some of the exploratory techniques that Tukey and others have developed,[2,3] let us discuss some guidelines about construction of good graphs. Graphic methods can be easy, powerful, and quick ways to present data. However, they can also be deceiving and misleading if improperly designed and implemented. Graphic presentations should be simple to read and interpret, yet portray the truth and the influence of each data point. As you prepare graphs, keep the following guidelines in mind.

Labeling

Label axes clearly. Do not clutter the axes with excess tic marks or excess tic mark labels. Graphs with axes on all four sides of the plot are easier to read than graphs that only have two axes presented. Tic marks can be placed inside the frame or outside, but must be consistent from graph to graph. Tic marks should be of varying lengths according to the magnitude of the significant figures of the data points they represent. This means that a tic mark that represents a significant figure in the 1s position should be smaller than 10s or 100s. Similarly, the tic mark that represents a significant figure in the 10s position should be smaller than the tic mark representing 100s or 1000s. Scaling of the axis is critical. The scale should not be so large that nothing appears significant, nor so small that everything appears significant. A good example of this is the amount of noise in a baseline of a chromatogram compared to the height of the chromatographic peak. If only the baseline is scaled to the

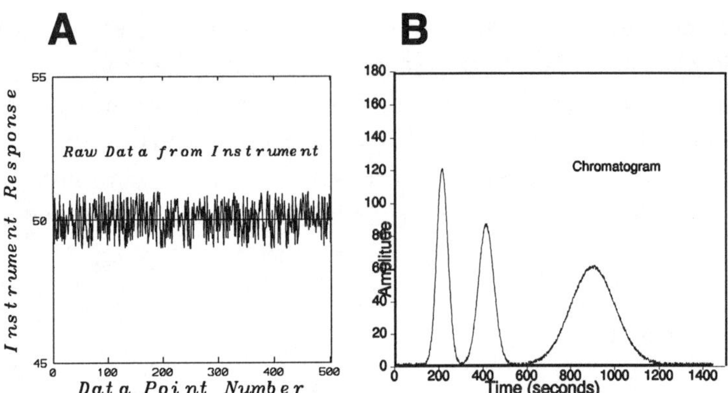

Figure 3-1. A) Chromatogram baseline B) Peaks The noise in the baseline is quite large, but against the peaks it is small.

magnitude of the data,[†] the variability in the signal appears huge (Figure 3-1A). Yet this same variability when viewed against the total signal of the chromatographic peak appears insignificant (Figure 3-1B).

Titles

Give a title to each axis of the graph. The title should be descriptive of the variable being plotted. The general convention is that the independent variable is graphed as the ordinate and the dependent variable is graphed as the abscissa. Almost every graph regardless of type (bar, scatter, etc.) will fit this mold. Some such as the pie chart will not. Titles on the axes should also contain the units of the variable, e.g. liters, milligrams/liter, atmospheres, frequency, etc. If abbreviations are used, use the proper SI abbreviations [e.g., L for liter, mL for milliliter, μL for microliter, m for meter (Table 3-2)].

The title of the graph should contain information about the graph and also describe the contents of the graph. Usually one line of text is not sufficient to adequately describe a graph. Do not be afraid to add more than one line of

---

[†] Computerized graphing procedures automatically account for the magnitude of the data. The automatic scaling of the data to account for differing magnitudes in different data sets is called autoscaling.

Table 3-2. Systemé Internationale (SI) Units (Also Referred to as Metric Units)

Fundamental Units

Quantity	unit	abbreviation	Definition
Mass	kilogram	kg	The mass of type kilogram held at Sèvres, France. This unit is different than all other SI units because it is not defined in terms of fundamental physical constants
Time	second	s	Duration of 9,192,631 770 periods of the radiation of the ground state of ^{113}Cs
Length	meter	m	Distance light travels in a vacuum during $(299\ 792\ \ 581)^{-1}$ of a second
Temperature	Kelvin	K	The temperature such that the triple point of water is 273.16 K and absolute zero is 0 K
Amount of substance	mole	mol	The number of atoms in exactly 0.012 g of ^{12}C
Electric current	Ampere	amp	The amount of current which will produce a force of $2 \cdot 10^{-7}$ N/m between two parallel, straight conductors of infinite length and negligible cross section that are separated by 1 m

Derived Units

Quantity	unit	abbreviation	In terms of fundamental SI units
Frequency	Hertz	Hz	1/s
Force	Newton	N	$m \cdot kg/s$
Pressure	Pascal	Pa	$kg/(m \cdot s^2)$
Electric charge	Coulomb	C	$s \cdot A$
Energy, work, heat	Joule	J	$m^2 \cdot kg/s^2$
Electric potential	Volt	V	$m^2 \cdot kg/(s^3 \cdot A)$
Electric resistance	Ohm	Ω	$m^2 \cdot kg/(s^3 \cdot A^2)$

text if it is needed. Contain information such as the experimental conditions in the title. Other information in the title could include date, time, and reactions. Labels for the variables should also be placed on the graph. The legend should show the symbol for the variable and the variable name. The names of the variable and associated data symbol should also be contained in the caption of the plot. The titles should convey sufficient information to the reader of the graph that reference to the text of the paper is not necessary.

Plotting of Data

Whenever possible, all of the data points should be included on a graph. The inclusion of the data points helps eliminate the questions asked by the reader of how many data points are represented in the data set. Whenever lines are used on the graph, they should convey the mathematical model that describes the data. If there is no model for the data, a line may be included if it is intended to lead the reader from one data point to the next. An example of this would be plotting of pH in rain at a sampling location over a several year period. There are mathematical models that could be applied to the data, but if it is desirous to show only the general trends in the data, a line that connects the data points should be sufficient. It should be stated in a caption, a legend, or as a subtitle that the line connects only the data points and does not convey mathematical relation.

Table 3-3 shows a set of integers generated by a RANDOM number function in a spreadsheet. These data were generated such that the probability of selecting any integer between 0 and 100 is identical. The probability is not unlike that of obtaining a 1, 2, 3, 4, 5 or 6 on the throw of a single die, i.e., the probabilities are equal.

Table 3-3. Unsorted Uniform Data

17	96	74	42	82	58	63	94	67	24	8	47	94	20	22	91
71	52	88	18	25	6	24	36	58	64	74	14	96	49	46	24
37	7	35	33	64	19	50	23	91	67	40	5	5	10	20	20
58	89	27	34	51	97	80	67	87	15	50	6	60	94	58	67
53	30	71	54	24	87	96	59	30	34	6	57	73	30	59	4
20	68	29	68	42	70	92	30	96	67	34	92	68	70	97	46
69	73	22	39	96	52	91	50	18	34	71	18	87	40	16	94
95	61	79	47	47	21	12	89	8	98	17	28	35	20	28	62
69	29	89	22	36	13	60	14	56	17	21	57	98	64	92	5
40	47	93	7	67	46	63	74	98	23	53	15	90	62	30	33
80	31	97	3	31	76	96	96	36	23	46	67	30	70	92	29
46	73	59	9	70	70	5	52	63	9	68	94	66	49	59	7
17	45	58	19	99	51	92	19								

$$P(0) = P(1) = \cdots = P(99) = P(100) \tag{5}$$

One goal of data analysis is to be able to easily recognize trends or similarities in data sets. Unfortunately data tables, such as Table 3-3, do little to quickly orient an individual to recognize trends or patterns in the data since these are just a grouping of numbers. Transformation of these numbers into a plot or some other visual representation allows an individual to recognize patterns much easier and quicker. The human eye can rapidly discern patterns in data if properly presented. We want to be able to construct representations for one-dimensional data such that extremes, middle, clusters, and spread of the data set may be easily seen. A variety of techniques of graphing univariate data are available.

Frequency Distribution Diagrams

Discrete probability distributions were introduced in a previous section. The probability distribution is presented either in the mathematical form or more conveniently in a graphic form. Most chemical data are not in discrete form, but are an expression of a continuous random variable. Rarely can we count the number of occurrences of a single value since there is a continual change from low to high values. It is, therefore, more often observed that the number of occurrences in an interval of the values of the random variable is counted.

The frequency distribution diagram (often called a histogram) is a plot of the frequency (number of data points) within an interval of the data set as a function of the intervals. The plot is most often graphically plotted as a bar plot — the height of the bar representing the number of occurrences of the data values in the interval. To manually construct a histogram, you must first sort data low to high (as shown in Table 3-4). Separate the data into a convenient number of equal intervals (not equal numbers of data points). The data in Table 3-3 suggests that the interval should be 10 data units (e.g., 0–10, 11–20, ..., 91–100). Plot the number of values in each interval versus the endpoint of the interval with the bar being centered on the midpoint of the interval. When manually generating a plot as a bar graph, it is easiest to have the intervals go from low to high and include the number of occurrences in that interval, rather than trying to calculate the midpoint of the interval. Alternatively, if a scatter diagram (XY plot) is being generated, then the midpoint of the interval is much

Table 3-4. Sorted Data from the Purported Uniform Distribution Data Set

1	11	21	31	42	51	61	71	82	91
3	12	21	31	42	51	62	71	87	91
4	13	22	33	45	51	62	71	87	91
5	14	22	33	46	52	63	73	87	92
5	14	22	34	46	52	63	73	88	92
5	15	23	34	46	52	63	73	89	92
5	15	23	34	46	53	64	73	89	92
6	16	23	34	46	53	64	74	89	92
6	17	24	34	46	54	64	74	90	93
6	17	24	35	47	56	66	74		94
7	17	24	35	47	57	67	76		94
7	18	24	36	47	57	67	79		94
7	18	25	36	47	58	67	80		94
7	18	27	36	49	58	67	80		94
8	19	28	37	49	58	67	80		95
8	19	28	39	50	58	67			96
9	19	29	40	50	58	67			96
9	20	29	40	50	59	68			96
9	20	29	40		59	68			96
10	20	30			59	68			96
	20	30			59	68			96
		30			59	69			96
		30			60	69			97
		30			60	70			97
		30				70			97
						70			97
						70			98
						70			98
									98
									99

more meaningful.

Although the bars are generally plotted as bars of constant width, it is not necessary, since the critical property of the frequency distribution diagram (as we shall see later in more detail) is the AREA of the bar and not the width. But if the bars are plotted with a constant width, then the height of the bar represents directly the number of data points in the interval.

Let us look at two different types of plots that can be generated using these two concepts. First the bar plot. To manually construct the bar chart, the data must be sorted from low to high. The sorted data (from the data in

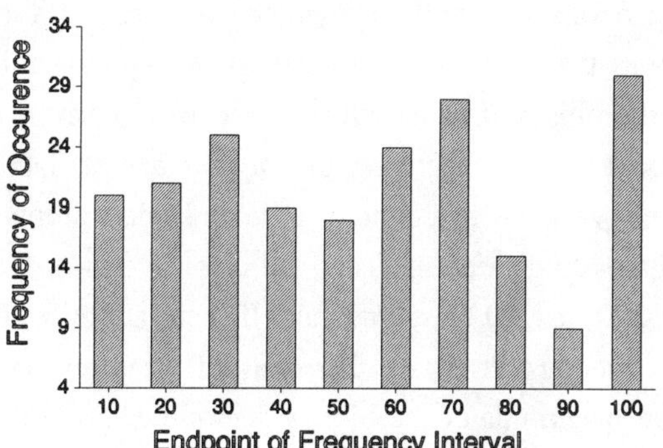

Figure 3-2. Histogram of the sorted uniform data.

Table 3-3) are shown in Table 3-4. The data have also been separated into 10 class intervals each 10 data units wide. The number of data points in each of the classes would then be counted and plotted as a bar 10 units wide as shown in Figure 3-2.

In this day when computers are so easily available, it makes little sense to manually graph any set of data. The better way to construct a frequency distribution diagram is to use the plotting capabilities of commercial software programs. Some of the more popular plotting programs are given in Appendix A. Spreadsheet programs (such as Lotus 123,[‡] Quattro-Pro,[§] or PlanPerfect[**]) are among the most popular to plot the data in Table 3-3 since the spreadsheet will not only do the plotting for you, but it will also count the number of occurrences in the class intervals. A variety of ways are available to get the data into the spreadsheet, such as importation of an ASCII text file, generating the data in the program using a model, or entering the data one cell at a time. Almost all of the manual tedium is eliminated with the use of a spreadsheet. For example, although one can have the spreadsheet program sort the data, it is not a necessary step. Before the frequency distribution can be plotted, it must be calculated. In finding the frequency distribution with the spreadsheet, the user

---

[‡] LOTUS 123 is a registered trademark of the Lotus Development Co.

[§] Quattro Pro is a registered trademark of Borland Inc.

[**] PlanPerfect is a registered trademark of the WordPerfect Corporation.

must define the intervals that will separate the data set. These intervals are known collectively as the bin block. Knowing how many bins to create and also the magnitude of the bins will come with experience. However, the number of bins should be about 10–20 depending on the number of data points with a minimum of five intervals. There should be at least 10 data points in an interval. If there are less than 100 data points use 5–10 class intervals. If there are over 100, but less than 200, use 10 class intervals. If there are over 200 data points, consider using 20 intervals. It is rare that one will need to use more than 20 class intervals. The magnitude of the bins is determined from the range of the data being divided by the number of bins.

$$\text{Bin Interval} = \frac{\text{Maximum} - \text{Minimum}}{\text{Number of intervals}} \tag{6}$$

The bin interval is then adjusted to break on an integer. If the maximum data value is less than 10, the bins should break on 1, 2, 3, etc. If the maximum data value is less than 100, the bins should break on 10, 20, 30, etc.

Although it is not necessary that the bins be of equal width, it is highly recommended. Before the sequence is initiated, the user should be aware that the bin block should first be set with the desired number of class intervals and filled with the desired class intervals. After the bin block has been set, the program is instructed to count the number of occurrences in each interval. The resulting frequencies are then placed in the bins. To plot the data, the ordered bins are plotted along the X-axis and the values in the bins are plotted along the Y-Axis as a bar graph. Figure 3-2 shows the results of plotting the data in Table 3-3 and Table 3-4 using spreadsheet software. Figure 3-3 shows the results of plotting the data as a frequency distribution diagram using STATGRAPHICS, a statistical software package.

The data may also be plotted as a scatter diagram or x–y plot. The frequencies are plotted as a function of the endpoint of the interval. The scatterplot may be graphed with a spreadsheet by choosing as the graph type, x–y plot (LOTUS or Quattro) or scatterplot (PlanPerfect) or it may be graphed using a graphing package such as ***Grapher*** (see Appendix A). The scatterplot of the frequency distribution graphed using ***Grapher*** is shown in Figure 3-4.

Box-Whisker Plot

There are other ways to visually inspect a data set besides histograms. Two that are popular are the box-whisker and stem-leaf diagrams. The box-whisker plot shows the spread of the data in terms of the median, the medians of the upper and lower halves of the data, and the spread between the medians of the upper and lower halves. This type of plot is very useful for the detection of outliers and also for rapid visual comparison of several different data sets.

The box-whisker plot is constructed as a box that has two whiskers protruding from it. The initial values that must be calculated for this plot are the median and the medians of the upper and lower halves of the data set. The medians of the upper and lower halves of the data set are called the upper and lower fourth, respectively. These medians, also called the hinges of the box plot, are then used to construct the box on the plot. First, draw a scale that is approximately the range of the data (maximum of the data set - minimum of the data set). Next indicate the median of the entire data set. The size of the box is the difference between the upper and lower fourths; the difference is called the fourth spread. The fourth spread (f_s) is then multiplied by 1.5 and also by 3.0. These two values are used to compute the limits of mild and extreme outliers. The values, 1.5 f_s and 3.0 f_s are added to the upper fourth and subtracted from

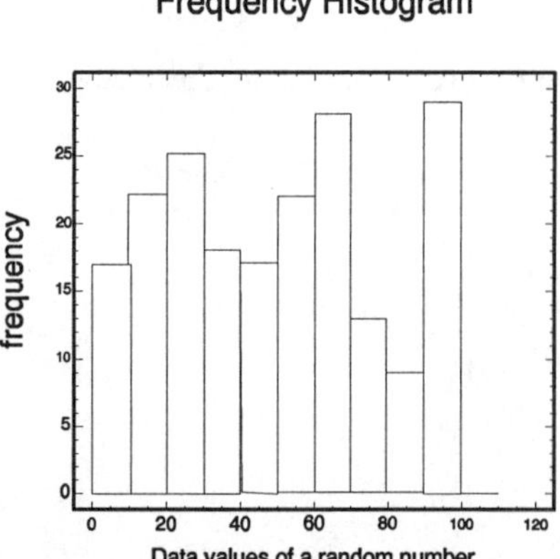

Figure 3-3. Frequency distribution of data in Table 3-3 using STATGRAPHICS statistical software.

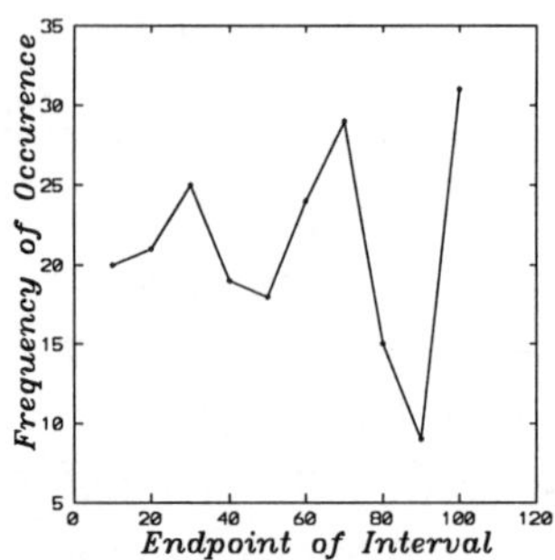

Figure 3-4. Frequency distribution diagram plotted as an X-Y Scatterplot using Grapher.

the lower fourth. The data set is then examined to determine the maximum value in the data set which is greater than the upper fourth and less than the upper fourth + 1.5 f_s. This is shown by a dot on the graph and a whisker is extended from the line indicating the upper fourth to that dot. Any values between upper fourth + 1.5 f_s and upper fourth + 3.0 f_s are indicated by an open circle; these are termed mild outliers. Any values greater than the upper fourth plus 3.0 f_s are termed extreme outliers and are indicated by a solid circle. Similar calculations and plotting are accomplished for the lower fourth, but subtracting the 1.5 f_s and 3.0 f_s. The box plot of the data in Table 3-4 is shown in Figure 3-5.

Stem-Leaf Diagram

The stem-leaf diagram is very similar to the histogram except that instead of

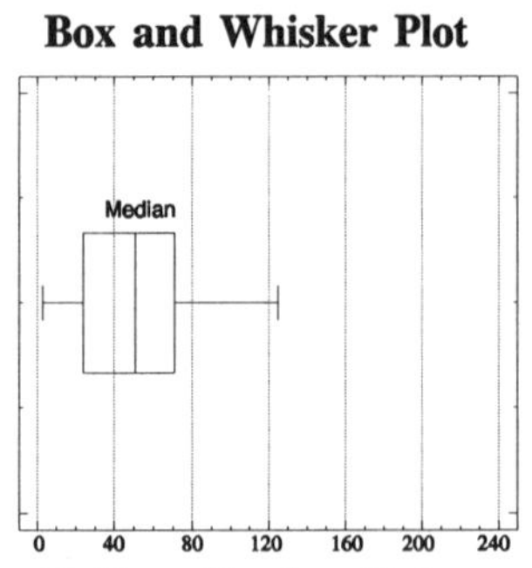

Figure 3-5. Box plot of random generated uniform distribution data.

using a solid bar to represent the frequency of the data values in the class or interval, the last digit of the data values is used. These values then visually also tell the story of the distribution within the data set. The stem-leaf diagram will also indicate high and low data values that may be outliers.

Although there are no hard and fast rules for constructing the stem-leaf diagrams, some general guidelines can be given. To construct the stem-leaf diagram, you should consider that the number of stems be between 10 and 20. To decide this, look at the sorted data to determine a convenient place to split the data. For example, if all the data were in the range 10 to 25, the convenient place in the number to split the data would be at the decimal point. However, if the data were in the range 15,000 to 30,000, the convenient place to split the data would be at the thousands (i.e., 15, 16, 17, etc.) for the stems. Once the number of stems has been determined and the location in the number is decided for the stems, a list of the stems should be made from the lowest stem to the highest stem. A stem should be written down for each set of leading values in the interval even if there are no leaves to go on the stem. This can give an indication of the presence of missing data. For each stem write down the digits after the split location. Indicate what the stem and leaf represent by giving a statement such as 9‖3 = 93.

After all the data have been entered on the stems, it must be decided whether there are some of the stems that are widely distant from the other stems (either high or low). If so, indicate these at the bottom of the stem-leaf display using HI=.... and LO=...., where .... represents the data values that are potential outliers.

When all of the above has been accomplished, count the number of leaves on each stem. A cumulative sum of number of leaves is written to the left of each stem from the lowest valued stem and from the highest valued stem. Each is accumulated until the numbers of data values in the accumulated sums are approximately equal and only one stem does not have an accumulated sum next to it. The number of leaves on that stem is written to the left in parentheses.

For example, to construct a stem–leaf display for the data in Table 3-3, the stems would likely be chosen to be the tens digits, and the leaves would then be the units digits. The stem–leaf diagram for those data is shown in Figure 3-6. As can be seen from this figure, the display resembles the frequency distribution diagram with the added advantage of seeing the data values.

```
          1||2 Represents 12

  16      0||3455556667778899
  34      1||023445567777888999
  58      2||000001122233344445788999
  79      3||000000113344445566679
  96      4||00022566666777799
(23)      5||00011222334677888889999
  81      6||0012233344467777777888899
  56      7||0000011133344469
  40      8||0027778999
  30      9||011122222344444566666667778889
```

Figure 3-6. Stem-leaf diagram for the random generated uniform distribution data.

Example

A set of radioactive decay rate data was taken and expressed as counts per hour. The data were plotted using a stem leaf diagram (Figure 3-7) and also a frequency

```
          1 |   2 represents 12

  26      0 |   00222 22224 66666 88888 88888 8
  46      1 |   00002 22244 46888 88888
  69      2 |   00000 22244 46666 66888 888
  90      3 |   00000 22444 44446 66688 8
 116      4 |   00000 02222 24444 44666 66888 8
(23)      5 |   00000 24444 44466 88888 888
 125      6 |   00000 00022 22222 22224 44466 66667 78888 88
  88      7 |   00000 00002 22222 24444 44444 44466 88888 88
  51      8 |   00222 22222 24444 44446 66668 899
  23      9 |   00002 24444 6668
   1     10 |   0
```

Figure 3-7. Stem-leaf diagram for a set of radioactive decay data.

distribution display (Figure 3-8). The frequency distribution diagram showed only the number of data points in a given interval. However, since the stem-leaf diagram also showed the actual data values for the leaves, examination of the data indicated that there were several data points that were possibly in error. It should be seen that all the data points, except 67 and 89, were multiples of two, likely explained by the taker counting the number of decays in an even amount of time, probably 30 minutes. How can you explain the presence of the 67 or the 89? Potentially, there are a couple of explanations, perhaps the taker interpolated the number of counts before multiplying; perhaps the digits are reversed and represent a transcription error; or perhaps the time interval was not 30 minutes for these two data points, but represent the actual number of counts in an hour. At any rate, the discrepancy is quite evident from the stem-leaf display.

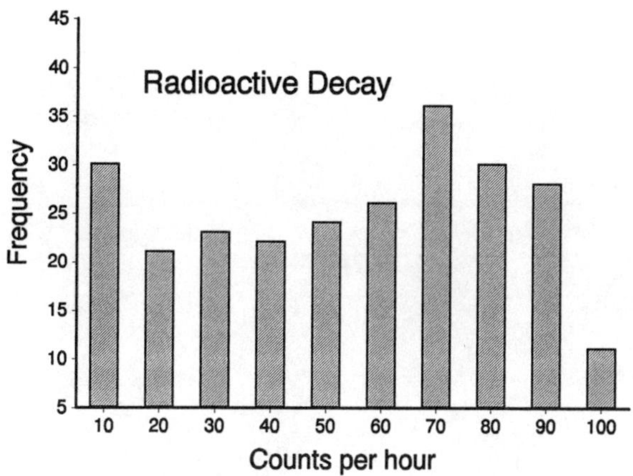

Figure 3-8. Frequency distribution for radioactive decay data.

References

1. Winkler, R.L. and Hays, ***W.L Statistics — Probability, Inference and Decision***, 2nd ed., Holt, Rinehart Winston, New York, 1975, p. 273.

2. McGill, R., Tukey, J.W., and Larsen, W.A. Variations of Box Plots, ***Amer. Stat.***, ***1978***, **32**, 12.

3. Velleman, P.F., and Hoaglin, D.C. ***Applications, Basics and Computing of Exploratory Data Analysis***, Duxbury Press, Boston, MA, 1981.

4. Cleveland, W.S. ***Elements of Graphing Data***, Wadsworth, Monterey, CA,1985.

5. Tufte, E. ***The Visual Display of Quantitative Information***. Graphics Press, Cheshire, CT, 1983.

4

CHARACTERIZATION OF DATA SETS

Introduction

Each data set may be characterized by two major measures, a measure of location and a measure of spread. The measure of location tells where the central portion of the data is. Several different measures of location will be discussed. The measure of spread tells us how far it is from one data point to another. Measures of location and spread will be discussed in this chapter.

Measures of Location

Mean or Expected Value

A large set of data is of limited utility because of the quantity of the data and the inability to rapidly discern measures of the data. It is more useful to present parameters or statistics that summarize the data set which are fewer in number than the entire set of data. One such useful characterization (the histogram) was presented in Chapter 3, Exploratory Data Analysis. A histogram is still rather bulky (although graphically it portrays a great deal about the data). Most data sets may be characterized by two general parameters, a measure of central tendency or location and a measure of dispersion. To understand the role that a measure of central tendency of a data set plays, turn again to the set of data presented in the previous chapter (reproduced as Table 4-1). The frequency distribution or histogram of the data indicated that the data were distributed fairly uniformly over the range of the data, yet there was a tendency for the data in the middle to have a higher frequency.

The measure of central tendency of this data set can be characterized by several different parameters, the most common being the mean. An alternative

Table 4-1. Unsorted uniform data

17	96	74	42	82	58	63	94	67	24	8	47	94	20	22
91	71	52	88	18	25	6	24	36	58	64	74	14	96	49
46	24	37	7	35	33	64	19	50	23	91	67	40	5	5
10	20	20	58	89	27	34	51	97	80	67	87	15	50	6
60	94	58	67	53	30	71	54	24	87	96	59	30	34	6
57	73	30	59	4	20	68	29	68	42	70	92	30	96	67
34	92	68	70	97	46	69	73	22	39	96	52	91	50	18
34	71	18	87	40	16	94	95	61	79	47	47	21	12	89
8	98	17	28	35	20	28	62	69	29	89	22	36	13	60
14	56	17	21	57	98	64	92	5	40	47	93	7	67	46
63	74	98	23	53	15	90	62	30	33	80	31	97	3	31
76	96	96	36	23	46	67	30	70	92	29	46	73	59	9
70	70	5	52	63	9	68	94	66	49	59	7	17	45	58
19	99	51	92	19										

terminology and symbology for the mean of X is the expected value of X [$E(X)$]. The mean of a data set is defined as

$$\bar{x} = \frac{\sum_{i=1}^{n} x_i}{n} \tag{1}$$

The frequency distribution introduced in the previous chapter may also be used to calculate the mean. A number that is used in many chemistry courses is the expected value. The expected value concept as taught in physical chemistry may be used to calculate the value of the momentum of a particle, the dipole moment of a particle, or the position of the particle. The student should realize that the expected values that are calculated can be thought of as average values of the property. One method of calculating these parameters is to use the wave function of the particles, Ψ and its complex conjugate, Ψ^* in the formula

$$\langle X \rangle = \int_{-\infty}^{\infty} \Psi \, x \, \Psi^* \, d\tau \tag{2}$$

where x is the quantity (momentum, velocity, dipole moment, etc) for which we wish to calculate an expected value.

Table 4-2. Frequency distribution of the uniform data by individual data value.

Value	Frequency	Value	Frequency	Value	Frequency	Value	Frequency
3	1	23	3	49	2	71	3
4	1	24	4	50	3	73	3
5	4	25	1	51	2	74	3
6	3	27	1	52	3	76	1
7	3	28	2	53	2	79	1
8	2	29	3	54	1	80	2
9	2	30	6	56	1	82	1
10	1	31	2	57	2	87	3
12	1	33	2	58	5	88	1
13	1	34	4	59	4	89	3
14	2	34	4	60	2	90	1
15	2	35	2	61	1	91	3
16	1	36	3	62	2	92	5
17	4	37	1	63	3	93	1
18	3	39	1	64	3	94	5
18	3	40	3	66	1	95	1
19	3	42	2	67	7	96	7
20	5	45	1	68	4	97	3
21	2	46	5	69	2	98	3
22	3	47	4	70	5	99	1

A concept similar to this can be used if a frequency distribution of the data set has been obtained. The frequency distribution must be obtained for the entire data set and the class intervals must be defined as each unique data value in the data set. To calculate the probability of obtaining a unique data value, one must obtain the fraction that represents the number of times the unique data value occurs in the population as a portion of the number of data values in the entire data set. The probability of obtaining any unique data value in the data set is the number of data points with the data value divided by the total number of data values in the set:

$$p(x_i) = \frac{n(x_i)}{\sum_{i=1}^{k} n(x_i)} \tag{3}$$

where k is the number of data points, and $n(x_i)$ is the number of data points with the value x_i. The mean is then calculated as

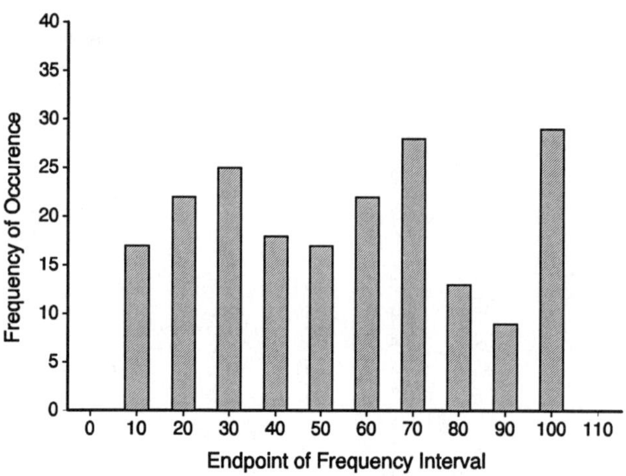

Figure 4-1. Histogram of data representing a uniform distribution.

$$\bar{x} = \sum_{i=1}^{k} p(x_i) \cdot x_i \tag{4}$$

Example

Table 4-2 shows the data from Table 4-1 for which a frequency distribution of each data value has been tabulated. The frequency distribution is plotted in Figure 4-1. To calculate the mean of the data set, each frequency is multiplied by the data value; all are then summed. The sum of the frequencies is then calculated. The sum of the products is divided by the sum of the frequencies to obtain the mean of the data set. The sum of the products is

$$\sum x_i \cdot n_i = 3 \cdot 1 + 4 \cdot 1 + \cdots = 10{,}168 \tag{5}$$

and the sum of the frequencies is

$$\sum n_i = 1 + +1 + 4 + 3 + \cdots = 200 \tag{6}$$

and the mean is obtained

$$\frac{\sum x_i \cdot n_i}{\sum n_i} = \frac{10{,}168}{200} = 50.84 \qquad (7)$$

One caution about using the mean to describe the central location of a data set is in order, though. The value of this parameter is often seriously affected by the presence of very large (or very small) numbers in the data set. The ability of a data analysis parameter to correctly portray the structure of the data set and not be unduly influenced by outliers is called robustness. The mean is a statistic that is not robust since the presence of outliers will often significantly bias the mean.

Median

There are several other measures of the central location of the data set that have proven useful and are also quite robust. The median is a central tendency measure, which is often calculated to overcome the influence of outliers. Quite simply it is eliminating the influence of the extremes of the data and generally includes one or at most two of the original data points in the calculation. The median is reported as the middle data point in a sorted data set. Of course, this is easy to calculate if the number of points in the set is odd, since it is

$$\tilde{x} = x_{(n+1)/2} \qquad (8)$$

the center data point such that half of the data points are greater in magnitude and half are less in magnitude.

But if n is even, the median is calculated as

$$\tilde{x} = \frac{x_{n/2} + x_{(n+2)/2}}{2} \qquad (9)$$

that is, the two center data values are averaged. The median is the ultimate in the eliminating of outliers and only depends on the center value(s).

Example

The median is calculated for the data in Table 4-1 using equation (9):

$$\tilde{x} = \frac{x_{100} + x_{101}}{2} = \frac{51 + 51}{2} = 51 \qquad (10)$$

Trimmed Mean

It is not always desirable to eliminate so many data points. Another measure of the central tendency that is gaining prominence is the trimmed mean. The trimmed mean is calculated by deleting a given percentage of the data points from the top and the bottom of the sorted data and using the equation for the mean to obtain a mean of the remaining data points. The two most common of the trimmed means in the literature are the 5 and 10 % trimmed means. Some data analysis packages (e.g., MINITAB) contain commands to calculate a trimmed mean from the data set. It is always possible to calculate a trimmed mean from any data set using the definition of the mean after deleting the percentage of the data set.

Example

Calculate the 5 and 10% trimmed mean for the data in Table 4-1. The trimmed mean is calculated by first sorting the data as in Table 4-2. For the 5% trimmed mean, the lowest 10 values and the highest 10 values are deleted and the mean calculated for the remaining 180 data values as above. For the 10% trimmed mean, the lowest and highest 20 values are deleted and the mean calculated for the remaining 160 data values. The 5% trimmed mean is 50.80 and the 10% trimmed mean is 50.6125.

Measures of Spread

Although the measure of location of a data set is important, of equal importance is a measure of the spread of the data. It is not enough to state where the data are centered, but you also must be able to specify something about the width or clustering of the data set. Several parameters are defined that can indicate how spread out or how close the data are. We would like to be able to say how far each data point is from the mean. Let us examine different ways of doing this.

Range

One measure of spread that is quite useful and is in fact used in the exploratory data techniques discussed in the previous chapter is range of the data. The range is defined as the difference between the maximum and the minimum of the data set

$$\text{Range} = \text{Maximum}(X) - \text{Minimum}(X) \tag{11}$$

A disadvantage of the range is that there no relation to the mean of the data set may be inferred.

Example

The range of the data set in Table 4-1 is calculated as

$$\text{RANGE} = x_{max} - x_{min} = 99 - 3 = 96 \tag{12}$$

Deviation from the Mean

A measure that is related to the mean is the average deviation from the mean

$$\overline{\Delta x_i} = \frac{\sum_{i=1}^{n} (x_i - \bar{x})}{n} \tag{13}$$

But unfortunately when the deviations are summed the sum is theoretically equal to zero, which in turn leads to the average deviation being zero. An alternative to this which eliminates the zero sum is to average the absolute value of the differences

$$\overline{\Delta x_i} = \frac{\sum_{i=1}^{n} |(x_i - \bar{x})|}{n} \tag{14}$$

Example

The sum of the differences is calculated to be zero. But the sum of the absolute value of the differences is calculated

$$\sum |x_i - \bar{x}| = 4928.3 \tag{15}$$

and when divided by 200 yields an average absolute difference of 24.64.

Variance and Standard Deviation

The most common of the measures of spread of the data, because of its prominence as a parameter of the normal distribution[††] is the variance. The parameter is the average of the square of the deviation from the mean:

$$\sigma^2 = V(X) = \frac{\sum_{i=1}^{n} (x_i - \bar{x})^2}{n} \tag{16}$$

The standard deviation is the square root of the variance:

$$\sigma = \sqrt{(V(X)} = \sqrt{\frac{\sum_{i=1}^{n} (x_i - \bar{x})}{n}} \tag{17}$$

The divisor in this equation must be modified if the data set does not represent a population but only represents a sample by subtracting 1 from n:

$$s = \sqrt{s^2} = \sqrt{\frac{\sum_{i=1}^{n} (x_i - \bar{x})}{n-1}} \tag{18}$$

The divisor must be modified because one degree of freedom has been eliminated by estimating the mean of the population with the mean of the sample. By so doing, only n-1 of the data values are free to be random values.

[††] The normal distribution is often referenced as the bell curve. Most students are familiar with this since grades usually follow a normal distribution; 95% of normally distributed will be within about $\pm$ 2 σ of the expected value.

The last value is fixed by the previous n-1 and the mean of the sample.

The variance may also be calculated as the expected value of the square of the difference between the actual data value and the expected value of the data value:

$$V(X) = E[(X-\mu)^2] \tag{19}$$

Alternative equations may be given that simplify the computation of the standard deviation and the variance are

$$V(X) = E(X^2) - [E(X)]^2 \tag{20}$$

or

$$s = \sqrt{\sum_{i=1}^{n} x_i^2 - \frac{\left[\left(\sum_{i=1}^{n} x_i\right)^2 / n\right]}{n-1}} \tag{21}$$

Coefficient of Variation

The variance and standard deviation, although indicating the amount of spread in the data, are quite sensitive to the scale of the numbers. Since they are calculated as the sum of the squares of the deviations from the mean, the values of the random variable that are further from the mean contribute greatly to the overall standard deviation. There is also no indication of the relation between the standard deviation and the mean unless both numbers are given. We can scale the standard deviation by dividing the standard deviation by the mean.

$$CV = \frac{s}{\bar{x}} \tag{22}$$

This calculation produces the coefficient of variation (CV) or the relative standard deviation.

If the coefficient of variation is multiplied by 100, the percent relative standard deviation ($\%rsd$) is obtained

$$\%rsd = CV \cdot 100 = \frac{s}{\bar{x}} \cdot 100 \tag{23}$$

This will give us an indication of the magnitude of the standard deviation as a portion of the mean.

Example

The measures of spread and central tendency for the set of random numbers in Table 4-1 are suumarized

mean	50.84	standard deviation	28.59
range	96	variance	817.25
average absolute deviation	24.6416		
coefficient of variation	1.778	average deviation	0.000
median	50.5	10% trimmed mean	50.6125

5
PROBABILITY DISTRIBUTIONS

Introduction

The concept of the random variable as being a relation between properties of a sample and the values that those properties may assume was discussed earlier. Examples were given such as the concentration of sulfate in rain water, the pH of a soil sample, and the mass of each beaker in a chemistry laboratory. Associated with each value of a random variable is a probability of obtaining that value. The set of all probabilities for the values of a random variable is given a special name, the *PROBABILITY DISTRIBUTION FUNCTION*. Each random variable, whether discrete or continuous, has a *PROBABILITY DISTRIBUTION FUNCTION* associated with it.

The manner of treatment of the probability distribution function is, however, very different for continuous and discrete variables. The discrete random variable uses the mathematics of summation, for the discrete random variable can take on only whole number values. The continuous random variable, on the other hand, is treated with differential and integral calculus. Each probability distribution function is characterized by one or more parameters that can be used to completely describe the mathematical form of the distribution. By stating the parameters for a probability distribution, one is able to reduce the data set from N elements to the number of parameters for the distribution. The mathematical form described by the parameters may then be used to calculate the probability of obtaining any value(s) or range of values of the random variable. This concept applies whether the distribution is discrete or continuous.

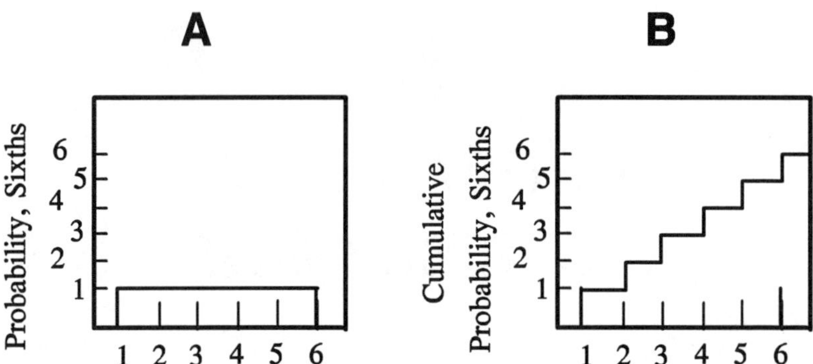

Figure 5-1. A. Probability distribution function B. Cumulative distribution function for the random variable *ROLL A SINGLE DIE*

Probability Distribution Functions

A probability distribution function may be obtained by knowing the probability of occurrence of each event in the event space. The probability distribution function is constructed by first identifying each possible outcome in the sample space and assigning a probability to each event in the sample space. The experiment that is performed is, of course, the random variable for the experiment. Associated with the random variable will be the values that the random variable may assume. The assignment of probabilities to the values of the random variable may be obtained either empirically or theoretically. Each of the possible values of the random variable will have a probability associated with it. The values of the random variable (the experiment) are the possible outcomes of the experiment. One must be able to identify all of the possible outcomes in the sample space before probabilities may be calculated. The hardest part of the composition of a probability distribution function is the delineation of all the possible events in the sample space.

Consider initially constructing a probability distribution function only for discrete random variables. Probability distribution functions for discrete random variables are often intuitive. A common example of a random variable for which a probability distribution function may be constructed is the tossing of a coin. First, identify all of the possible outcomes for a given experiment.

The experiment must be laid out in advance and must be definitive; EXPERIMENT— ***toss a coin once***. At the same time that we have defined the experiment, we have also defined the random variable, *THE SINGLE TOSS OF A COIN*. There are only two possible outcomes for the coin toss, a *HEAD* or a *TAIL*. These are the values of the random variable. The probability of obtaining either a head or a tail is ½ if we have a fair coin. The probability distribution function for a single toss of a fair coin is

Event	**Toss a head**	**Toss a tail**
Probability	½	½

Another example of an intuitive experiment is to roll a single die for which the associated random variable is *THE NUMBER FROM A ROLL OF A SINGLE DIE*. The sample space may be determined to be the outcomes that are possible from rolling this die, namely the numbers 1 through 6. The numbers 1 through 6 are the values of the random variable. The probability for each outcome is the same, 1/6, and the probability distribution function is shown as

Event	1	2	3	4	5	6
Probability	1/6	1/6	1/6	1/6	1/6	1/6

Cumulative Distribution Function

Associated with each probability distribution function is a cumulative distribution function that is, as the name implies, the probabilities of successive events added together. The manner of addition is quite unique and very specific. The probability and the associated value of the random variable are sorted in ascending sequence. The probabilities of the two lowest values of the random variable are added. To this sum is added the probability of the third lowest value of the random variable. To that sum is added the probability of the fourth lowest random variable. The process is continued until the probability of each value of the random variable has been added to the sum. The interpretation of the cumulative distribution function is that it is the probability of obtaining a value of the random variable that is less than or equal to the given value:

$$P(X \leq x_n) = \sum_{j=0}^{n} P(X=x_j) \tag{1}$$

The cumulative distribution function for the random variable *ROLL A SINGLE DICE* may be calculated by cumulatively adding each probability. The cumulative probabilities may be appended to the table for the probability distribution function and tabularly shown as follows

Event	**1**	**2**	**3**	**4**	**5**	**6**
Probability	1/6	1/6	1/6	1/6	1/6	1/6
Cumulative probability	1/6	2/6	3/6	4/6	5/6	6/6

The probability and cumulative distribution functions are shown graphically in Figure 5-1.

Properties of Probability Distribution Functions

There are several axioms that a function must follow if the function is to be termed a probability distribution function. The first of these is that the value of the probability distribution function must always be positive in the domain of interest. If the probability distribution function is plotted as an *XY* plot with origin at (0,0), the values of the function must exist only above the x axis. This is mathematically stated as

$$p(x) \geq 0 \tag{2}$$

The second requirement of the distribution function depends on whether the probability distribution function is for discrete or continuous random variables. If the probability distribution function is for a discrete random variable, then the sum of the probabilities must be 1:

$$\sum_{i=1}^{N} p(x_i) = 1 \tag{3}$$

If the probability distribution function is for a continuous random variable, then the integral over the domain must equal 1:

$$\int_{-\infty}^{+\infty} f(x)\,dx = 1 \tag{4}$$

This leads to the full statement of the probability that an event will occur:

$$0 \leq p(x) \leq 1 \tag{5}$$

Most often, the concern is not whether the total probability is 1, but what the probability is that a value of a random variable may be less than a given value. An immediate example that comes to mind is the situation of an action limit for a pollutant. Manufacturing facilities, chemical waste treatment plants, power plants, etc. are able to discharge pollutants only below levels designated in their emission permits. It is required to know if the analyte of interest is present in a sample at a concentration less than or equal to that permitted level. We must have a means available to determine the probability that the value for the concentration found in the analysis is less than or equal to the permitted level.

To find the probability that a value of a random variable will be less than or equal to an action level, we can sum the probabilities of all values less than or equal to the action level. The probability that the value of a random variable is less than or equal to a given value may be calculated from the probability distribution function. The manner of calculation of the probability is different depending whether the random variable is discrete or continuous. For the discrete random variable, the cumulative probability depends on the inclusion of the value of the random variable or not.

For example, if it is desired to calculate the probability that the sum of the two numbers from the simultaneous roll of two dice is less than or equal to seven, the entries in Table 5-1 may be used. The rows in the table indicate the relative expected number of occurrences of each sum [the row labeled f(x)], the relative cumulative number of occurrences [the row labeled F(x)], the relative probability of obtaining any given sum [the row labeled $p(X=x)$], and the cumulative probability [the row labeled $P(X \leq x)$]. The probability that a given sum has been obtained is calculated by first determining all of the potential, unique outcomes of throwing two dice and adding the two numbers together that are face up on the two die. After determining all of the unique combinations, the number of combinations that yields 2 as a sum is counted (1), the number of

Table 5-1. Probability that the Value of a Variable is Less Than or Equal to a Given Value for Throwing Two Dice

x	2	3	4	5	6	7	8	9	10	11	12
$f(x)$	1	2	3	4	5	6	5	4	3	2	1
$F(x)$	1	3	6	10	15	21	26	30	33	35	36
$p(X=x)$	1/36	2/36	3/36	4/36	5/36	6/36	5/36	4/36	3/36	2/36	1/36
$P(X\leq x)$	1/36	3/36	6/36	10/36	15/36	21/36	26/36	30/36	33/36	35/36	36/36

occurrences that the sum is three is counted (2), and so forth. These values are the relative number of occurrences of that sum. The cumulative relative number of occurrences is found by adding the number of occurrences that the sum is two (1) to the number of occurrences that the sum is three (2) to obtain (3); to this sum is added the number of occurrences that the sum is four (3) to obtain (7) and so forth. Because each throw is independent, the number of unique occurrences may be found by multiplying the number of outcomes (6) on the first die by the number of outcomes on the second die (6) to obtain 36. The cumulative number of occurrences of unique sums must equal this number.

The probability of occurrence of any sum is found by dividing the relative number of occurrences by the total number of occurrences (36):

$$p(X=x) = \frac{n_x}{\sum n_x} = \frac{n_x}{36} \tag{6}$$

The entire set of probabilities, $p(X=x)$, is the probability distribution function for the random variable defined by the experiment ***throw two dice; add the resulting numbers.*** The set of cumulative probabilities, $P(X \leq x)$, is the cumulative distribution function for the random variable defined by the experiment ***throw two dice; add the resulting numbers.***

The probability distribution function and the cumulative distribution function may be graphed and are shown in Figure 5-2 and Figure 5-3.

The cumulative distribution function may be used to obtain the probability of occurrence of a value less than or equal to a given sum by reading the cumulative probability directly from the chart. The cumulative probability is really the sum of probabilities of obtaining any of the values less than or equal to a desired sum. Let us examine an example; if we desire to know the probability of obtaining a sum less than or equal to 7, we would read the probability directly from the table as 21/36. This number was obtained from the sum:

$$P(X \le 7) = p(X{=}7) + p(X{=}6) + p(X{=}5) + p(X{=}4) + p(X{=}3) + p(X{=}2)$$
$$\frac{21}{36} = \frac{6}{36} + \frac{5}{36} + \frac{4}{36} + \frac{3}{36} + \frac{2}{36} + \frac{1}{36} \tag{7}$$

However, for discrete random variables, the probability that the sum is less than a given value does not include the endpoint. For example, if we desire to know what the probability of obtaining a sum that is less than 7, we would read the cumulative probability directly from the table as 15/36, i.e., the probability of obtaining a sum that is less than or equal to 6:

$$P(X<7) = p(x{=}6) + p(x{=}5) + p(x{=}4) + p(x{=}3) + p(x{=}2)$$
$$\frac{15}{36} = \frac{5}{36} + \frac{4}{36} + \frac{3}{36} + \frac{2}{36} + \frac{1}{36} \tag{8}$$

The cumulative distribution function may also be used to determine the probability of obtaining an outcome in an interval of the random variable. The probability, as with the probability of obtaining an outcome less than a given value of the random variable, depends on whether the end points are included or not. The probability to find a value in the interval *a* to *b*, where *a* and *b* are values of the random variable is described by four different situations:

$$p(a<x<b) = P(x \le (b-1)) - P(x \le (a+1)) \tag{9}$$

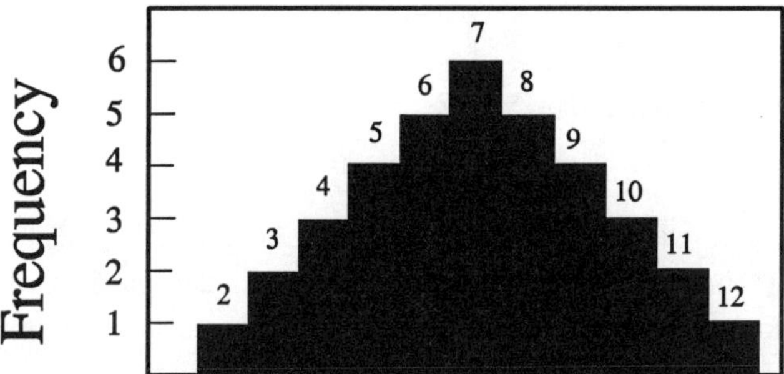

Figure 5-2. Probability distribution function for the sum of two dice.

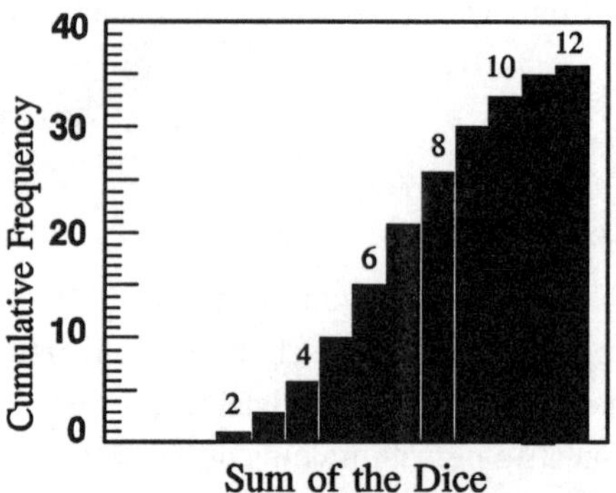

Figure 5-3. Cumulative frequency distribution for the sum of two dice.

$$p(a \leq x < b) = P(x \leq (b-1)) - P(x \leq a) \quad (10)$$

$$p(a < x \leq b) = P(x \leq b) - P(x \leq (a+1)) \quad (11)$$

$$p(a \leq x \leq b) = P(x \leq b) - P(x \leq a) \quad (12)$$

where a and b are assumed to be integers and X is a discrete random variable. We can now calculate the probability that the value of the random variable lies in the interval, $4 \leq X \leq 9$ as

$$p(4 \leq x \leq 9) = P(x \leq 9) - P(x \leq 4) = \frac{30}{36} - \frac{6}{36} = \frac{24}{36} \quad (13)$$

Continuous Random Variables

Continuous random variables are also governed by probability distribution functions. The probability distribution functions for continuous random variables must also conform to the axioms given for the discrete random variable. The integral of the function in the domain of interest must equal 1.

$$\int_{-\infty}^{+\infty} f(x)\, dx = 1 \tag{14}$$

The function must also be positive in the domain of interest. These two conditions lead to the full statement of the probability of the occurrence of a continuous random variable:

$$0 \le P(X) \le 1 \tag{15}$$

Similarly, a cumulative distribution function exists for the continuous probability distribution function. Instead of a summation, the cumulative distribution function for the continuous random variable is an integral. Let $f(x)$ denote a general continuous probability distribution function. Then the probability to obtain a value of the random variable less than or equal to a given number is given by the integral:

$$p(x \le a) = \int_{-\infty}^{a} f(x)\, dx \tag{16}$$

We may also find the probability of finding a value in a given interval as the integral from a to b:

$$p(a \le x \le b) = \int_{a}^{b} f(x)\, dx \tag{17}$$

However, unlike the discrete distribution, all of the situations given in Equations (9) to (12) are equivalent. The probability that a value of a random variable lies in an interval is not dependent on the inclusion of the limits of the interval. The interpretation of the evaluated integral is that the probability of occurrence of a value of a random variable in a given interval is the area under the probability distribution function in the domain of interest. An adjunct consequence of this integral is that the probability of finding a given value for a random variable that is distributed as a continuous function is zero.

There are several *PROBABILITY DISTRIBUTION FUNCTIONS* that have special significance in chemical applications. We will discuss the following probability distribution functions in this chapter: the uniform, the binomial, the Poisson, the normal, the log-normal, the F, Student's t, and the χ^2 distributions.

Uniform Distribution

Discrete Random Variables

A rather simple distribution, but one from which many of the intuitive concepts of probability distribution functions are easily obtained, is the uniform distribution that was illustrated earlier in the chapter on exploratory data analysis techniques. If the random variable is discrete and follows the uniform distribution, the probability of every integer in the interval

$$A \leq X \leq B \tag{18}$$

is equally likely and is denoted by the sequence:

$$P(A) = P(A+1) = P(A+2) = \cdots = P(B-2) = P(B-1) = P(B) \tag{19}$$

Mathematically, the formula for the probability distribution function is given by:

$$f(x) = \begin{cases} \dfrac{1}{A - B} & A \leq x \leq B \\ \\ 0 & \text{otherwise} \end{cases} \tag{20}$$

which is shown graphically in Figure 5-4.

The uniform distribution is useful since many intuitive examples are described by the distribution: the throwing of dice, the drawing of a card from

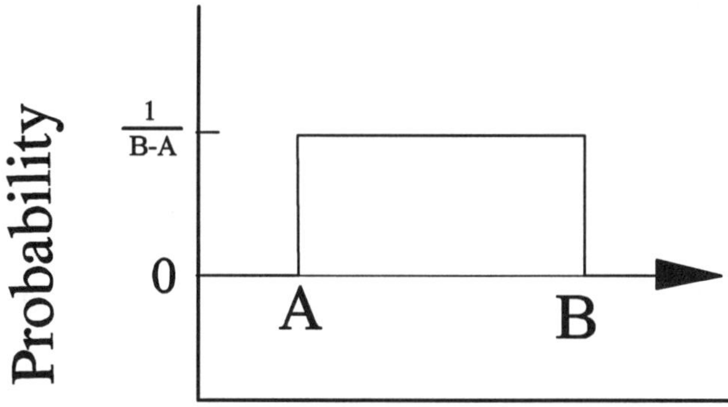

Figure 5-4. Graphic representation of the uniform probability distribution.

a deck of cards, dialing a telephone number at random, the numbers drawn in the lottery, or the generation of random integers on an interval. Each of these is described by the uniform probability distribution. The probability of occurrence of any event described by a uniform probability distribution function is given by the probability distribution function and is merely the inverse of the number of elements in the set:

(21)

Example

If a die is thrown, there are six possible outcomes. Intuitively, we know that the probability of obtaining any one of the values of the random variable is 1/6. If instead of throwing a six sided die, we throw a tetrahedron, the number of possible outcomes is 4, and the probability of obtaining any one of the sides of the tetrahedron is ¼. If we throw a dodecahedron, the number of possible outcomes is 12 and the probability of any one of the faces of the dodecahedron showing is 1/12. Many problems of probability are easily solved by recognizing that the probability distribution function that describes the random variable is the uniform probability distribution function.

The discrete random variable described by a uniform probability distribution function is of particular interest when sampling. Consider the situation in a chemical plant where the product is being conveyed from one process to another. Quality control must be performed at each stage of the synthesis to ensure that no particular portion of the overall process may adversely affect the general quality of the product. At each stage, samples are obtained that are then analyzed. Each sample must be given an equal chance of being selected. Great care is taken to ensure that the samples are truly random and that each of the samples that is taken is representative of the whole. Recognize, however, that only the probability of obtaining any given sample is uniform. The values of the purity or the content of the sample may vary widely and probably will not be described by a uniform distribution. This argument holds for any sampling process; obtain samples that are uniformly distributed in the sense that each sample has an equal likelihood of being selected.

Continuous Random Variables

The uniform probability distribution may also describe continuous random variables. For the continuous random variable, the probability distribution function is also

$$f(X) = \begin{cases} \dfrac{1}{A - B} & A \le x \le B \\ 0 & \text{otherwise} \end{cases} \tag{22}$$

but with this distribution, ANY value in the interval, $A \le x \le B$ is possible, whereas with the discrete variable, only certain values would be possible. The probability distribution function for the continuous uniform random variable is shown in Figure 5-4.

Binomial Distribution

Let us consider a random variable that can have only two values, A or B, although there may be more than one element of the sample space that has the value A or B. The important thing to note is that each member of the population may have only the value A or B. The fraction of A in the population is designated by P_A and the fraction of B in the population is designated by P_B, which indicates probability of A or probability of B. It should be noted that P_B is $(1 - P_A)$. These two probabilities may be calculated using the formulas

$$P_A = \frac{N_A}{N} \qquad P_B = \frac{N_B}{N} \tag{23}$$

where:

$$N = N_A + N_B \tag{24}$$

Note that the sum of the probability that event A will occur and the probability that event B will occur must be one. Note also that the fraction of occurences of event A is exactly the probability that event A will occur. A similar conclusion is drawn for event B.

Example

Consider a bag which has two types of marbles in it, 250 green and 500 red. The probability of obtaining a green marble from the bag is

$$P_{green} = \frac{N_{green}}{N_{green} + N_{red}} = \frac{250}{250 + 500} = \frac{250}{750} = 0.33333 \quad (25)$$

and the probability of obtaining a red marble is:

$$P_{red} = 1 - P_{green} = 0.66667 \quad (26)$$

The variety of experiments that can be conducted that have only two outcomes is very broad. Since the number of experiments is so large, it was early recognized by Bernoulli who developed the theory of binomial variables. Random variables that have only two outcomes are often denoted as Bernoulli variables. Experiments conducted with random variables which have only two values are considered Bernoulli or binomial experiments. However, four conditions must be met if the experiment is to be termed a binomial experiment.

1. The number of trials must be set in advance.
2. There may be only two outcomes for each trial.
3. The probability of success must be constant from trial to trial.
4. Each of the trials must be independent according to the condition of independence:

$$P(T_j | T_n) = P(T_j)$$

If the experiment meets all of these conditions, then the probability of x successes in n trials will be shown to be the relative number of combinations of n items (trials) taken x at a time. Each of these conditions is quite important. These conditions may be applied to chemical situations with ease. The conditions are generally met for situations in chemistry that have only two outcomes. In some chemical situations, there may be more than two possible outcomes. The following discussion is easily modified to include that possibility.

The binomial experiment can be thought of as being mathematically described by the expansion of the binomial $(A+B)^N$ with the coefficients of the resulting polynomial being related to the number of combinations for the given number of successes. The ratio of the coefficients is the ratio of probabilities for obtaining each of the distinct outcomes in the experiment. The probability for obtaining each of the outcomes is, as explained before, the ratio of the number of outcomes for each event divided by the total number of outcomes. The binomial probability and cumulative probability distribution are graphically shown in Figure 5-5. If there are more than two possible outcomes, a polynomial $(A+B+C+...)^N$ will be expanded rather than a binomial.

Example

If two marbles are drawn from a bucket that contains only black and white marbles, four possibilities for the sample exists: WW, BW, WB, and BB. After the first marble is drawn, it is replaced; then the second marble is drawn. If the probability of drawing a white marble is P_A and the probability of drawing a black marble is P_B, then the probability of obtaining two white marbles is $P_W \bullet P_W$; the probability for obtaining white followed by black is $P_W \bullet (1-P_W)$; the probability for black followed by white is $(1-P_W) \bullet P_W$; and the probability for two black marbles is $(1-P_W) \bullet (1-P_W)$. However, if we are concerned only with the number of white marbles drawn and not the order in which the marbles are drawn, then the probabilities are

$$P(X = 2 \text{ white marbles}) = P_W \cdot P_W$$

$$P(X = 1 \text{ white marble}) = 2 \cdot P_W \cdot (1 - P_W) \tag{27}$$

$$P(X = 0 \text{ white marbles}) = (1 - P_W) \cdot (1 - P_W)$$

The concept can be extended to 3, 4, ..., n with the results for 4 being

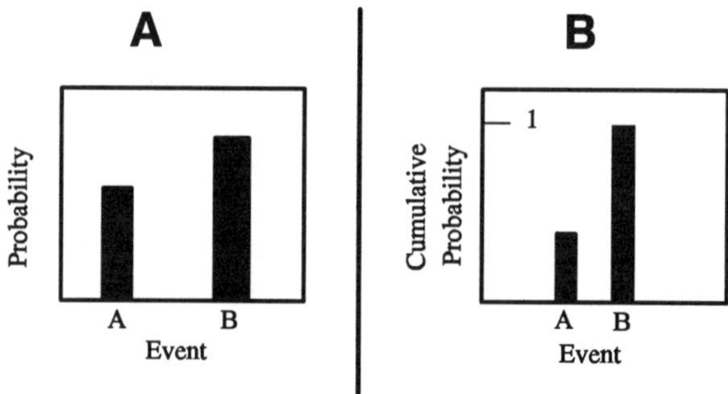

Figure 5-5. **A.**Probability distribution and **B.** Cumulative probability distribution for binomial distribution.

$$
\begin{aligned}
P(X = 4 \text{ white marbles}) &= P_W^4 \\
P(X = 3 \text{ white marbles}) &= 4 \cdot P_W^3 \cdot (1\ P_W) \\
P(X = 2 \text{ white marbles}) &= 6 \cdot P_W^2 \cdot (1 - P_W)^2 \\
P(X = 1 \text{ white marble}) &= 4 \cdot P_W \cdot (1 - P_W)^3 \\
P(X = 0 \text{ white marbles}) &= P(X = 1 \text{ black marble}) = (1 - P_W)^4
\end{aligned}
\tag{28}
$$

One should be able to see a pattern developing which results in the general formula for the number of white marbles drawn in a binomial experiment:

$$b(x; n, p) = \binom{n}{x} p^x \cdot (1 - p)^{n-x} \tag{29}$$

where n is the number of trials, x is the number of white marbles drawn, and p is the probability of drawing a white marble in any trial. Notice that the binomial probability distribution function is described by two parameters, n and p. This same formula may be applied to any Bernoulli experiment where it is desirous to calculate the probability of a number of events occurring.

Before we apply a binomial experiment to a chemical example, we need to know two other important properties of the distribution, the expected value and the standard deviation. As was given earlier, the expected value is calculated as the sum of the products of the probability of a value times the value:

$$E(X) = \sum x \cdot p(x) \tag{30}$$

for the simple case of a binomial random variable having the two values, 0 and 1, the expected value for a binomial experiment of one trial is merely p, the value of the probability of the success. If there is more than one trial, and the values of the random variable are still 1 and 0, then the expected value is merely $n \bullet p$, where n is the number of trials.

The standard deviation of a binomial probability is given as the square root of the variance of the random variable:

$$\sigma = \sqrt{V(X)} = \sqrt{E(X^2) - [E(X)]^2} \tag{31}$$

which reduces to $[p \bullet (1-p)]^{1/2}$ for one trial or $[n \bullet p(1-p)]^{1/2}$ for more than one trial.

Let us now apply these principles to several chemical examples.

Example

ATOMIC MASS BROMINE

Let us first calculate the atomic mass of bromine. Bromine has two predominant isotopes with atomic masses of 79 and 81 with natural abundances of 50.69 and 49.31% (approximately 50% each). The molecular weight of bromine is found by using the formula for the expected value of a binomial distribution. That is, sum the products of the atomic mass of the isotope (79 and 81) multiplied by the frequency of natural occurrence (the percentage expressed as a fraction .5069 and .4931) to give the mean mass (79.99).

$$\textit{Molecular mass}_{\text{Bromine}} = 79.00 \cdot 0.5069 + 81 \cdot 0.4931 = 79.99 \tag{32}$$

Similar calculations may be done for any of the elements using the formula for finding the expected value. The calculation of atomic mass by this method is not dependent on the element only having two isotopes, but may be extended to any number of isotopes using the expression for the expected value of any random variable:

$$E(X) = \sum_{j=1}^{n} x \cdot p(x) \tag{33}$$

where x is the atomic mass of the isotope and $p(x)$ is the natural abundance of the isotope.

Example

MASS SPECTRA OF ORGANIC COMPOUNDS

Another application of the binomial distribution is to predict mass spectral patterns. Let's examine several polysubstituted halocarbons. Bromine first. The expected mass spectral pattern (Figure 5-6) for a monobromo-alkane, a dibromo-alkane, or a tribromo-alkane may be calculated from the expansion of the binomial $(A+B)^n$; where n is the number

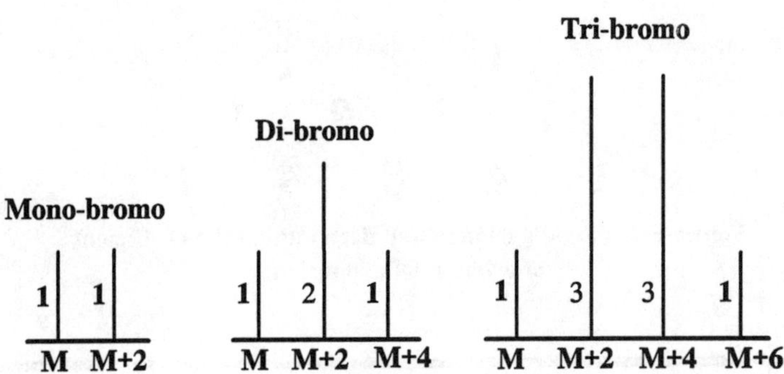

Figure 5-6. The expected peak intensities in a mass spectrum of bromo-substituted alkanes.

of bromine atoms in the molecule and A and B are the probabilities of obtaining a Br^{79} or a Br^{81}. The coefficients are easily obtained from PASCAL's triangle (Figure 5-7).

Using PASCAL's triangle to predict what the coefficients should be if there are but a few bromines is an easy matter, since it is so regular. But for a molecule that might contain many bromines, using the formula given in Equation (29) for binomial probabilities is much easier. Approximating the fraction of bromine atoms by 0.5 for each of the 79 and the 81 isotopes, the probabilities for tribromo–substituted bromoalkane are given in Equation (34). The calculated coefficients reduce to an integer ratio of 1:3:3:1, which are the relative peak heights for the grouping of peaks in the mass spectrum of a molecule that contains bromine.

$$
\begin{aligned}
P(3\,\mathrm{Br}\ ^{79}) &= \binom{3}{3} \cdot (0.5)^3 \cdot (1-0.5)^0 \\
P(2\,\mathrm{Br}\ ^{79}) &= \binom{3}{2} \cdot (0.5)^2 \cdot (1-0.5)^1 \\
P(1\,\mathrm{Br}\ ^{79}) &= \binom{3}{1} \cdot (0.5)^1 \cdot (1-0.5)^2 \\
P(0\,\mathrm{Br}\ ^{79}) &= \binom{3}{0} \cdot (0.5)^0 \cdot (1-0.5)^3
\end{aligned}
\qquad (34)
$$

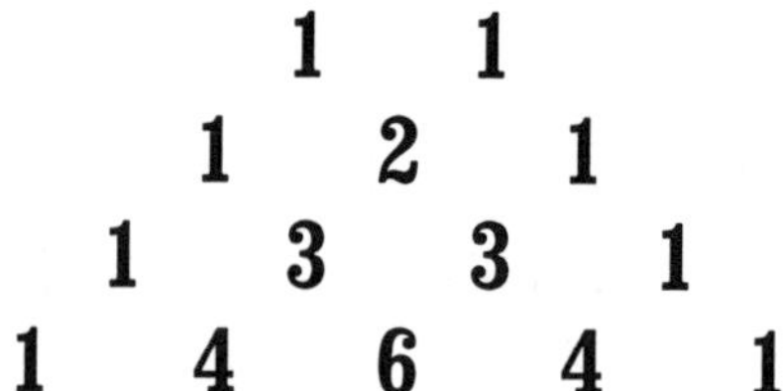

Figure 5-7. Pascal's triangle for determining the coefficients of a binomial expansion.

Example

MASS SPECTRUM OF CHLORINE SUBSTITUTED HYDROCARBON

The probabilities are rather easily worked for the case where isotope abundances are equal. But what about the situation where they are not equal, e.g., $C_nH_{2n+2-x}Cl_x$? The ratios for the mass spectral patterns for mono–, di–, and trisubstituted chlorohydrocarbons may be calculated from the expansion of the binomial $[P(Cl^{35})+P(Cl^{37})]^N$. But since the natural abundance of the Cl^{35} isotope is about three times that of the Cl^{37} isotope, the binomial which is expanded is in effect $(3A+B)^N$. The coefficients that are found from the expansion are

Mono	3:1
Di	9:6:1
Tri	27:27:9:1

The probabilities for a trichloroalkane are more accurately obtained from the binomial probability formula and the actual abundances [$P(Cl^{35})$ = **0.66** and $P(Cl^{37})$ = **0.34**] of the two isotopes

$$P(3\ Cl^{35}) = \binom{3}{3} \cdot (0.66)^3 \cdot (1-0.66)^0$$

$$P(2\ Cl^{35}) = \binom{3}{2} \cdot (0.66)^2 \cdot (1-0.66)^1$$

$$P(1\ Cl^{35}) = \binom{3}{1} \cdot (0.66)^1 \cdot (1-0.66)^2 \quad (35)$$

$$P(0\ Cl^{35}) = \binom{3}{0} \cdot (0.66)^0 \cdot (1-0.66)^3$$

which when reduced to integer coefficients are 27:27:9:1.

Poisson Distribution

There is no simple experiment that can lead to the Poisson distribution since it is the limiting situation of a complex argument. Simeon Denis Poisson was a French mathematician living in early nineteenth–century France. He dabbled not only in mathematics (publishing over 300 papers) but was a highly respected educator and politician. For his work in deriving the distribution, it has since come to be known as the Poisson distribution. The Poisson distribution is used to find the probability of events occurring which happen very infrequently. The Poisson distribution is only described by one parameter (unlike the binomial, which was described by two.) This parameter, λ, should be thought as a rate (number of occurrences per unit time) or as a density (number of occurrences per unit area or unit volume). The Poisson probability distribution function is given as

$$p(x; \lambda) = \frac{e^{-\lambda} \cdot \lambda^x}{x!} \tag{36}$$

where λ is the rate and x is the value of the random variable whose probability is being calculated. The expected value and the variance of the Poisson distribution are both equal to λ. Appendix C gives the results of calculation of several sets of Poisson probabilities for a few values of the parameter, λ, and the desired number of occurrences. The probability distribution function and the cumulative distribution function are shown in Figure 5-8.

Some examples of processes that are described by a Poisson probability distribution function are the number of trees in a forest, the number of traffic accidents that occur on a street corner, the number of incoming telephone calls in an office, the number of radioactive decays measured in a laboratory, the number of samples that contain particulate contamination, and the number of gold nuggets found in a stream bed.

An interesting aspect of the Poisson distribution is its relation to the binomial distribution. If the probability of one of the outcomes of a binomial

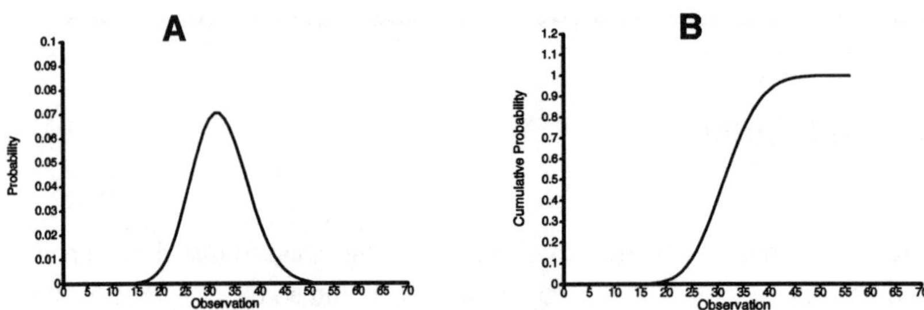

Figure 5-8. Poisson distribution. A. Probability distribution. B. Cumulative probability distribution.

random variable is quite small ($p<0.01$), the number of trials is quite large ($N>100$), and the product of the probability and the number of trials is constant and less than about 20 ($np \leq 20$), the probability is more easily calculated using the Poisson distribution.

Let us examine an example of the Poisson distribution as it relates to chemistry.

Example

Radioactive decay or the measurement of a cosmic ray follows the Poisson distribution. If the rate of decay of a radioactive element has been measured, then that rate of decay could be designated as α. The number of decays that would be expected in a preset time period would then be $\lambda=\alpha t$, where t is the time period of interest. The probability of an observed number of decays occurring in the time period may then be calculated as

$$p(n;\alpha \cdot t) = \frac{e^{-(\alpha \cdot t)} \cdot (\alpha \cdot t)^n}{n!} \quad (37)$$

What is the probability that exactly 9 observances of radioactive decay will be observed in 1 hr if the average rate of decay is 8 decays per 15 minutes? $\alpha=8$ decays per 15 min and $t=60$ min, thus $\alpha t=32$; $n=9$. These values are substituted in the expression for the Poisson distribution:

$$p(9,32) = \frac{e^{-32} \cdot 32^9}{9!} = 1.23 \cdot 10^{-6} \quad (38)$$

The previous probability functions were all examples of discrete probability functions. However, there are also continuous probability distribution functions which are extremely important. The most important of the continuous probability distribution functions is the normal probability function. This particular distribution is so important because of the number of random variables that may be described by the distribution. Another reason why it is so important is that the distribution of the means of samples of any random variable (no matter what its probability distribution function) may be described by the normal probability distribution function. More on this property of the normal probability distribution function will be presented later.

Normal Probability Distribution

Perhaps the best known distribution because of its use in grading is the normal or Gaussian probability distribution function. The function is known to accurately describe the frequency distribution of many phenomenon. An additional interesting aspect of the normal probability distribution function is that it can be used to approximate (sometimes quite accurately) the distribution of many random variables that do not exactly follow the normal distribution function. The normal probability distribution is the limit of many of the other probability distributions. For example, compare the graphic representation of the Poisson distribution given in Figure 5-8 with the normal distribution shown in Figure 5-9. As can be seen the shapes of the distributions are very close indeed. Similar observations can be made for the binomial distribution when the number of trials is very large.

The normal distribution, derived by Gauss, a French mathematician living in the eighteenth century, is used to describe many chemical analyses and processes. The concentration of a solution, if determined many times, the errors associated with any calibration process, the random deviations from a baseline in a chromatogram, the hydrogen ion concentration in rain at a given location, and the mass of a beaker if weighed many times may all be described by the normal probability distribution function.

The normal probability distribution function is described by two parameters: μ the mean, describes the central location of the distribution, and σ, the standard deviation, describes the spread of the data. The distribution is

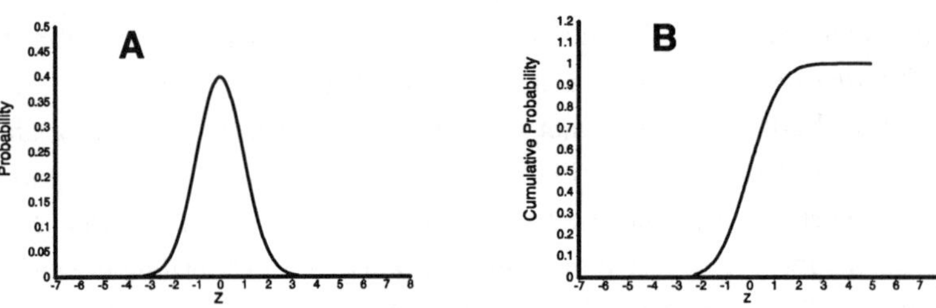

Figure 5-9. Normal distribution: A. Probability distribution. B. Cumulative probability distribution.

symmetric about the mean. The normal probability distribution function is

$$f(x) = \frac{1}{\sigma\sqrt{2\pi}} e^{-\frac{(x-\mu)^2}{2\cdot\sigma^2}} \tag{39}$$

Unfortunately, this expression is not easily integrated so as to obtain the cumulative distribution function. In fact, an analytical solution does not exist, so we must rely on numerical integration to obtain values of the distribution. As with discrete probability functions, the integral of all probabilities of the normal probability distribution function must also sum to 1. However, since we are interested in a continuous random variable, the sum is found by integrating the probability distribution function over the interval, $-\infty$ to $+\infty$:

$$\frac{1}{\sigma\sqrt{2\pi}} \int_{-\infty}^{+\infty} e^{-\frac{(x-\mu)^2}{2\cdot\sigma^2}} dx = 1 \tag{40}$$

The normal probability distribution function may also be integrated from $-\infty$ to a desired value to obtain the probability that X is less than any given value x

$$P(X \leq x) = \frac{1}{\sigma\sqrt{2\pi}} \int_{-\infty}^{x} e^{-\frac{(X-\mu)^2}{2\sigma^2}} dx \tag{41}$$

Since, this is not easily integrated, the random variable is generally transformed to a standard normal distribution by subtracting μ and dividing by σ.
This has the effect of creating a distribution that is centered at zero (the expected

$$Z = \frac{X - \mu}{\sigma} \tag{42}$$

value of Z) and a standard deviation of 1. By making this transformation, the integral becomes

$$P(Z \le z) = \frac{1}{\sqrt{2\pi}} \int_{-\infty}^{z} e^{-\frac{z^2}{2}} dz \tag{43}$$

which is easily tabulated. Every normal distribution can be transformed to the standard normal distribution for the ease of calculating probabilities. Probabilities of the standard normal distribution have been tabulated in Appendix C. The probabilities that are given in this table are the probabilities

$$P(Z \le z) = \Phi(Z) \tag{44}$$

The normal probability and cumulative probability distributions are shown graphically in Figure 5-9.

The standard Z value is a measure of the number of standard deviations that the value is from the mean of the random variable. If $E(X) = \mu = 10.0$, $\sigma = 2.0$ and the value of X is 8, we can calculate a Z value:

$$Z = \frac{X-\mu}{\sigma} = \frac{8-10}{2} = -1.0 \tag{45}$$

which indicates that the value of X is one standard deviation less than the mean. The probability that a value less than or equal to 8 would be obtained is read from the table in Appendix C as

$$P(X \le 8) = \Phi(Z) = \Phi\left(\frac{8-\mu}{\sigma}\right) = \Phi\left(\frac{8-10}{2}\right) = \Phi(-1) = 0.1587 \tag{46}$$

The probability that a value of a random variable may lie in an interval may be calculated using the following:

$$P(A \le Z \le B) = \frac{1}{\sqrt{2\pi}} \int_A^B e^{-\frac{z^2}{2}} dz \tag{47}$$

which can be interpreted as being

$$P(A \le Z \le B) = \Phi(Z_B) - \Phi(Z_A) \tag{48}$$

where Z_A and Z_B are the Z values for A and B.

Example

Calculate the probability that the analyzed concentration of chloride ion from a sample of a groundwater system will be in the range 28 to 38.5. The mean of the chloride concentration in the ground water system is known to be 35.00 mg/L with a standard deviation of 3.5 mg/L. The Z-values are calculated as

$$Z_A = \frac{28-35}{3.5} = -2.0$$

$$Z_B = \frac{38.5-35}{3.5} = 1.0 \tag{49}$$

and the probability is then calculated as the difference

$$P((X=28) \le Z \le (X=38.5)) = \Phi(Z_{38.5}) - \Phi(Z_{28})$$

$$= 0.8413 - 0.0228 = 0.8185 \tag{50}$$

Log–normal Distribution

Some individuals may think that when sampling a system such as rain or a lake or the atmosphere, that the values obtained would be normally distributed. In actuality, they are not! An example of data obtained from sampling atmospheric precipitation is the data set of the concentrations of sulfate ion at a single sampling site. The frequency distribution of the data is shown in Figure 5-10. Data sets that exhibit such asymmetry may be transformed to a normal

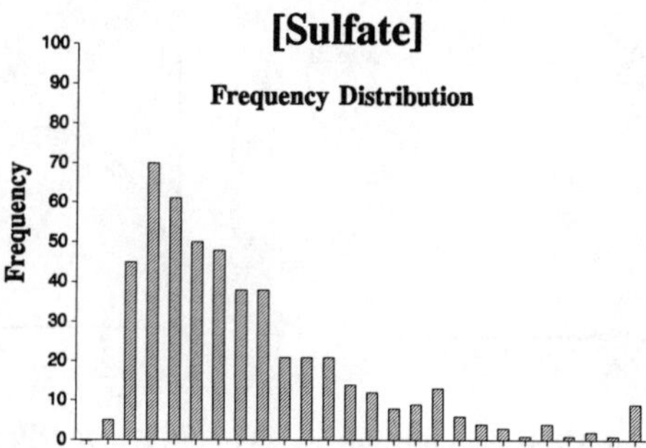

Figure 5-10. Frequency distribution of concentration of sulfate ion in atmospheric precipitation.

distribution if the logarithms of the values of the random variable are obtained,

$$Y = \log(X) \tag{51}$$

the resulting distribution will be normal or Gaussian.

$$f(Y) = \frac{1}{\sigma_Y\sqrt{2\pi}} e^{-\frac{(Y-\mu_Y)}{2\sigma_Y^2}} \tag{52}$$

The lognormal distribution[1] is described by the equation

$$f(x) = \frac{1}{x\sqrt{2\pi}\,\sigma} e^{-\frac{(\ln x - \mu)^2}{2\sigma^2}} \tag{53}$$

The distribution is graphically portrayed in Figure 5-11.

The mean of the log-normal distribution is described by the geometric mean rather than the arithmetic mean. The geometric mean is the arithmetic mean of the logarithms of the data values and may also be calculated from the distribution equation and the following:

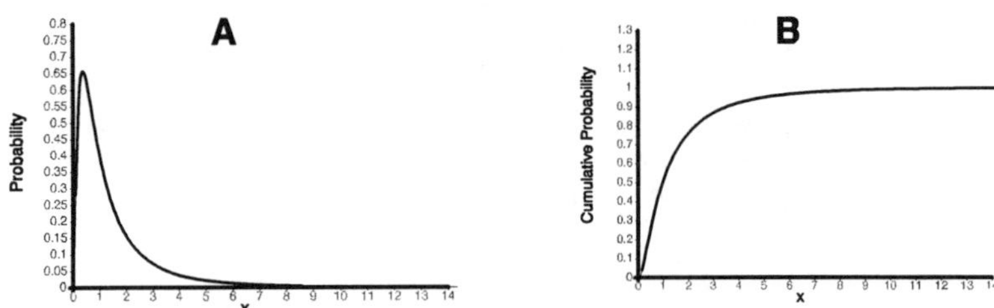

Figure 5-11. Log-normal distribution A. Probability distribution function B. Cumulative probability distribution function.

$$E(Y) = \frac{\sum_{i=1}^{n} Y_i}{n} = \sqrt[n]{\prod_{i=1}^{n} X_i} \tag{54}$$

The variance of this distribution is calculated as

$$V(x) = e^{(2\mu + \sigma^2)}\left(e^{\sigma^2} - 1\right) \tag{55}$$

where μ and σ are calculated as given previously, i.e. the arithmetic mean and the standard deviation of the x values.

Student's *t*–Distribution

William Gossett, a late nineteenth–century chemist, worked in an English brewery and wrote under the pseudonym, STUDENT. As he was performing analyses, he became concerned about the distribution of means of small samples. His work led to the discovery of the distribution that bears his nom de plume, Student. The name of the distribution does not derive from the fact that it is used by students. This distribution is quite important because it does not depend on having estimates for the parameters of a population but depends only on the value for the number of members in the sample. Student's *t*–distribution does not describe a population but rather describes a sampling distribution. The main advantage of the Student *t*-distribution is that small samples may be used. The sample size, *n*, if less than 30, will qualify the sample to be small. This is not a hard and fast rule, but it is sufficient in most instances. As the number of elements in the sample increases, the *t*–distribution approaches the normal distribution described earlier. In fact, the form of the *t*–statistic that will be introduced in greater detail later is of the same form as the *Z*–variable of the standardized normal distribution:

$$t = \frac{(X - \bar{X})}{s / \sqrt{n}} \quad (56)$$

The mathematical form of the *t*–distribution is

$$f(x) = \frac{\Gamma\left[\frac{(\nu + 1)}{2}\right]}{\Gamma\left[\frac{\nu}{2}\sqrt{\nu \cdot \pi}\right]} \frac{1}{\left(1 + \frac{x^2}{\nu}\right)^{(\nu + ½)}} \quad (57)$$

where ν is equal to the number of elements in the sample minus 1. The Γ function is defined as

$$\Gamma[x] = x \cdot \Gamma[x - 1] \quad (58)$$

thus for x, an integer, the Γ function is merely the factorial. Student's *t*–distribution is shown graphically for several different values of the number of elements in the sample in Figure 5-12.

Although introduced here, the application of Student's *t*–distribution will be deferred until a later chapter. In the later chapter, the *t*–distribution will be used to compare means of samples.

F-Distribution

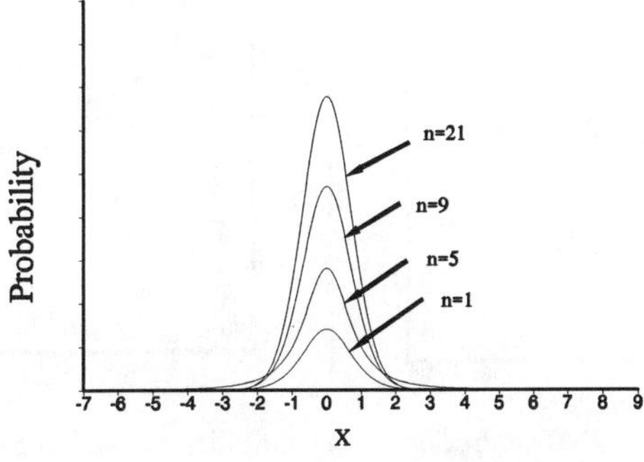

Figure 5-12. Student's *t*–distribution for sample sizes of 1, 5, 9, and 21.

Of considerable utility is the *F*-distribution because it allows statistical tests to be performed on random variables that do not follow the normal or similar distributions. *F*-tests are designed to allow comparisons between random variables that follow the χ^2 distribution and depend on the ratio of two such random variables. The most common of the random variables that follow the χ^2 distribution is the ratio of the variance of a sample to the variance of the population:

$$\frac{(n-1)\cdot s^2}{\sigma^2} \tag{59}$$

The *F*-distribution is described by the equation

$$f(x) = \frac{\Gamma[(\nu_n + \nu_d)/2]}{\Gamma[\nu_n/2]\Gamma[\nu_d/2]}\left(\frac{\nu_n}{\nu_d}\right)^{\frac{\nu_n}{2}} \frac{x^{\frac{(\nu_n - 2)}{2}}}{\left[1 + \left(\frac{\nu_n}{\nu_d}\right)\cdot x\right]^{\frac{(\nu_n + \nu_d)}{2}}} \tag{60}$$

where ν_n = **number of elements in the numerator - 1** and ν_n = **number of elements in the numerator - 1** and $x>0$; $\nu_n>0$ and $\nu_d>0$. The *F*–distribution probability and cumulative probability are shown in Figure 5-13.

As will be shown in the next chapter, the *F*–statistic will be introduced as a way to ascertain the equivalence of two sample variances. The values of the *F*– distribution are tabulated in Appendix C.

A compendium of probability distribution functions, associated equations

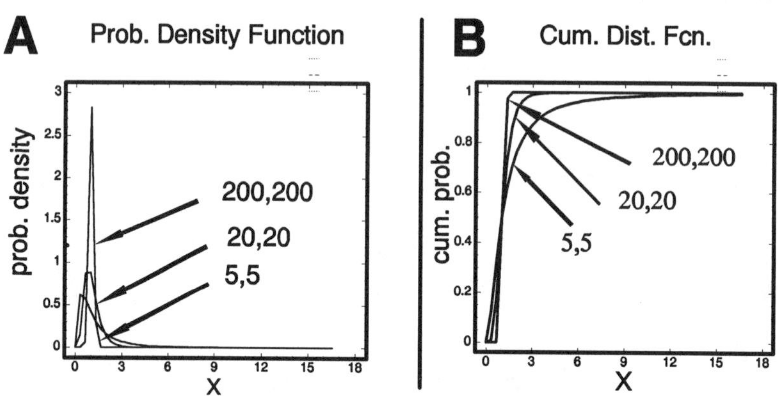

Figure 5-13. ***F*–Distribution A. Probability density function. B. Cumulative probability function.**

for expected value and standard deviation may be found in Hastings and Peacock[2].

References

1. Aitchison, J., and Brown, J.A.C. ***The Lognormal Distribution***. Cambridge University Press, New York, 1957.

2. Hastings, N.A.J.and Peacock, J.B. ***Statistical Distributions***. Butterworth's, London, 1974.

6

FITTING FREQUENCY DISTRIBUTIONS

The description of data sets by a frequency distribution leads one to question how to determine which of the classical distribution functions should be used to describe the values of a random variable that comprise a data set. Values of the random variable have been obtained through chemical experiments such as analyses, syntheses, or weighing of products. As the amount of data (values of random variables) grows, the data may be represented graphically by a frequency distribution diagram such as described in Chapter 4. However, the graphic representation is not suitable to determine intermediate values of a function or finding theoretical or expected frequencies of values of a random variable. Additionally the entire data set must be used to construct the histogram. It is desirous to be able to represent a set of data by a smaller number of values if possible. A way to do this is to describe the data set by a frequency distribution function. Of course, the volume of data should be such that representation by a smaller number makes legitimate sense. The ensuing discussion will assume that a large amount of data is available.

One should examine critically the need for your data set to be described by a distribution function. Can the salient features be described by a graphic representation in a histogram? Is it necessary to reduce the size of the data set by a mathematical description?

The type of distribution to describe the data will be greatly influenced by the type of measurement process and the type of data to be represented. As an example, multiple determinations of the concentration of sulfate ion in a single sample of rainwater by ion chromatography could be hypothesized to be described by a normal distribution. However, the recording of the sulfate concentration in rain water samples at a single location spread over time would probably best be described by a log-normal distribution as shown in Figure 6-1.

Let us expand on these two examples to see more fully what these two settings describe. Sampling of rain is accomplished at many locations throughout the United States, a major effort that has been underway for more than a decade

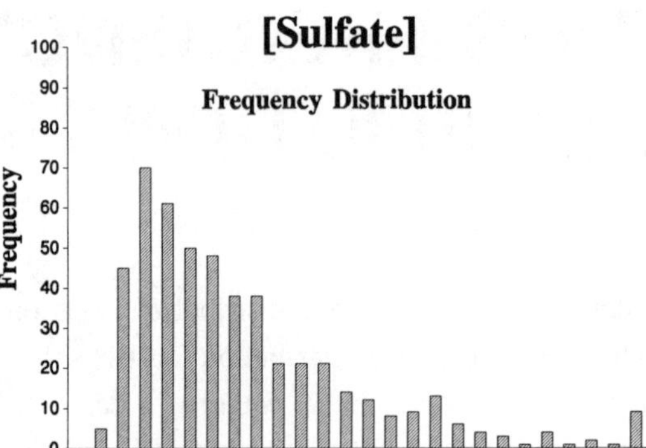

Figure 6-1. Frequency distribution for sulfate ion concentration at the NY99 site of the National Atmospheric Deposition Program.

to ascertain spatial and temporal trends in concentration of chemical species in the rain. Each sample container, that contains the rain that falls during the week, is collected. If the rain that is collected in that sampler is analyzed many times, a normal distribution would be expected. The distribution would be centered on the expected value for the concentration of the sulfate ion in the sample ($E($ $[SO_4^{2-}]$)) as calculated by

$$E\left(\left[SO_4^{2-}\right]\right) = \frac{\sum x_i}{n} \tag{1}$$

The spread of the data can be described by the sample standard deviation:

$$s = \sqrt{\frac{\sum x_i^2 - \frac{\left(\sum x_i\right)^2}{n}}{n-1}} \tag{2}$$

These two values ($E($ $[SO_4^{2-}]$) and s) are the estimates of the parameters of the normal distribution and will be used in the equation for the probability distribution function:

$$f(X) = \frac{1}{s \cdot \sqrt{2\pi}} e^{-\frac{(x_i - \bar{x})}{2 \cdot s^2}} \tag{3}$$

or the cumulative distribution function:
to describe the data.

This distribution describes only the average concentration for a week of

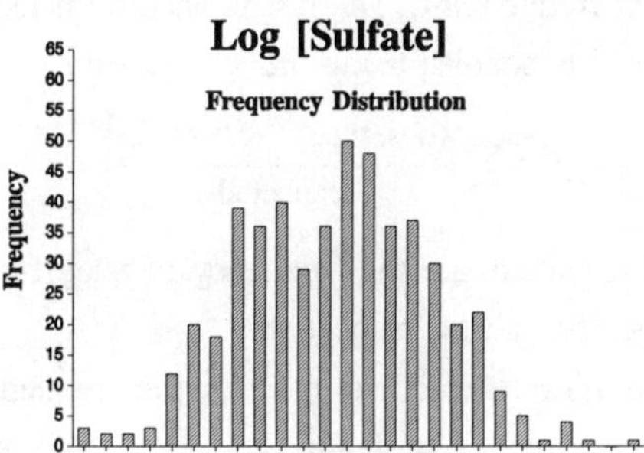

Figure 6-2. Frequency distribution of the logarithm of the concentration of sulfate ion at the NY99 site of the National Atmospheric Deposition Program.

$$F(X) = P(X<x) = \frac{1}{\sqrt{2\cdot\pi}} \int_{-\infty}^{X} e^{-\frac{(x-\bar{x})^2}{2\cdot s^2}} dx \tag{4}$$

rain. We can ascertain virtually nothing about the week–to–week variation of the concentration (Figure 6-1). Data from week–to–week analyses show long tails at higher concentration and the distribution shows a marked peak at lower concentration (Figure 6-1).

If, however, the frequency of the logarithm of the concentration is plotted, the familiar bell-shaped normal curve is obtained (Figure 6-2). This type of distribution is noted in many systems involving the environment.

We now need to discuss different ways to determine how well a given frequency histogram may be described by a probability distribution function. Two classical methods are the χ^2 and the Kolmogorov–Smirnov tests.

χ^2 and Kolmogorov-Smirnov Tests for Distribution

χ^2 Test for Distribution

The χ^2 test[‡‡] to determine how well a set of data is described by a probability

---

[‡‡] The χ^2 tests were first proposed by Karl Pearson.

distribution function depends upon being able to calculate or arrive at a set of expected and observed frequencies. The test is accomplished by comparing the observed and expected frequencies using the χ^2 equation:

$$\chi^2 = \sum \frac{(\text{observed}_i - \text{expected}_i)^2}{\text{expected}_i} \tag{5}$$

Where observed_i and expected_i are the frequency of occurrence of the observed values and of the expected values, respectively.

Of question is how to calculate each of the frequencies, observed and expected. To use this equation will almost always require that the data be discrete; at least discrete in the sense that not all data values are observed even though they may be possible. Calculation of the observed frequencies is easy since we can rely on the data from which our graphic representation of the frequency, the histogram, was formed. To obtain the expected frequencies, however, requires that we propose a probability distribution function to describe the data set.

Let us examine a set of data that purportedly comes from a uniform distribution. The data are meant to represent the mass of 250–mL beakers that are found in a laboratory. Five hundred beakers were weighed and the masses tabulated. A frequency distribution was obtained and is shown in Table 6-1. We will formulate the null and alternative hypotheses for this example as

H_0: the distribution is uniform

H_a: the distribution is not uniform

Recall that the probability distribution function for the uniform distribution is

$$f(X) = \begin{cases} \dfrac{1}{B - A} & A \le X \le B \\ 0 & X < A \text{ or } X > B \end{cases} \tag{6}$$

The beaker mass data supposedly had an average value of 100 and ranged from 98 to 102. Thus in Equation (6), $A=98$ and $B=102$. Then the probability of obtaining any mass between 98 and 102 is equally likely. This can be extended to say that the probability of obtaining a set of masses between any two limits is equal to the probability of obtaining a set of masses with two other limits provided that the ranges of the two limits are equal. Mathematically this is described as

Table 6-1. Observed and Expected Frequencies for a Purported Uniform Distribution.

Interval	Observed	Expected	χ_i^2
98.0			
98.4	55	50	0.50
98.8	51	50	0.02
99.2	46	50	0.32
99.6	52	50	0.08
100.0	52	50	0.08
100.4	46	50	0.32
100.8	56	50	0.72
101.2	40	50	2.00
101.6	56	50	0.72
102.0	46	50	0.32

$$P(A_1<X<B_1) = P(A_2<X<B_2) = \frac{1}{B_1 - A_1} = \frac{1}{B_2 - A_2} \tag{7}$$

provided that

$$B_1 - A_1 = B_2 - A_2 \tag{8}$$

We can now divide the interval $98<X<102$ into an equal number of subintervals. Let us divide this set into 10 equal subintervals of length 0.4. The number of elements in any one of the subintervals will then be the total number of elements in the sample (500) divided by the number of subintervals (10). The expected frequency (number of beakers) in any one subinterval is 50. The observed frequencies (number of beakers) are shown in Table 6-1 together with the expected frequency (number of beakers) for the interval. The expression in Equation (5) is then evaluated for each of the observed and expected frequencies. All of the χ_i^2 are added together to obtain the test statistic, χ^2. For this example, $\sum\chi_i^2$ is 5.08. The test statistic is then compared to the table value of χ^2 for the desired level of significance and the number of degrees of freedom. If we desire not to reject the hypothesis at a significance level of 5% then $\alpha=0.05$. The number of degrees of freedom is

$$\nu = \text{number of intervals} - 1 - \text{number of values estimated} \tag{9}$$

The number of values estimated is 1 since we estimated the expected number of beakers for a given mass in each interval. Thus for this example, ν=8. The table value for χ^2 is 15.51. Thus we would not reject the null hypothesis that the distribution is uniform.

Let us examine whether the distribution of the mass of the beakers would adequately be described by the normal distribution. For this, we formulate the null and alternative hypotheses as

H_o: the distribution is normal

H_a: the distribution is not normal

We must now calculate the expected frequencies for the intervals. To do this, transform each of the limits of the intervals to a Z-value as introduced in Chapter 5:

$$Z = \frac{(X - \bar{X})}{S} \tag{10}$$

Then calculate the expected fraction of the population that would be less than the calculated Z value. For example, the P_i for a Z value of -1.333 is 0.092 and is the $\Phi(Z)$ discussed in Chapter 5. To calculate the value of ΔP_i, 2 consecutive values of P_i are subtracted from each other. For example, to find ΔP_i for the interval between 99.6 and 100.0, 0.396 is subtracted from 0.500 giving a ΔP_i value of 0.104 that represents the proportion of the population of beakers that should have masses between 99.6 and 100.0. Multiplying this number (0.104) times the size of the population (500) will give the expected number of beakers whose masses should lie between 99.6 and 100.0 (51.87). The χ^2 value is then calculated using Equation (5). The results of the calculation are shown in Table 6-2. As can be seen, $\sum\chi_i^2$ is 114.765. The number of degrees of freedom for this example is not the same as for the uniform distribution because we have estimated two parameters of the distribution, μ and σ. Thus the number of degrees of freedom will be "number of classes - 1 - 2" or seven degrees of freedom. The table χ^2 value is 14.07 for α=0.05. For this example, we would reject the null hypothesis that the distribution is normal.

There are some limitations of the χ^2 test, though. The test is mostly of historical importance and significance, especially if the full set of ungrouped sample data exists. If only the summary or the grouped data are available then the χ^2 test is the most desirable. The difference between ungrouped and grouped data is that with grouped data only the summary number of observations in each

Table 6-2. Expected and observed frequencies for mass of beakers assuming a normal distribution

Class Interval	obs f_i	Z_i	P_i	ΔP_i	exp f_i	χ_i^2
98.0	23	-1.333	0.092	0.092	45.900	11.425
98.4	55	-1.067	0.143	0.051	25.700	33.404
98.8	51	-0.800	0.212	0.069	34.350	8.071
99.2	46	-0.533	0.297	0.085	42.575	0.276
99.6	52	-0.267	0.396	0.099	49.605	0.116
100.0	52	0.000	0.500	0.104	51.870	0.000
100.4	46	0.267	0.604	0.104	51.870	0.664
100.8	56	0.533	0.703	0.099	49.605	0.824
101.2	40	0.800	0.788	0.085	42.575	0.156
101.6	56	1.067	0.857	0.069	34.350	13.645
102.0	23	1.333	0.908	0.051	25.700	0.284
			1.000	0.092	45.900	45.900
					$\sum \chi_i^2$	114.765

group or class is reported and not the value of each element of the group. If the full set of ungrouped data is available, other tests should be used.

Kolmogorov-Smirnov Test for Distribution

The Kolmogorov-Smirnov[1,2] test statistic to determine if a set of data is described by a particular distribution that is distribution free. This test is best used to detect differences in departure from the shape of the distribution, unlike the χ^2 test that is more likely to detect irregularities when comparing the observed distribution to an expected distribution. Another difference between the two tests is that the Kolmogorov-Smirnov relies on the cumulative frequency distribution rather than the frequency distribution used for the χ^2 test.

Since the full set of ungrouped data for this example is available, these data may be examined by the Kolmogorov–Smirnov test. This test calculates the difference between the observed cumulative frequency and the expected cumulative frequency for each class:

$$D_x = F_n(x) - F(x) \tag{11}$$

where $F_n(x)$ is the observed value of cumulative frequency of the ungrouped data set:

$$F_n(x) = \frac{n(X<x)}{n} \tag{12}$$

and $F(x)$ is the expected cumulative frequency calculated from the proposed probability distribution function:

$$F(x) = \int_{-\infty}^{x} f(x)\, dx \tag{13}$$

Which for a normal distribution becomes

$$F(x) = \frac{1}{s \cdot \sqrt{2\pi}} \int_{-\infty}^{x} e^{-\frac{(x-\bar{x})^2}{2s^2}}\, dx \tag{14}$$

and for the uniform distribution

$$F(x) = \int_{A}^{B} \frac{1}{b-a} dx \tag{15}$$

Each of the D_xs that is calculated is compared to find the maximum deviation from the expected cumulative frequencies. The test statistic is then the maximum of the absolute value of the positive or negative deviation from the expected cumulative frequency from observed cumulative frequency:

$$D = \frac{\max|D_x|}{n} = \frac{\max\left(|D_x^+|, |D_x^-|\right)}{n} \tag{16}$$

Example

The results of calculating the set of D_x s using Equation (11) for the set of beaker weights is shown in Table 6-3. for both the normal and the uniform distributions. Aseen in Table 6-3A, for the test of the uniform distribution, D^+ is 0.062 and D^- is nonexistent since all of the observed cumulative frequencies are greater than the expected cumulative frequencies, therefore the Kolmogorov–Smirnov test statistic for the uniformly distributed beaker masses is calculated to be 0.062, the maximum of the two. If 0.05 is chosen as the desired significance level and the number in the sample is 500, then the table value for the Kolmogorov-Smirnov test statistic is 0.061. If 0.01 is chosen as the desired

Table 6-3. Results of Calculating the Kolmogorov Test Statistic for Normal and Uniform Distributions

Part A, the uniform distribution

f_i	exp f_i	Σf_i	Σ exp f_i	$\Delta f_i/n$
23	0	23	0	0.046
55	50	78	50	0.056
51	50	129	100	0.058
46	50	175	150	0.050
52	50	227	200	0.054
52	50	279	250	0.058
46	50	325	300	0.050
56	50	381	350	0.062
40	50	421	400	0.042
56	50	477	450	0.054
23	50	500	500	0.000

Part B, the normal distribution

f_i	exp f_i	Σf_i	Σ exp f_i	$\Delta f_i/n$
23	45.90	23	45.90	-0.046
55	25.70	78	71.60	0.013
51	34.35	129	105.95	0.046
46	42.58	175	148.53	0.053
52	49.61	227	198.13	0.058
52	51.87	279	250.00	0.058
46	51.87	325	301.87	0.046
56	49.61	381	351.48	0.059
40	42.58	421	394.05	0.054
56	34.35	477	428.40	0.097
23	25.70	500	454.10	0.092
	45.90	500	500.00	0.000

significance level and the number in the sample is 500, then the table value for the Kolmogorov-Smirnov test statistic is 0.0728. Thus we would not reject the null hypothesis that the distribution is uniform at the $\alpha=0.01$ significance level, but would reject at the $\alpha=0.05$ significance levels. As seen in Table 6-3B, D^+ is 0.097and D^- is

-0.046, therefore, the test statistic is 0.097. The critical values remain the same as for those given previously for the uniform distribution, as 0.0728 for $\alpha=0.01$ and 0.062 for $\alpha=0.05$.

Quadratic Empirical Distribution Function Statistics

Although, the χ^2 and the Kolmogorov–Smirnov are useful, they are not the most powerful of the techniques for determining the probability distribution function of a data set. A second class of statistics is more useful and more powerful. The general class is the Cramér–von Mises family of statistics. The general mathematical description of the Cramér–von Mises family is

$$C = n \int_{-\infty}^{\infty} [F_n(x) - F(x)]^2 \, \psi(x) \, dx \tag{17}$$

where $\psi(x)$ is a weighting function for the observed square. Two main statistics are proposed in the literature and depend on the weighting function. If $\psi(x)$ is 1, the statistic is the Cramér–von Mises[3] statistic, usually denoted by W^2. If the weighting function is

$$\psi(x) = \frac{1}{F(x) \cdot [1 - F(x)]} \tag{18}$$

the statistic is known as the Anderson–Darling[4,5] statistic, A^2. Operationally, these are not easy to calculate since we usually do not have sufficient data to approximate the continuous function that is required. However, we can integrate each of the functions and obtain the following as calculational models for the Anderson–Darling and the Cramér–von Mises statistics. First, the data are scaled by an appropriate transform such as the Z-transform in the case of a normal distribution. The Z-values are then ordered from lowest to highest. The statistics are then calculated using

$$W^2 = \sum_i \left(Z_i - \frac{(2 \cdot i - 1)}{2\,n} \right)^2 + \frac{1}{12\,n} \tag{19}$$

for the Cramér–von Mises and

$$A^2 = -n - \frac{1}{n}\sum_i (2 \cdot i - 1) \cdot \left[\ln Z_i + \ln (1 - Z_{n+1-i}) \right] \tag{20}$$

for the Anderson–Darling.

To use the Cramér–von Mises or Anderson–Darling, one must know or at least be able to estimate all the parameters of the population. The power of these tests increases, the more fully the parameters are known. The general theory of using these two statistics is quite well developed and tables of their values are published and available (see, for example, paper by Stephens [6]). All of these tables depend on the sample being finite (which is, of course, why we must estimate the "closeness" of describing a population distribution by knowing something about a portion of the population). Even if we do not know all of the parameters of the population, these statistics may be used if estimates of the parameters can be calculated and the values of the parameters replaced in the probability distribution function by the estimate of the parameter. Of the distributions that may be tested by these statistics, it is interesting to note that the distribution of the empirical distribution function statistics does not depend on the true value of the parameter for several distributions including normal and exponential, but depends only on the distribution being tested and the size of the sample being tested.

Stephens[7] gives a summary of the use of empirical distribution function statistics when distributions such as normal or exponential are being tested or the values of all population parameters are fully specified.

1. EDF statistics are usually much more powerful than the Pearson chi-square statistic..
2. The most well-known EDF statistic is D[§§], but it is often much less powerful than the quadratic statistics, W^2 and A^2.
3. A^2 often behaves similarly to W^2, but is on the whole more powerful for tests when $F(x;\theta)$ departs from the true distribution in the tails, especially when there appears to be too many outlying X-values for the $F(x;\theta)$ as specified. In goodness-of-fit, departure in the tails is often important to detect, and A^2 is the recommended statistic.

Graphic Technique

Although one can calculate fit statistics to determine if the data are adequately

---

§§ D is the Kolmogorov-Smirnov statistic.

described by a given distribution, an alternative graphic method may be examined. One of the ways to graphically determine if a set of data is described by a given distribution is to construct a *probability plot*. If the data are adequately described by the distribution such a plot should be close to linear. If the distribution of the data is something other than the one hypothesized, such a plot should deviate significantly from the straight line. One of the advantages of using the following method of constructing probability plots is that the plot may be constructed for any of the distributions we have discussed or others that may be found in the literature[8,9,10].

To construct the probability plot first requires one to calculate the percentiles of the data and plot them against the theoretical percentiles of the distribution. To calculate the percentiles of the data, recall that a percentile is the number on the measurement scale of the data set such that $100 \bullet p$ percent of area is under the cumulative probability distribution curve. Thus the value of the data set that gives a p value of 0.5 would represent the median of the data and 50% of the data would be less than the data value and 50% of the data would be greater than the data value. Percentiles may be determined rather easily from the cumulative distribution tables such as the cumulative distribution table for the normal distribution given in Appendix C. For example, since the numbers in the center of the table represent the portion of the area to the left of the z-value in the left–hand column, if one wanted to find the z value that gave the median, one would find 0.500 in the center under a z-value of 0.00. To find the z-value that corresponded to the 25th percentile, locate the value in the center of the table that is closest to and brackets 0.2500 on the line headed by z=-0.6. The value of 0.2500 would lie between the column headings of 0.07 and 0.08 (corresponding to -0.67 and -0.68) and, by interpolation, it can be seen that the z-value would be approximately 0.675.

The percentiles of data from a sample should, at least, crudely follow the percentiles of a population, i.e., the 50th percentile should be the value of the data set that has roughly 50% of the data less than the value and 50% greater than the data value. Likewise, the 80th percentile should represent the value of the data set for which roughly 80% of the data set is less in magnitude. To find the percentiles generally requires that the sample size be quite large.

However, there are means to calculate the percentiles of any reasonable sample size. The procedure is to first sort the data from low magnitude to the highest magnitude of the data set. If there are n values of the random variable in the sample, then there will be n percentiles that can be calculated from

$$i\text{-th percentile} = \left[\frac{100 \cdot (i - 0.5)}{n}\right] \tag{21}$$

where i ranges from 1 to n, the number of values of the random variable in the sample. For example, if $n=20$, then the values of the percentiles would be 2.5, 5.0, 7.5, ... , 92.5, 95, and 97.5%. For $n=10$, the values of the percentiles would be 5, 10, 15, ... , 85, 90, and 95%. These values represent the theoretical percentiles of the distribution. Thus the actual percentiles that are calculated from the sample should not differ too much from the theoretical percentiles. If the z-values for the actual percentiles are plotted against the z-values for the theoretical percentiles, a plot that has a slope of 1 should be obtained.

Example

Consider a set of data taken as the concentration of sulfate ion in several lakes in the northeastern United States. The data set has a mean of 5 and a standard deviation of 1.5 and consists of the following:

Percentile	5	15	25	35	45	55	65	75	85	95
z-value	-1.645	-1.037	-0.675	-0.385	-0.126	0.126	0.385	0.675	1.037	1.645
x-value	-1.85	-1.17	-0.75	-0.49	-0.15	0.15	0.68	0.81	1.25	1.65

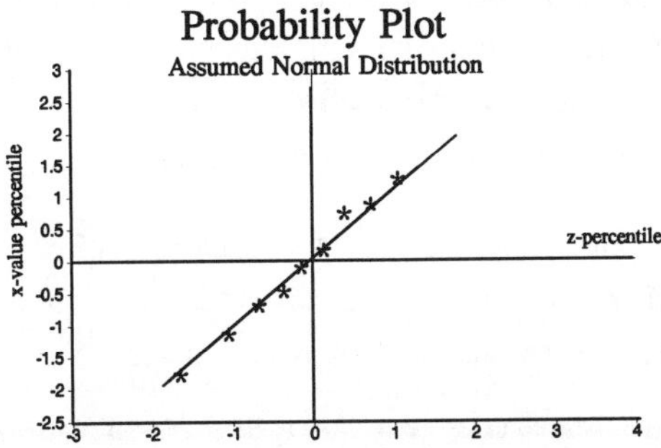

Figure 6-3. A probability plot of sulfate ion in lakes data. This plots the z-values. Note that the plot is nearly linear.

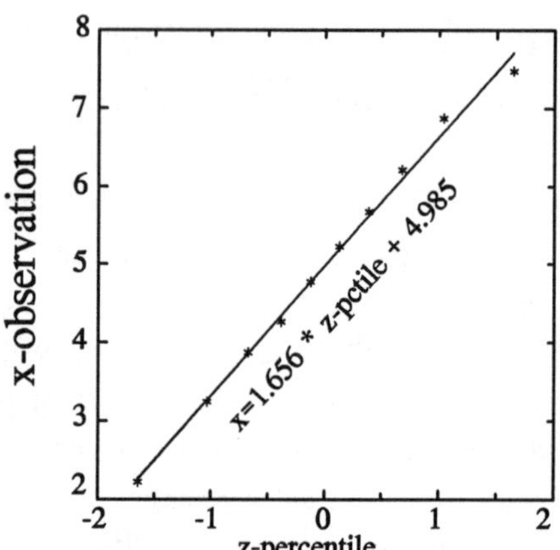

Figure 6-4. Probability plot of benzene in water data supposedly described by a normal distribution. Note that the line is nearly linear.

The data is plotted as the x-values (z-values from actual data values) against the z-values (theoretical z-values for percentiles) as shown in Figure 6-3.

A second set of data that could be examined to show the probability plot of a sample is the following where x-value represents the concentration of benzene found in water expressed in μg/L and the z-values are calculated as before:

z-value	-1.645	-1.037	-0.675	-0.385	-0.126	0.126	0.385	0.675	1.037	1.645
x-value	2.225	3.245	3.875	4.265	4.775	5.225	6.02	6.215	6.875	7.475

The plot is shown in Figure 6-4.

Note that in both cases the curve is nearly linear and the data points do not show much deviation from the straight line. Note also in the second plot that the graph may be described by the relation

$$x - \text{observation} = \bar{x} + s \cdot z - \text{percentile} \tag{22}$$

Thus a probability plot can also lead to a determination of estimates of the parameters μ and σ of the normal distribution. Similar plots for other distributions will lead to estimates of their parameters.

References

1. Kolmogorov, A. ***Ann. Math. Statist.***, 1941, **12**, 461-463.

2.Smirnov, N. ***Ann. Math. Statist.***, 1948, **19**, 279-281.

3. Durbin, J., Knott, M., and Taylor, C.C. Components of Cramér-von Mises statistics II. ***J. Roy. Statist. Soc., B***, 1975, **37**, 290-307.

4. Anderson, T.W.; Darling, D.A. Asymptotic theory of certain goodness-of-fit criteria based on stochastic processes, ***Ann. Math. Statist.***, 1952 **23**, 193-212.

5. Anderson, T.W., Darling, D.A. A Test of Goodness-of-fit, ***J. Amer. Statist. Assoc.***, 1954, **49**, 765-769.

6. Stephens, M.A. "Use of the Kolmogorov-Smirnov, Cramér-von Mises and related statistics without extensive tables", ***J. Roy. Statist. Soc.***, 1970, **B32**, 115-122.

7. Stephens, M.A. Tests Based on EDF Statistics in ***Goodness of Fit Techniques*** D'Agostino, R.B., and Stephens, M.A., eds., Marcell–Dekker, New York, 1988, p. 110.

8. See, for example, Hastings, N.A.J., and Peacock, J.B. **Statistical Distributions**, Butterworths, London, 1974, which describes several distributions including the normal, log–normal, Student's *t*, and *F*.

9. Derman, C., Gleser, L., and Olkin, I. **Probability Models and Applications**. Macmillan, New York, 1980.

10. Johnson, N., and Kotz, S., **Distributions in Statistics: Continuous Distributions**. Vols 1 and 2, Houghton Mifflin, Boston, MA, 1970.

7

POINT ESTIMATORS, CONFIDENCE INTERVALS, SIGNIFICANCE LEVELS AND HYPOTHESIS TESTS

Introduction

Although the measures of a central location of a data set and the spread of a data set have been discussed and equations have been given with which to calculate the quantities for the population, the discussion has not been extended to the calculation of these measures for a sample. The measures of central location and spread for a sample provide estimates of the measures for the population. Estimators for populations come in two forms, point and interval. Procedures to estimate the parameters of a population are the aims of this chapter. Estimators of the population parameters must be obtained because the parameters of the population are usually not obtainable. The estimators calculated from samples are called *statistics.* The concepts of *statistics* of a sample from which the population parameters may be estimated will be presented. Rarely, if ever, will the point estimator give the "exact" value of the population parameter. Therefore, a point estimator needs a way to indicate the strength or confidence of the point estimate. This confidence will be indicated by calculating an interval that will contain the population parameter to some degree of confidence.

Statistics

It is difficult, if not impossible in most situations, to calculate population parameters. In Chapter 2, a relation was given between population and samples and the use of statistics and probability to be able to tell something about a sample given a population and something about a population given a sample. The parameters of a population could be inferred from the data contained in the sample. To make the inference the random variables of the sample must be united in specific combinations. The allusion was made that the parameters could be estimated with statistics. However, the term *statistic* was never defined. Random variables can be combined to a set of numbers termed the *statistics* of the sample. For now let us be content to examine the *statistics* that can be used to infer the values of the *parameters* of the population.

Point Estimators

A population that is composed of a number of random variables may be sampled to obtain a subset of the population. Let X denote a general random variable in the population, then a subset or sample of the X_is $\{X_1, X_2, X_3, \ldots, X_n\}$ is drawn from the population. These n random variables comprise the sample. The random variables are then combined to form new random variables through linear combinations:

$$Y = a_1 \cdot X_1 + a_2 \cdot X_2 + a_3 \cdot X_3 + \cdots + a_n \cdot X_n \quad (1)$$

The operator for the expected value of a random variable is a linear commutative operator, therefore the expected value of the linear combination is given by

$$E(Y) = a_1 \cdot E(X_1) + a_2 \cdot E(X_2) + \cdots + a_n \cdot E(X_n) \quad (2)$$

If each of the a_is is equal to 1, then the expected value of the linear combination is equal to

$$E(Y) = n \cdot E(X) \quad (3)$$

dividing both sides by n results in an expression for the mean of Y as being

$$\bar{Y} = \frac{n \cdot E(X)}{n} = E(X) \quad (4)$$

If a value for each of the random variables is measured, the mean of the sample may be calculated:

$$\bar{x} = \frac{\sum_{j=1}^{n} x_i}{n} \tag{5}$$

The quality of a point estimator is determined by whether the resulting statistic that is calculated is an unbiased estimator of the corresponding population parameter. An unbiased estimator of the population parameter is one that will not introduce a measure of personal interpretation in the statistic but will fairly represent the population. Bias can often be introduced in a sample if the researcher uses a method of picking the sample that does not allow each element in the population an opportunity to be selected. The sample obtained under conditions of bias can never represent the population without a degree of bias. The point estimator just given for the expected value of the random variable, *X,* is unbiased. Several other estimators could also be obtained that would also be unbiased, that leads to the second criterion for the goodness of an estimator and that is the best estimator will be one which will give the lowest variance of several estimators.

The criterion of lowest variance derives from the desire to have the error about the estimator (Y-Ŷ) to be as small as possible.

Example

Consider the normal distribution that often describes the noise about the baseline of a chromatogram. If only one value (x_1) for the baseline is available, the value for the baseline must be estimated as that one value. However, if there is more than one value for the baseline, several combinations could be contrived that would fill the criterion of unbiasedness.

A baseline average value (*B*) can be computed using

$$\bar{B} = \sum_{i=1}^{n} a_i \cdot X_i \tag{6}$$

which if all the coefficients, a_i, are equal to 1 over the number of points in the baseline reduces to an average value of the baseline being

$$\bar{B} = \frac{\sum_{i=1}^{n} x_i}{n} \tag{7}$$

There is no constraining reason why the coefficients need be equal, however. If the constraint that the sum of the coefficients is 1,

$$\sum_{i=1}^{n} a_i = 1 \tag{8}$$

then any set of a_i that sums to 1 may be used.

Consider that a baseline is estimated from the initial and final values of the chromatogram. If there were two values, the following combinations could be used,

$$\begin{aligned} baseline &= \frac{X_{initial}}{2} + \frac{X_{final}}{2} \\ baseline &= \frac{3 \cdot X_{initial}}{4} + \frac{X_{final}}{4} \end{aligned} \tag{9}$$

that have as the expected value of the baseline, μ, the following:

$$\begin{aligned} E(baseline) &= E\left(\frac{X_1}{2} + \frac{X_2}{2}\right) = E(X) = \mu \\ E(baseline) &= E\left(\frac{X_1 \cdot 3}{4} + \frac{X_2}{4}\right) = E(X) = \mu \end{aligned} \tag{10}$$

The variances of the two, however, are quite different from the following:

$$V(baseline) = a_1^2 \cdot V(x_{initial}) + a_2^2 \cdot V(x_{final}) \tag{11}$$

where the coefficients a_1 and a_2 are the coefficients in the linear combination:

$$\bar{B} = a_1 \cdot X_{initial} + a_2 \cdot X_{final} \tag{12}$$

then the variances for the two combinations given in Equation (9) are

$$\begin{aligned} V\left(\frac{X_1 \cdot 3}{4} + \frac{X_2}{4}\right) &= \frac{9}{16} V(X_1) + \frac{1}{16} V(X_2) = \frac{10}{16} V(X) \\ V\left(\frac{X_1}{2} + \frac{X_2}{2}\right) &= \frac{V(X_1)}{4} + \frac{V(X_2)}{4} = \frac{V(X)}{2} \end{aligned} \tag{13}$$

The variance of the linear combination that has as coefficients, ¾ and ¼, is larger than the linear combination that has ½ for both coefficients. Thus the minimum variance unbiased estimator is the arithmetic mean of the data values.

Similar calculations could be made for each of the other population distributions discussed in Chapter 5 to find the best estimator with unbiasedness and minimum variance. The results of the calculations indicate that the best unbiased estimator for the expected value of the distribution is the arithmetic mean for all distributions.

The best estimator for the variance of the normal distribution is somewhat different as shown in the following. Let us propose that the unbiased estimator is the variance of the sample as calculated by

$$S^2 = \frac{\sum (X_i - \bar{X})^2}{n-1} \tag{14}$$

for which the expected value is

$$E(S^2) = E\left[\frac{\sum (X_i - \bar{X})^2}{n-1}\right] = \frac{1}{n-1} E\left[\sum (X_i - \bar{X})^2\right] \tag{15}$$

which reduces to

$$E(S^2) = \frac{n}{n-1}\left[E(X^2) - E(\bar{X})^2\right] \tag{16}$$

since $V(X) = E(X^2) - [E(X)^2]$, it follows that $E(X^2) = V(X) - [E(X)^2]$ and

$$V(\bar{X}) = \frac{V(X)}{n} \tag{17}$$

which results in the value for the expected value of the square of X being

$$E(\bar{X}^2) = V(\bar{X}) - \left[E(\bar{X})^2\right] = \frac{V(X)}{n} - \left[E(X)^2\right] \tag{18}$$

which when substituted in equation (16) yields as the expected value of the sample standard deviation, the variance of X. This explains why $n-1$ is used in the denominator of the definition of s^2 rather than simply n, i.e., the best unbiased estimator of the variance of X is the sample variance.

An alternative explanation is often propounded that $n-1$ is used because of the number of degrees of freedom that are available. Recall the Phase Rule, that relates the number of independent phases which may be present in a system to the number of variables, such as temperature and pressure, that can independently be altered. The statistician uses a similar concept to describe the number of degrees of freedom in a statistical sample. The degrees of freedom are related to the number of the random variables in the statistical sample which can be independently varied. For the standard deviation, one of the degrees of freedom is lost because of the use of $\bar{x}$ in the equation. Since $\bar{x}$ is set, only $n-1$ of the random variables are truly independent. The value of the nth variable is set by the other $n-1$ and the value of $\bar{x}$. More will said about degrees of freedom in a later chapter.

Central Limit Theorem

The Central Limit Theorem is a vital concept that helps to explain the general applicability of the equation of the average value of a data set as

$$\bar{x} = \frac{\sum_{j=1}^{n} x_j}{n} \tag{19}$$

Similarly, it was previously shown that the best estimator of the population variance is the sample variance:

$$s^2 = \frac{\sum_{j=1}^{n} (x_j - \bar{x})^2}{n-1} \tag{20}$$

Both of these were derived as being the best estimator without regard to the underlying distribution of the population. As random samples are obtained, the relation between the distribution of the means of the samples, the mean of a sample, the variance of a sample, the variance of the population, and the expected value or mean of the population should be understood.

Consider the random sample to be a set of random variables that has been chosen from the population of all such variables. An example of this would be the concentration of sulfate in atmospheric deposition that is obtained at many different locations on the same day. Since the random sample is a set of random variables, linear combinations of the random variables can be formed such as an expression for the mean of the elements of the random sample:

$$\bar{X} = \frac{\sum_{j=1}^{n} X_j}{n} \tag{21}$$

But just what is the expected value of this linear combination? It has already been shown that the expected value of the mean of the sample is μ. Thus the central limit theorem says that regardless of the underlying distribution of the population, the mean of the sample approximates or estimates the mean of the population. The variance or standard deviation of the means must also be determined, i.e., $V(\bar{X})$. This can be obtained by expanding the expression of X within the variance:

$$V(\bar{X}) = V\left(\frac{\sum_{j=1}^{n} X_J}{n}\right) = \frac{V(X_1 + X_2 + \cdots + X_j)}{n} \tag{22}$$

but since the variance operator is linear and commutative:

$$V\left(\frac{X_1 + X_2 + \cdots + X_n}{n}\right) = \frac{V(X_1) + V(X_2) + \cdots + V(X_n)}{n \cdot n} = \frac{n \cdot \sigma^2}{n^2} = \frac{\sigma^2}{n} \tag{23}$$

where σ^2 is the variance of the population. This is an extremely important formula to remember for it relates the variance of the sample means to the variance of the population.

Consider now what the distribution of the means of the samples is. Consider a set of beakers (N=125,000). 500 samples of size n=1, n=5, n=10, and n=25 are acquired. That is we sample 500 times from the pool of 125,000 pulling 1 out at a time; then pull 500 sets of 5, then 500 sets of 10; then 500 sets of 25. A frequency distribution is generated from the means calculated of each sample. The frequency distributions of the means are shown in Figure 7-1. As the size of the random sample increases, it can be seen that the distribution becomes more peaked. The symmetry of the frequency plots leads one to theorize that the distribution may be normal. As the sample size increases, the goodness of fit to a normal distribution becomes better. This particular sampling distribution is expected to fit the normal distribution very well because the distribution of the underlying population is normal. A linear combination of independent and identically distributed random variables from any population is expected to be distributed the same as the random variables.

Is it possible to show a similar conclusion for a distribution that is not normal? An example that has a uniform distribution would be one that is very dissimilar to a normal distribution. Consider that instead of the mass of the beakers being normally distributed, there is a population that is uniformly distributed between 98 and 102 g. Draw samples of size n=5, 10, and 25. The histogram of the population from which the samples are derived is shown in Figure 7-2A. As can be seen the population *appears* to be somewhat uniform. The histograms that are derived from sample sizes of n=1, 5, 10, and 25 (Figure 7-2) show that the central limit theorem applies to the uniform distribution. As the sample size increases, the distribution of the means looks *normal*.

The proof that the distribution of the means is normal can be accomplished with the tests given in Chapter 6. However, some guidelines

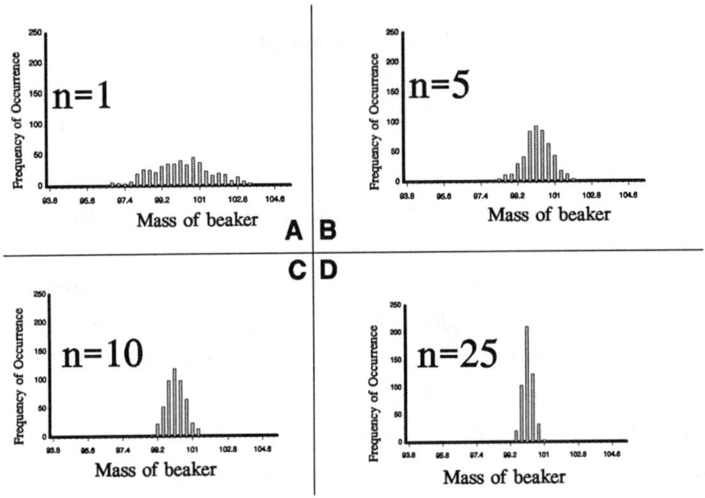

Figure 7-1. Frequency distributions of 500 means from a normally distributed population. **A.** n=1 **B.** n=5 **C.** n=10 **D.** n=25.

should be given that can be used in the approximation that the distribution of means from a population with *ANY* underlying distribution, continuous or discrete, is normal. The main guideline that must be adhered to is the size of the random sample. For this approximation, the size of the random sample should be at least 30. The rigidity of this guideline can be determined only by the sampling situation. If the distribution of the underlying population is known, then a smaller sample size may be used. If the distribution is not known, a larger sample size may be required. The best situation and one that technically does not depend on the central limit theorem is that the distribution of the underlying population is normal. The distribution of the sample means will automatically be normally distributed if the population is normally distributed. The distribution of sample means is a random variable that is a linear combination of random variables. The probability distribution of each sample, if taken from the same population, must be identical to the distribution of the parent distribution. This is true no matter what the underlying distribution of the

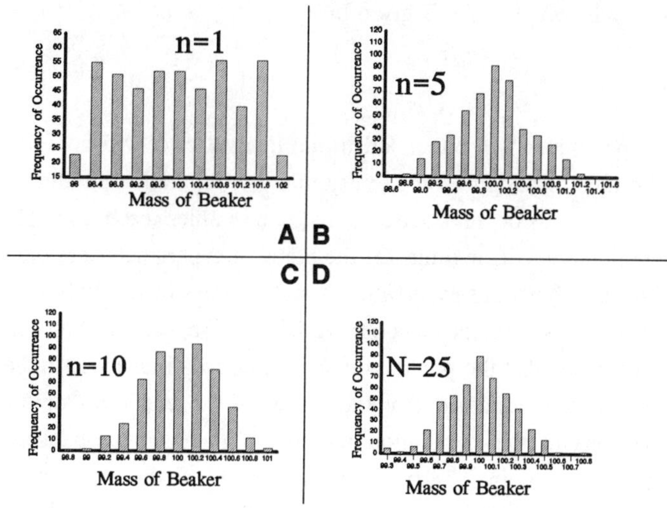

Figure 7-2. Frequency distributions of 500 samples from a uniform population. A. n=1 B. n=5 C. n=10 D. n=25.

parent population is. The distribution of a random variable derived as a linear combination of random variables that are independent and identically distributed is identically distributed as the random variables from which it was derived.

Interval Estimators

Having an estimate of a population parameter is often not sufficient for the needs of the chemical investigator, for something must also be known about the error bars on that point estimate. The error bars will indicate something about the confidence that the true value of the population parameter lies within the limits indicated by the error bars. If the lower and upper limits of the interval are denoted by e_l and e_u and the unknown parameter is symbolized by θ, then the interval denoted by the limits is $e_l < \theta < e_u$. The limits of the interval may be estimated from the values of a random sample $(x_1, x_2, \cdots, x_n)$ of the random variable X by some function or functions of the values of the random sample. If the function that leads to the limits is given by

$$e_l = g_l(x_1, x_2, \cdots, x_n)$$
$$e_u = g_u(x_1, x_2, \cdots, x_n) \qquad (24)$$

then the population parameter will either fall in the interval or it will not. But we would like to be able to assign a probability that the population parameter does lie in the interval. Let us obtain a series of random samples, each a sample of size n from the population. Error limits for the population parameter are then calculated using the same functions each time, then the values of the endpoints of the intervals themselves become random variables. We can then speak rigorously of the probability that the population parameter does indeed lie in the interval. If, in general, it is assumed that the values of the endpoints of the interval do not depend on the population parameter, then the probability, $P\{e_l < \theta < e_u\}$, that the endpoints encompass the value of the population parameter lies between 0 and 1. This probability will be a measure of the chemist's certainty that the statement, $e_l < \theta < e_u$, is true. Since the probability must lie between 0 and 1, this probability is $1\text{-}\alpha$, where this quantity is some desired level of confidence. For example, if the desired level of confidence is 95%, then the probability must be 0.95 and α is 0.05.

Consider a situation in which the true average weight of all the 250 mL beakers in a chemical laboratory must be estimated. The population parameter being estimated is the true average weight of the beaker. The expected value of the true average weight could be determined by weighing all of the beakers, summing the weights, and dividing by the number of beakers. This could be an extremely long and time consuming process. A more efficient manner of estimation would be to sample the beakers, weighing n of the beakers. As indicated above, the best estimator of the population parameter (true average weight of the beakers) is the point estimator, $\bar{x}$. Although knowing an estimate of a population parameter is important, how close the estimator lies to the actual value of the population parameter must also be determined. If the value of the population parameter is known, the value of the parameter could simply be subtracted from the estimator. Since the value of the population parameter, μ, is usually not known, we must rely on the confidence intervals that can be constructed to tell us how close the estimator is to the parameter.

For any situation that might be encountered, the interval that can be constructed will fall in one of five categories:

σ of population known; n any size; $X \sim N(\mu, \sigma^2)$
σ of population unknown; n large; $X \sim N(\mu, \sigma^2)$
σ of population unknown; n large; $X \nsim N(\mu, \sigma^2)$
σ of population unknown; n small; $X \sim N(\mu, \sigma^2)$
σ of population unknown; n small; $X \nsim N(\mu, \sigma^2)$

where $X \sim N(\text{mu}, \sigma^2)$ indicates that the random variable X is normally distributed with a mean of μ and a variance of σ^2.

Let us now construct confidence intervals for each of these situations and also examine the properties of the confidence intervals.

Confidence Interval Construction

Normal Population, μ Unknown, σ Known

Assume that the weights of the beakers are normally distributed and that σ of the population is 1.5 g. The mean of the population is not, however, known. Construct the interval based upon a sample size of 50. To construct the confidence interval consider two statistics, Z_u and Z_l, of the sample:

$$Z_{u,l} = \frac{\bar{X} \pm \mu}{\sigma/\sqrt{n}} \tag{25}$$

the endpoints of the interval can be estimated if the mean of the sample, $\bar{X}$, is estimated. μ is isolated between the two inequalities by multiplying the statistics by $-\sigma/\sqrt{n}$, and adding μ to obtain:

$$\bar{X} - \frac{Z \cdot \sigma}{\sqrt{n}} \leq \mu \leq \bar{X} + \frac{Z \cdot \sigma}{\sqrt{n}} \tag{26}$$

The values of Z are obtained from the table of normal deviates given in Appendix C and are based on the desired level of confidence, α. It should be recognized that if the desired level of confidence is α, then the Z must be obtained from the table using $\alpha/2$ because the confidence is the area under the probability curve between the two confidence limits. Half of the area outside the confidence limits will be contained above the upper limit and half of the area will be contained below the lower limit. The $Z_{\alpha/2}$ value for a confidence level of 90% (α=0.10) is ± 1.645. If the calculated mean of the sample is 98.45, the confidence limits are

$$98.45 - \frac{1.645 \cdot 1.5}{\sqrt{50}} \leq \mu \leq 98.45 + \frac{1.645 \cdot 1.5}{\sqrt{50}} \qquad (27)$$

To learn what the confidence interval really means, construct confidence intervals from 50 different samples of size $n = 5$, 10 and 25 with the assumption that the standard deviation of the population is known and equal to 1.5 g.. To ascertain if the confidence interval is the probability that the mean of the population will lie in the interval, a value for the population mean must be assumed. For this example, let us assume that the mean of the population is 100.00 g, i.e., $\mu=100.00$. The accompanying figure (Figure 7-3) shows fifty 95% confidence intervals for each of $n=5$, 10, and 25. The subfigures show that the width of the confidence intervals is very dependent on the size of the sample. As n increases, the confidence interval decreases dramatically. This is in consonance with the Central Limit Theorem, which indicates that the mean of the population will be approximated more closely by the mean of a sample if the sample size is large. The standard deviation of the means is also affected by the sample size. As the sample size increases, the standard deviation of the mean decreases. Another thing to note from the figures is that all of the confidence intervals do not have the same endpoints, but are quite varied. The width of the confidence interval, however, is constant. The only statistic that affects the placement of the confidence interval is the mean of the sample from which the confidence interval is derived. The solid line at 100.00 represents the true mean of the population. Other aspects about these confidence intervals should be noted. Each of the confidence intervals for a given sample size is the same

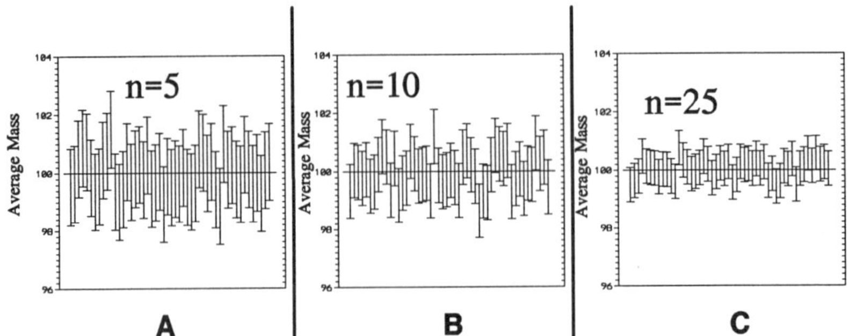

Figure 7-3. Fifty confidence intervals for samples from a normal population where σ=1.5 g. A. n=5 B. n=10 C n=25.

width. It should also be noted that as the sample size increases, the width of the interval decreases as the square root of the sample size. This has practical importance if a sampling program is being planned and the variance of the population is known. The sampling program can be planned so as to be confident to any desired level that the true mean will be encompassed by conducting that sampling program. Even if the variance of the population is not known, an estimate of the variance can be used with confidence to plan the sampling program.

As can be seen, most of the confidence intervals include the mean. In fact, if enough confidence intervals were constructed in this manner, 95% of the confidence intervals would encompass the mean of the population. The confidence interval does not represent the probability that the mean of the population will be contained in the interval, but only that (1-α)·100% of all the possible confidence intervals will include the mean. This is a fine point, but one that should be carefully noted.

Normal Distribution or Unknown Distribution; Unknown μ and σ

This last example indicated how confidence intervals would appear if the standard deviation of the population were known. What if neither the standard deviation nor the underlying distribution of the population were known? Rarely will the standard deviation of the population be known and the population distribution will be normal. However, no matter, the underlying distribution of the data, the best point estimator for the variance of a population, σ^2, is the sample variance, s^2 calculated from

$$s^2 = \frac{\sum_{j=1}^{n} (x_i - \bar{x})^2}{n-1} \tag{28}$$

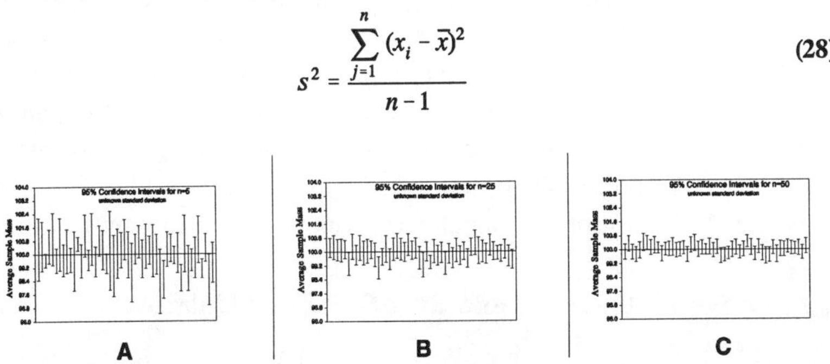

Figure 7-4. 50 confidence intervals for an unknown distribution and σ^2 also unknown. A. n=5; B n=25; C n=50.

The square root of this estimator for the variance can be used in the confidence interval equation given above:

$$\bar{x} - \frac{Z_{\alpha/2} \cdot s}{\sqrt{n}} \leq \mu \leq \bar{x} + \frac{Z_{\alpha/2} \cdot s}{\sqrt{n}} \qquad (29)$$

A major limitation applies to the use of this equation, however; the size of the sample must be greater than 30. This allows us to invoke the conditions of the Central Limit Theorem regardless of the distribution of the underlying population. Ideally, of course, we would like to use this formula if the distribution of the underlying population is normal.

The appearance of the confidence intervals constructed from these conditions, i.e., standard deviation and mean of the population are not known, is quite different from those confidence intervals constructed from the assumption that the population standard deviation is known. The appearance of the confidence intervals will be the same regardless of the underlying distribution if the sample size is large (i.e., $n>30$) and the standard deviation of the population is unknown.

Fifty confidence intervals for $n=5$, 10, and 25 are shown in the accompanying figure (Figure 7-4). Some things should be noted on examination of these confidence intervals. The first is that the width of the "error bars" gets smaller as the sample size increases. In fact the width of the interval decreases as the square root of the sample size. This is not too surprising considering the equation from which they were calculated. This has some practical consequences that we can put to use in the construction of confidence limits. The Central Limit Theorem allows us to estimate the size of the confidence interval about a mean as being dependent on the sample size. As we increase the sample size, one can be more confident that the true mean will lie in a smaller interval about the mean of the sample. This is a very practical application of the Centrail Limit Theorem. If one is constrained to estimate the true expected value of a population to a given interval, the only way to do so with rigor is to increase the sample size. The practicality of this equation is that as the statistical sample size is increased, the cost of the sampling and analysis program grows.

Normal Population; Sample Size Small; σ^2 Unknown

Chemists, though, will not often have the luxury of having a sample size of $n>30$, but will often be able to say that the population That is being sampled is normal. If distribution of the underlying population is normal (or at least near normal), confidence intervals may still be constructed. For this situation, a chemist working in an English brewery, William Gossett, derived a sampling distribution that is similar in graphic appearance to the normal distribution. He published under the pseudonym of Student, and thus the distribution is most often called Student's *t*-distribution. The form of the confidence intervals constructed from Student's *t*-distribution values is similar to the confidence intervals constructed using the *Z*-distribution:

$$\bar{x} - \frac{t \cdot s}{\sqrt{n}} \leq \mu \leq \bar{x} + \frac{t \cdot s}{\sqrt{n}} \qquad \textbf{(30)}$$

These confidence intervals are constructed for those samples (statistical) that have less than 30 elements and the underlying distribution of the population is normal.

Student's *t*-distribution is dependent on only one parameter, the number of degrees of freedom of the sample, ν. As the discussion above indicated, the degrees of freedom are calculated by subtracting one from the number of data points, n-1. Student's t value is found from the Student's t table in Appendix C, by calculating the significance level from the desired confidence level using the relation

$$\text{Confidence level} = 100 \cdot (1 - \alpha) \qquad \textbf{(31)}$$

The significance level, α, must be divided by 2 to enter the table, however, since the confidence level represents the area under the distribution between the two confidence limits. The remainder of the area under the distribution is represented by α. Half of that area will be under the curve that lies to the left of the lower confidence limit and half of the area will be under the curve that lies to the right of the upper confidence limit. If the statistical sample size is 10, ν=9. If the desired confidence level is 90%, α=0.10 and $\alpha/2$=0.05. The table would be entered with these two values, ν=9 and $\alpha/2$=0.05. The value which is obtained is t=1.833.

What types of situations might be applicable to use the *t*-distribution to construct confidence intervals? One example that comes to mind are situations where a sample is analyzed multiple times. The underlying distribution of the values of the concentration of an analyte in a single sample would be expected to be normal. This presupposes that the same sample can be analyzed an infinite number of times, which is generally not possible since most analytical techniques are destructive. If it were possible to analyze the sample an infinite number of times, then the distribution would be normal. A second example would be the values obtained for a given analytical standard from which a calibration curve is constructed.

Example

Analysis of samples of rain is often accomplished by ion chromatography as introduced by Small, et al.[1] The analysis will yield values for the concentration of the common anions and many of the common cations in separate determinations. Before quantitation can be accomplished, however, analytical standards for which the concentration is known must be analyzed.[***] The values for the response of the ion chromatograph for 10 analyses of an analytical reference standard solution of chloride ion in aqueous solution whose concentration is 5 mg/L are 0.675, 0.672, 0.670, 0.679, 0.675, 0.676, 0.677, 0.674, 0.675, and 0.676 cm^2. The average value of the response is 0.6752 and the standard deviation is 0.002821. The Student's t value for a confidence level of 95% is 2.262. Note that Student's t value is obtained for $\alpha=0.025$, not for $\alpha=0.05$. Substitution of these values in Equation (30) gives as the confidence interval

$$\bar{x} \pm \frac{t \cdot s}{\sqrt{n}} = 0.6752 \pm \frac{2.262 \cdot 2.821 \cdot 10^{-3}}{\sqrt{10}} \tag{32}$$

$$0.6730 \le \mu \le 0.6772$$

Unknown Distribution; Unknown σ^2; Small Sample Size

The cases given previously have all been covered except the case where σ is not known, n is small, and the underlying distribution of the population is not known and it is suspected that the distribution is not normal. For the purposes of this text, the confidence intervals for this situation *cannot* be calculated.

[***] Quantitation will be covered in detail in the chapter on calibration curves.

Confidence Intervals for the Standard Deviation

Confidence intervals may also be obtained for the standard deviations that are calculated. Consider the situation where the random experiment is normally distributed but has an unknown expected value (mean) and an unknown variance or standard deviation. Several values of the random variable are obtained by sampling the population and calculating the standard deviation using the equation

$$s = \sqrt{\frac{\sum x_i^2 - \frac{\left(\sum x_i\right)^2}{n}}{n-1}} \tag{33}$$

as the best unbiased estimator for the standard deviation. The confidence interval can be estimated from the probability statement

$$P\left[\frac{\chi^2 \cdot \left(1 - \frac{\alpha}{2}, \nu\right) \cdot \sigma^2}{\nu} \leq S^2 \leq \frac{\chi^2\left(\frac{\alpha}{2}, \nu\right)}{\nu}\right] = 1 - \alpha \tag{34}$$

by rearranging the expression within the brackets to isolate σ^2,

$$\frac{\nu \cdot s^2}{\chi^2\left(\frac{\alpha}{2}, \nu\right)} \leq \sigma^2 \leq \frac{\nu \cdot s^2}{\chi^2\left(1 - \frac{\alpha}{2}, \nu\right)} \tag{35}$$

ν in this equation is the symbol for the number of degrees of freedom and is equal to n-1. This set of inequalities gives the required equation to calculate the confidence interval for a variance. The distribution that describes the probabilities is the χ^2 distribution. Values of χ^2 to use are found in the table in Appendix C. Note that this distribution is not symmetric and that the value of the lower $\chi^2_{\alpha/2,\nu}$ and $\chi^2_{1-\alpha/2,\nu}$ are not equal. Both values must be found from the table.

To find the confidence interval for the standard deviation, take the square root of the expression. One major limitation to the use of this equation to find the confidence limits on a standard deviation is that the underlying distribution of the population must be normal.

Hypothesis Testing

The concepts of estimators to predict values of population parameters, such as the mean and the standard deviation and confidence intervals, have been

The concepts of estimators to predict values of population parameters, such as the mean and the standard deviation and confidence intervals, have been developed for one purpose— to be able to decide whether the point estimator and associated confidence interval are inclusive of a population parameter. To be able to make that decision, we must formulate a **hypothesis** about the value of the population parameter.,e.g., we might hypothesize that the true mean of the population of the mass of the beakers was 100 g, i.e., $\mu = 100.00$ g. Alternatively, we could have hypothesized that $\mu < 100$ g or $\mu \geq 100.00$ g.

As intimated above, in every hypothesis testing situation, there will always be at least two hypotheses to choose from. The objective of sampling and subsequent calculation of point estimators and confidence intervals will be to decide which of the hypotheses is valid. Formulation of hypotheses should be accomplished in terms of the favored outcome. Generally, we will not reject a hypothesis without an overwhelming reason to so do. This favored hypothesis is given the name of *null hypothesis* and is characterized by the symbol, H_o. The hypothesis that will be accepted only with convincing evidence is termed the *alternative hypothesis* and is denoted by the symbol, H_a. The null hypothesis for the mass of beakers might be $H_o{:}\mu = 100.00$ and the alternative hypothesis would be $H_a{:}\mu \neq 100.00$. For this situation, if the estimators calculated from the sampling program strongly indicated that the mean was either significantly less than 100.00 or significantly greater than 100.00 would the null hypothesis be rejected.

Formulation of the null and alternative hypotheses will always depend on the sampling situation. The experimenter must closely scrutinize what is trying to be proved (or disproved) and the hypotheses should be formulated accordingly. In the situation above, the alternative hypothesis was formulated for the sample mean being either significantly less or significantly greater than the population mean. This can be related to the confidence interval that has been calculated about the sample mean (or other population parameter). We indicated that the confidence interval could be compared to the probability that a particular confidence interval includes the population parameter or not.

We might also formulate the alternative hypothesis in terms of a simple inequality, rather than the implied double inequality of the previous example. If the objective of sampling was to ensure that a new manufacturing process would significantly alter the profit margin of a chemical, we would want to adopt that the new process is more efficient. If the profit of the old process is \$1.25/kg, then the new, proposed process must yield a profit of greater than \$1.25/kg. The null hypothesis would be $H_o\!:\mu=\$1.25$ and the alternative hypothesis would be $H_a\!:\mu>\$1.25$.

A new gasoline additive will supposedly significantly increase the mileage of vehicles that use the fuel. The cost of the additive is such that addition will be justified only if the mileage is greatly increased. The average mileage without the additive is 30 miles per gallon (mpg). The null hypothesis would be $H_o\!:\mu=30$ mpg and the alternative hypothesis would be $H_a\!:\mu>30$ mpg.

Null hypotheses need not always be formulated for the mean or expected value. Often the variability of a process is of concern. Here the confidence interval about the standard deviation or variance would be calculated. A sampler to obtain aliquots of a reagent in a robotic analysis scheme is required to have a variability of performance of less than 0.01 cm. The null hypothesis would be $H_o\!:\sigma=0.01$ cm and the alternative hypothesis would be $H_a\!:\sigma<0.01$ cm. Only if the robotic sampler had a variability that was significantly less than 0.01 cm would we accept the alternative hypothesis.

A manufacturing process has a defective rate of 100 blemishes per 50,000 tires manufactured. The analytical chemist feels that addition of a different curing agent at one stage of the manufacturing process will decrease the number of blemishes. The null hypothesis would be $H_o\!:\mu=100$ blemishes and the alternative hypothesis would be $H_a\!:\mu<100$ blemishes.

In each of these situations, data must be taken and point estimators and confidence intervals must be calculated. On the basis of these estimators and intervals, the validity of the null or alternative hypothesis will be decided.

The reader should have noticed some concepts about the formulation of null and alternative hypotheses from the foregoing examples. Foremost among the principles is that the null hypothesis is ***most easily*** formulated as an equality; $H_o\!:\mu=$value; $H_o\!:\sigma=$value. However, the alternative hypothesis is not so constrained and may be formulated as either a simple inequality, $H_a\!:\mu<100.00$ or $H_a\!:\sigma>0.01$ or as an implied double inequality, $H_a\!:\mu\neq100.00$ or $H_a\!:\sigma\neq0.01$. The last inequality implies that $\mu>100.00$ ***OR*** $\mu<100.00$ would be accepted if the evidence from the sampling were convincing enough.

Philosophical Discussion about Significance Tests, Hypotheses, and Confidence Intervals

As the astute reader will have noticed two ways to test the validity of a hypothesis have been presented. Both involve the use of the standard normal distribution. However, it is the application of the standard normal distribution that makes a difference. On the one hand, confidence intervals were calculated by knowing the mean, standard deviation, and size of the statistical sample (*n*) through the equation

$$\bar{x} - \frac{Z_\alpha \cdot s}{\sqrt{n}} \leq \mu \leq \bar{x} + \frac{Z_\alpha \cdot s}{\sqrt{n}} \tag{36}$$

To reject the hypothesis that the mean calculated from another sample represents another population, the new mean must fall outside of the confidence limit. However, it is possible to also determine a Z-value for the new mean compared to a previous mean:

$$Z = \frac{(X - \bar{X})}{s/\sqrt{n}} \tag{37}$$

which must be compared to a table Z value to determine if the null hypothesis should be rejected. These type of calculations are given the name *tests of significance*. The confidence interval is always built using a predetermined significance level, e.g., 5, 10, or 1%. Either of these is strictly an accept or reject situation. The test of significance, on the other hand, gives a definitive probability of the percentage of the time one would expect a mean to exceed the calculated Z-value. The confidence interval approach also gives this information, but with more subtlety. One can immediately compare the mean with the confidence interval to see how close (or how far a result) is from the limits. If the mean exceeds the limits by a small amount, one should be cautious about declaring that the null hypothesis is rejected. However, if the limit is exceeded by a large amount, one should be ready to declare loudly that the limits are exceeded and the null hypothesis is rejected. Another advantage of the confidence interval approach is that the confidence interval is very dependent on the size of the statistical sample. For example, suppose that tests are being conducted to determine if a manufacturing process of a plastic produces batches of the required purity. Assume that only batches that pass the purity test will be used in further manufacturing, then if the number of samples that are taken from the batch is increased, the confidence interval for batch acceptance will be smaller. The results must be looked at carefully to decide which should be used. Through the rest of this book, both methods will generally be presented.

Reference

1. Small, H., Stevens, T.S., and Bauman, W.C. *Anal.Chem.*, 1975, **47(11)**, 1801.

8

TESTS FOR COMPARISON OF MEANS

General

Now that the major distributions and hypothesis testing have been discussed, application of these concepts to problems that are encountered in chemistry needs to be presented. A task often faced is to determine if two statistical samples that have been taken are the same (i.e., whether the two samples come from the same population). Examples such as deciding if the set of lead concentrations in a set of samples as determined by atomic absorption spectrophotometry (AAS) or the lead concentrations in the same set of samples as determined by differential pulsed anodic stripping voltammetry (DP–ASV) are the same; that a contaminant is effectively removed by a treatment system (implying a before and after comparison); or the yields of dibenzo–18–Crown–6 as synthesized by Jones are the same as the yields of dibenzo-18-Crown-6 as synthesized by Young; or that the Cl^- content of a water sample as analyzed by ion chromatography (IC) and ion specific electrode (ISE) yields the same concentration. Data sets are shown in Tables 8-1 though 8-4 for each of these situations and each will be examined in subsequent sections.

The first calculations that will be made are means and standard deviations for each data set. Although not necessary[##], normality will be assumed and the equations for mean and standard deviation of the normal distribution will be used. The means, standard deviations, and number of samples in the data sets are given in Table 8-5.

[##]Recall that the best estimator of the variance is the square of the standard deviation and the best estimate of the expected value of any distribution is the arithmetic mean.

Table 8-1. **Yields from Synthesis of 18-Crown-6 by Two Different People. Yields Are in Percent**

Jones	Young	Jones	Young
98.21	95.55	90.58	98.68
97.91	90.80	90.27	94.80
97.88	91.91	98.34	97.01
91.42	95.26	97.14	94.39
92.96	98.56	94.17	98.97
98.07	95.70	90.53	95.21
98.23	97.67	97.82	90.97
97.51	97.00	98.49	92.50
92.55	93.93	93.54	97.59
97.96	98.97	95.36	93.93
97.87	97.07	97.86	98.85
97.35	94.04	97.56	96.11
96.14	92.56	98.35	93.78
94.80	97.09	97.29	95.83
94.68	90.87	90.64	92.13
95.39	91.70	94.14	94.15
95.41	96.00	91.93	96.82
98.46	94.70	90.47	98.88
94.28	97.22	92.55	97.47
98.17	93.16	98.81	97.25

$\text{Mean}_{\text{Jones}} = 95.53$ $\qquad s_{\text{Jones}} = 2.80$

$\text{Mean}_{\text{Young}} = 95.38$ $\qquad s_{\text{Young}} = 2.43$

Table 8-2. Values Obtained by Multiple Analyses on a Rainwater Sample by Dual Column Suppressed Ion Chromatography and by Ion Specific Electrode

IC	ISE	ΔMethod	IC	ISE	ΔMethod
15.02	18.01	2.99	15.01	18.00	2.99
15.04	18.00	2.96	15.01	18.01	3.00
15.01	18.01	3.00	15.00	18.01	3.01
15.01	18.00	2.99	15.04	18.02	2.98
15.01	18.01	3.00	15.01	18.01	3.00
15.02	18.00	2.98	15.01	18.01	3.00
15.01	18.02	3.01	15.01	18.00	2.99
15.00	18.01	3.01	15.02	18.02	3.00
15.03	18.00	2.97	15.04	18.01	2.97
15.03	18.01	2.98			

$Mean_{IC} = 15.018$ $s_{IC} = 0.0125$
$Mean_{ISE} = 18.009$ $s_{ISE} = 0.0072$

Table 8-3. Analysis for Lead by Atomic Absorption Spectrophotometry (AAS) and Anodic Stripping Voltammetry (DP–ASV). Units mg/L for All Determinations

$[Pb^{2+}]$ AAS	$[Pb^{2+}]$ DP-ASV	Diff	$[Pb^{2+}]$ AAS	$[Pb^{2+}]$ DP-ASV	Diff
18.0	18.1	0.1	17.5	17.4	-0.1
58.1	57.9	-0.2	43.1	43.2	0.1
153.6	153.9	0.3	17.4	17.4	0.0
32.8	33.1	0.3	87.3	87.4	0.1
28.2	28.4	0.2	19.4	19.45	0.05
75.3	76.0	0.7	51.3	52.0	0.7
51.8	52.0	0.2	72.1	72.0	-0.1
19.5	19.4	-0.1	42.1	42.2	0.1
37.6	37.8	0.2	98.4	98.4	0.0
5.3	4.9	-0.4	251.8	251.9	0.1

$Mean_{AAS} = 59.03$ $s_{AAS} = 59.15$
$Mean_{DP\text{-}ASV} = 57.49$ $s_{DP\text{-}ASV} = 57.54$

Table 8-4. -- **Removal of Dicyclopentadiene by carbon adsorption filtration from groundwater samples. DCPD Determined by hexane extraction followed by GC-MS.**

Before	After	Diff	Before	After	Diff
62.27	13.10	49.17	66.97	17.78	49.19
22.02	4.95	17.07	50.60	11.42	39.18
32.64	6.86	25.78	48.38	10.66	37.72
76.01	18.32	57.69	40.99	8.34	32.65
68.98	16.87	52.11	50.88	13.60	37.28
69.36	18.00	51.36	72.88	17.59	55.92
70.56	17.77	52.79	78.60	18.32	60.28
67.64	14.61	53.03	28.84	7.14	21.70
59.28	15.86	43.42	67.17	15.05	52.12
78.96	19.53	59.43	37.37	8.65	28.72
22.86	6.14	16.72	60.37	15.41	44.96
51.53	10.59	40.94	47.37	11.27	36.10
24.16	6.57	17.59			

$\text{Mean}_{before} = 54.27$ $s_{before} = 18.24$
$\text{Mean}_{after} = 12.98$ $s_{after} = 4.61$
$\text{Mean}_{diff} = 41.32$ $s_{diff} = 13.83$

As we examine these numbers, we see some striking revelations. For example, the mean and standard deviation of the lead determinations are nearly

Table 8-5. Summary of mean, standard deviation, and number of samples for four sets of data used to illustrate tests for means.

Data set	mean	s	n
AAS	59.03	59.15	20
DP-ASV	57.49	57.54	20
DCPD Before	54.27	18.24	25
DCPD After	12.98	4.61	25
Jones	95.53	2.80	40
Young	95.38	2.43	40
IC	15.018	.0125	20
ISE	18.009	.0072	20

equal for both techniques. We also see that the standard deviations are quite large, except for the analysis of the chloride by ion chromatography and ion specific electrode. The standard deviations for the synthesis of 18-Crown-6 are also quite low, indicating that the range of the yields is quite small. One of the stated goals of this text was to be able to determine if two data sets are distinctly different or if they represent different statistical samples from the same population. As we recall from our discussion of significance level, the significance level of a statistical test is related to the confidence interval of the test statistic. We must then be able to construct confidence intervals about each of the data sets above. Before we can construct confidence intervals, however, we must first formulate hypotheses to test for significance. How would you formulate hypotheses for each of the data sets?

Let us formulate the hypothesis for the yield data set (Table 8-1). This data set represents the results of 40 different syntheses of a compound by two different individuals, Jones and Young. Only of concern is whether the average yield of either chemist is larger than the other's average yield. The null hypothesis to be tested for this data set is that the average yield of Jones equals the average yield of Young.

$$H_0\text{: } \overline{yield}_{Jones} = \overline{yield}_{Young} \quad (1)$$

The alternative hypotheses are that the yield of Jones is less than or greater than the yield of Young:

$$H_a\text{: } \begin{cases} \overline{yield}_{Jones} > \overline{yield}_{Young} \\ \quad OR \\ \overline{yield}_{Jones} < \overline{yield}_{Young} \end{cases} \quad (2)$$

We must now calculate the critical region, that is, the region for which we can reject the hypothesis of equal yields. The average yield of the synthesis for Jones was 95.53 with a standard deviation of 2.80; the average yield of the synthesis by Young was 95.38 with a standard deviation of 2.43. The Z-transform was given earlier as

$$Z = \frac{\bar{x} - \mu}{\sigma} \quad (3)$$

and it implied that the Z value corresponds to a significance level (α) for the data set compared to a known mean. However, since the true mean, μ, is not known, the true mean must be assumed to lie in a range about the sample mean. Mathematically this is shown by the range of inequalities:

$$\bar{x} - Z \cdot \sigma \leq \mu \leq \bar{x} + Z \cdot \sigma \tag{4}$$

This inequality may be used to calculate confidence intervals using different *Z*-values for several levels of significance. Critical values for several different α values are shown for a one–tailed rejection region, i.e., the alternative hypothesis is that the sample mean is significantly larger than the true mean:

α-Level	Critical Region	Decision
0.10	$z \geq 1.282$	Reject H_0
0.05	$z \geq 1.645$	Reject H_0
0.025	$z \geq 1.96$	Reject H_0
0.01	$z \geq 2.33$	Don't reject H_0
0.005	$z \geq 2.58$	Don't reject H_0

This table also shows the effect on accepting or rejecting a null hypothesis based on a calculated *Z*-value of 2.20. If the calculated *Z*-value is greater than the table value (column labeled *critical region* in the table), the null hypothesis is rejected. As the α value gets smaller, the area under the probability distribution function gets smaller and the probability of a sample mean lying outside the confidence interval decreases.

One would intuitively think that the larger the confidence interval (compare a 95% confidence interval to a 99% confidence interval), the higher the confidence of making a correct decision of not rejecting a true hypothesis. This is actually true, but only because the confidence interval is constructed on the percentage of time that a true mean would be contained in the interval. But, on the other hand, the larger the confidence interval, the farther one could be from the true mean and not reject the null hypothesis. As can be seen, the decision whether to reject or fail to reject the null hypothesis is very dependent on the acceptable significance level. The true mean should lie in the confidence interval that has been calculated. From this calculated confidence interval, it can be inferred that $(1-\alpha/2) \bullet 100\%$ of the time, the true mean for the population will lie in this confidence interval. If the two means do not lie in the same confidence interval, the null hypothesis is rejected. The comparison is made by calculating one confidence interval about one mean. If the other mean does not lie in this confidence interval, the null hypothesis is rejected.

P-Values

Although it is comforting to be able to say that a null hypothesis was rejected at an α significance level, something is lacking. One does not know by how far the hypothesis was rejected. That is, was the alternate mean just inside or a considerable distance into the rejection region? One would like to be able to indicate the distance from the true (or hypothesized) mean that the tested mean is. A significant disadvantage of reporting only that a null hypothesis was rejected at a specific significance level is that statisticians impose their analysis on the managers, thus not allowing the managers to assess the impact of accepting a false hypothesis or failing to accept a true hypothesis. The manager should be given the latitude to determine the effect of the critical region on decisions and not be limited to a specified significance level. The calculation of *P*-value allows the decision maker that latitude.

Alternatively, a *Z*-value can be calculated that compares the two means. To do so will require that one of the means is assumed to be the true mean. From the *Z*-value, an α is obtained from the table of values of the normal cumulative distribution function in Appendix C. This α-value is called the *P*-value and is much more useful than limiting our statistical argument to just a single significance level. Although this sounds simple, there are some subtleties that must be considered when calculating the *P*-value. The method of calculating the *P*-value will depend on the null and alternative hypotheses.

If an upper one-tailed test is considered, the critical region for rejection is that area under the probability distribution function curve to the right of the critical value (as shown in Figure 8-1A). The table above showed that the rejection region is very dependent on the significance level. The *P*-value will then be the area contained under the probability distribution function to the right of the critical value. The *P*-value will then be the significance level, α, corresponding to the calculated *Z*-value. If a lower one-tailed test is considered, the critical region for rejection is that area under the probability distribution function curve to the left of the critical value (as shown by the shaded area in

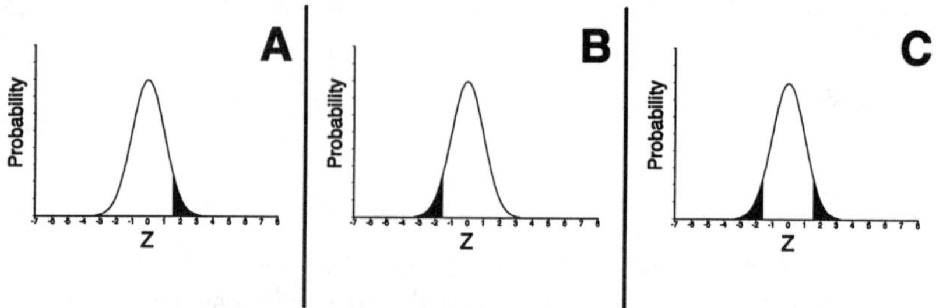

Figure 8-1. Depiction of rejection (critical) regions for P-values. The shaded area is that portion attributable to chance.

Figure 8-1B). The area to the right of the critical region has an area, α, therefore the area under the curve to the left of the critical value will be 1-α. If a two–tailed test is considered, then the area of rejection will be the sum of the areas to the left and to the right of the critical values (as shown by the shaded area in Figure 8-1C). If the areas are equal (which for a normal distribution they will be), then the area to the right of the upper critical value will be $\alpha/2$ and the area to the left of the lower critical value will be 1-$\alpha/2$ and the *P*-value will be α. Z-tables that are in print generally give cumulative probabilities that are denoted by $\Phi(Z)$ and is interpreted as the cumulative area such that $\Phi(Z)=P(Z\leq z)$. Then the *P*-values in terms of the cumulative probabilities are (1) lower tailed, $P=\Phi(Z)$ (2) upper tailed, $P=1-\Phi(Z)$, and (3) for a two–tailed test, $P=2[1-\Phi(Z)]$.

Example

The confidence interval about the mean of Jones is

$$95.53 \pm \frac{1.96 \cdot 2.80}{\sqrt{40}}$$

The 1.96 represents the *Z*-value for a significance level of $\alpha=0.05$ with half of the rejection area greater and half of the rejection area less than the critical value and is found in the table of values of the normal cumulative distribution function in Appendix C as $\alpha/2$ ($P=0.05$). The two critical values are then 96.39 and 94.66. This implies that 95% of the confidence intervals so obtained will contain the true mean. On the other hand, the interval for α=0.01 is 94.39–96.67. (The Z-value to use for this is 2.576.) Since the average yield for Young lies in both of these intervals, we cannot reject the hypothesis that the yields are equal.

The *P*-value can also be calculated based on either the average yield of Jones or the average yield of Young being the true average yield. Let us calculate the *P*-value based on the average yield of Jones being the true average yield.

$$Z = \frac{95.38 - 95.53}{2.80 / \sqrt{(40)}} = -0.339 \quad (5)$$

The standard deviation in the denominator is divided by the square root of *n* to account for the sample size. Φ(Z–0.338)=0.3702, i.e., the area under the *Z* probability distribution function to the left of *Z* of -0.338 is 0.3702. This is $\alpha/2$, therefore the *P*-value is 2•0.3702 or 0.7404. This is a highly significant result which indicates that only about 26% of the time would one expect to have the means separated by a greater distance.

F-Test for Equality of Variance

The other method of calculation is to use the difference of the two means to calculate a Z-value based on the Z-transform given earlier. However, to calculate a Z-value, the population variance must be known, for which we now have only an estimate. We can obtain a better estimate of the population variance if we can show that the variances of the two data sets are homogeneous. To do this, we will use the *F*- est (named so by George Snedecor in honor of R.A. Fisher, both prominent statisticians). *F*-values calculated using the equation for the *F*-distribution have been tabulated in the table of values for the cumulative distribution function for the *F*-distribution in Appendix C. The test statistic is obtained as the ratio of the variances:

$$F_{calc} = \frac{s^2_{Jones}}{s^2_{Young}} = \frac{7.84}{5.90} = 1.3277 \quad (6)$$

ν_n and ν_d are the degrees of freedom are the parameters required to obtain the critical value of *F* from the table of values of the cumulative distribution function for the *F*-distribution in Appendix C. ν_n is the number of degrees of freedom for the variance of Young and ν_d is the number of degrees of freedom for Jones. The hypothesis is that the variances are equal.

$$H_0\colon s^2_{Jones} = s^2_{Young}$$
$$H_a\colon s^2_{Jones} \neq s^2_{Young} \tag{7}$$

F_{calc} for the two yield variances is calculated as 1.328. $F_{critical}$ is obtained from the table for α=0.05, and 39 degrees of freedom for each the numerator and the denominator; the value is 1.925. Since the calculated F is less than $F_{critical}$, we fail to reject the hypothesis of homogeneous variances. The P-value for the test is 0.380. Since the variances are homogeneous, the variances can be pooled or combined. Variances are pooled using the following formula:

$$s_p = \sqrt{\frac{\sum_{i=1}^{2} s_i^2 \cdot \nu_i}{\sum_{i=1}^{2} \nu_i}} \tag{8}$$

The pooled variance is calculated as

$$s_p = \sqrt{\frac{7.84 \cdot 39 + 5.905 \cdot 39}{39 + 39}} = 2.62 \tag{9}$$

The pooled variance is calculated to be 2.62. The Z-value is calculated using the difference of the means:

$$Z_{test} = \frac{\bar{x}_{Jones} - \bar{x}_{Young}}{s_p} = \frac{95.53 - 95.38}{2.62} = 0.0573 \tag{10}$$

which indicates that the mean of Young is only 0.0573 standard deviations from the mean of Jones. The area to the right of the deviate is 0.4792. The area to the left of the negative of the deviate is also 0.4792. The total percentage is 95.84%. This is a highly significant result indicating that there is a confidence interval of only about 4%, which means that about 96% of the time such a variation would be explained by random probability. If the confidence interval was built using this P-value, we would expect that 96% of the time that the means would be separated by a larger difference.

This particular test is really appropriate to use only when the number of samples is greater than about 30. If the number of samples is less than 30, the t-test(s) must be used, which will be done for the three remaining data sets.

If the underlying distribution of the population is approximately normal and the size of the statistical sample is less than 30, the Student's t-distribution may be used to calculate significance tests.

t Tests

Four general formulas are given for t-tests — (1) the mean of the population is known, and the variances are homogeneous, (2) mean of population is not known and the variances are homogeneous, (3) the mean of the population is not known and the variances are not homogeneous, and (4) paired sampling and analysis.

Mean of Population Known; Variances Homogeneous

This test is used when comparing one mean against a standard. The test statistic is given by

$$t_\nu = \frac{(\bar{x} - \mu)}{s/\sqrt{n}} \tag{11}$$

where ν is the number of degrees of freedom, n-1. The test statistic can then be compared to either the t-value from the table of values for the Student's t distribution in Appendix C for a set significance level of α or a P-value can be determined from the same table to give the actual significance level.

Mean of Population Not Known; Variances Homogeneous

Used to compare two means that have been estimated and the variance of the means is the same. The test statistic is given by

$$t_\nu = \frac{(\bar{x}_1 - \bar{x}_2)}{s_p/\sqrt{n}} \tag{12}$$

where s_p is the pooled standard deviation as calculated by Equation (8) and where ν is the number of degrees of freedom, n-1. The test statistic can then be compared to either the t-value from the table of values of the Student's t distribution in Appendix C for a set significance level of α or a P-value can be determined from the same table to give the actual significance level.

Mean of Population Not Known; Variances Not Homogeneous

This test is used when comparing two means that have been estimated and the variances of the two means are not the same. The equality of the two variances has been tested using the F-test. The test statistic is given by

$$t_\nu = \frac{(\bar{x}_1 - \bar{x}_2)}{\sqrt{s_1^2/\nu_1 + s_2^2/\nu_2}} \quad (13)$$

where ν is calculated from

$$\nu = \frac{\left[s_1^2/\nu_1 + s_2^2/\nu_2\right]^2}{\left(s_1^2/\nu_1\right)^2(1/(\nu_1 - 1)) + \left(s_2^2/\nu_2\right)^2(1/(\nu_2 - 1))} - 2 \quad (14)$$

Paired Samples

A t-test for a situation where two similar treatments on a sample need to be compared can be accomplished if the treatments are accomplished on the same sample. This statistical test is primarily used for situations, such as before/after on the same sample or analysis of the same sample by two methods. This statistical test is not particularly appropriate to compare multiple methods. The analysis of variance should be used to determine if differences exist between more than two methods. The test statistic is given by

$$t_\nu = \frac{(\bar{x} - 0)}{s/\sqrt{n}} \quad (15)$$

where n is the number of pairs of samples that have been analyzed.

Example

Consider the data set for analysis of chloride ion by ion chromatography (IC) and by ion–specific electrode (ISE) next. The data are shown in Table 8-2. In this instance, a single analytical sample has been analyzed multiple times by two different techniques, suppressed, dual column IC and ISE. In practice, an analytical chemist probably would not analyze a sample so many times, but for illustrative purposes it is shown. For this example, the null and alternative hypotheses are formulated:

$$H_0\colon \hat{\mu}_{IC} = \hat{\mu}_{ISE} \qquad H_a\colon \hat{\mu}_{IC} \neq \hat{\mu}_{ISE} \tag{16}$$

As with the previous example, first determine whether the variances are homogeneous. The calculated F-statistic is 3.014; $F_{.05,18,18}$ is 2.222 at $\alpha = 0.05$. The P-value is 0.0121, a highly significant result. We must thus reject the hypothesis of homogeneous variance at the 0.05 significance level. More will be said about this result when "precision" of analytical results is discussed. Equations (13) and (14) must be used to calculate whether the two means are equal since the variances are shown to be nonhomogeneous. ν, the degrees of freedom for the critical t-value, is calculated from Equation (14). The number of degrees of freedom is calculated for this example as 15.2, which is rounded to the nearest integer, 15. The t-statistic is calculated to be 927.4, which is considerably larger than t_{critical} of 2.13 for α=0.05 (two-tailed) with 15 degrees of freedom. The P-value is 0.000. However, this result should have been obvious from looking at the data that the ISE method gives much higher results than the IC method. This statistical test gives a quantitative measure to show that a difference exists and that investigation needs to be completed to determine where the problem lies.

Example

Examine now the data set for lead analysis by atomic absorption spectrophotometry (AAS) and by differential pulsed anodic stripping voltammetry (DP-ASV) (Table 8-3). In this instance, ascertain if a difference exists in the analysis of samples using the two different analytical techniques. Twenty different samples were analyzed. Each sample is analyzed by both AAS and by DP-ASV. If no difference exists, then the mean of the difference of results obtained by each technique should be zero. Thus the null hypothesis is formulated as

$$H_0\colon \hat{\mu}_{\text{AAS}} - \hat{\mu}_{\text{DP-}ASV} = 0 \tag{17}$$

and the alternative hypothesis may be formulated as either

$$H_a\colon \begin{cases} \hat{\mu}_{\text{AAS}} - \hat{\mu}_{\text{DP-}ASV} > 0 \\ \hat{\mu}_{\text{AAS}} - \hat{\mu}_{\text{DP-}ASV} < 0 \end{cases} \tag{18}$$

or

$$H_a\colon \hat{\mu}_{\text{AAS}} - \hat{\mu}_{\text{DP-}ASV} \neq 0 \tag{19}$$

The mean of the differences is 0.116 and the standard deviation of the differences is 0.256. The t-statistic is calculated from the t-test for a known mean:

$$t_v = \frac{(\bar{x} - \mu)}{s/\sqrt{n}} = \frac{(0.116 - 0)}{0.256/\sqrt{20}} = 1.975 \qquad \textbf{(20)}$$

to be 1.975. $t_{critical}$ for α=0.05 (two-tail) and 19 degrees of freedom is 2.09. The P-value is 0.067. At a significance level of 5%, the hypothesis of equal means cannot be rejected. The significance level of the test is 6.7% from the P-value. Thus for this example, the two analytical techniques give the same results. The confidence interval for the mean of the differences is calculated from the range inequality:

$$\bar{x} - \frac{t_v \cdot s}{\sqrt{n}} \leq \mu \leq \bar{x} + \frac{t_v \cdot s}{\sqrt{n}} \qquad \textbf{(21)}$$

where, as indicated, the true mean lies between the addition and subtraction of $t \cdot s/\sqrt{n}$ from $\bar{x}$. If the intent of the experiment was to demonstrate comparability of two analytical techniques, the true mean of the difference should be zero. The confidence interval is calculated as -0.004 $\leq \mu \leq$ 0.236. So, although the test indicates that there is no difference between the two techniques, the difference borders on being rejected at the α = 0.05 significance level.

Example

The last set of data (Table 8-4), removal of dicyclopentadiene (DCPD) by carbon adsorption filtration, could also be considered a paired test since the results are obtained on the same sample of water before and after treatment. The question whether the treatment is effective or not is formulated by the hypothesis

$$H_0\text{: } \mu_{before} = \mu_{after}$$
$$H_a\text{: } \mu_{before} > \mu_{after} \qquad \textbf{(22)}$$

This test will then be a one-tailed test since we want to know only if the DCPD has effectively been removed.

To decide which of the three formulas to use, we must first calculate the F-ratio to ascertain whether variances are homogeneous. The F-ratio is 7.93 and $F_{0.05,24,24}$ = 2.27 for α=0.05 (two-tailed) with 24,24 degrees of freedom. The P-value for the test is 0.000, which would indicate that the results could not have occurred strictly by chance. Therefore, the hypothesis of equal variances is rejected. The equations for unequal variances must be used to first calculate the number of degrees of freedom to be 29. A critical region may be constructed by rearranging the equation to isolate one of the means in an inequality

$$\bar{x}_1 - t_\nu \sqrt{\left(\frac{s_1^2}{\nu_1}\right) + \left(\frac{s_2^2}{\nu_2}\right)} \leq \bar{x}_2 \leq \bar{x}_1 + t_\nu \sqrt{\left(\frac{s_1^2}{\nu_1}\right) + \left(\frac{s_2^2}{\nu_2}\right)} \tag{23}$$

where the subscripts 1 and 2 refer to before and after, respectively. In this inequality, the t value that is used (1.70) is obtained from the table of values of the Student's t distribution in Appendix C. The number of degrees of freedom has been calculated to be 29. However, we are really interested only in the lower portion of that inequality. The critical region is then calculated as that portion of the inequality < 47.87. The after average is definitely less than 47.87. Thus we can conclude that the treatment method is highly effective.

As with the other methods of calculation a t-value can also be calculated to compare to a critical t. The calculated t is 10.97 and the critical t value for 29 degrees of freedom is 1.70. The P-value is 0.000 which indicates that there is an extremely low probability that the results occurred by chance. The null hypothesis that the $\text{mean}_{\text{before}}$ is less than the $\text{mean}_{\text{after}}$ can be rejected.

Type I and Type II Errors

All of the significance tests that have been explained so far have been for the purpose of estimating what the probability is for rejecting a hypothesis when it is actually true. These are called Type I errors. The probability of committing Type I errors is the significance level, α. However, concern also needs to be taken with regard to Type II errors. These errors are of the nature that a false hypothesis is accepted when it should have been rejected. Table 8-6 shows the relation between the decision and the consequence of accepting that decision. As an example, a confidence interval can be calculated about any mean. If a second mean (whose variance is equal to the first) falls in that confidence interval, the two means would be accepted as being equal, or more correctly statistically indistinguishable. This is based on one of the means being the true mean of the population. However, if the true mean were in actuality different than the two sample means, the hypothesis of equality should possibly have been rejected. The probability that we would be able to correctly reject a false hypothesis based on a set of assumed true means can be calculated. The probabilities thus

Table 8-6. Decision Table for Determination of Type of Error

ACTUAL SITUATION	DECISION	
	Accept	Reject
H_0 is True	Correct probability$=1-\alpha$	**Type I Error** probability$=\alpha$
H_0 is False	**Type II Error** probability$=\beta$	Correct probability$=1-\beta$

calculated are known as the power of the significance test. The usual designation of the significance level of this test is β and each β that is calculated is dependent on a unique assumed mean.

The rejection region for the acceptance of a false hypothesis is related to the acceptance region of the calculated mean. The acceptance region for a calculated mean, μ_c, is shown in Figure 8-2A. The sampling distribution is centered on μ_c. If the true sampling distribution is, however, centered on μ_0, as shown in Figure 8-2B, then the shaded area in Figure 8-2B is the rejection region for the false hypothesis based on a mean of μ_0. The sum of the two shaded areas is β.

Example

The calculation of β is best explained by an example. Use as the data set to calculate the α and β limits, six retention times from the analysis of pyrene

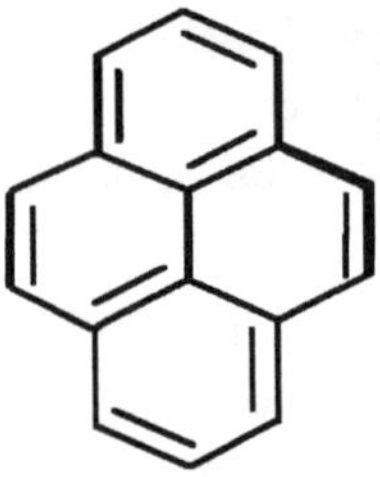

PYRENE

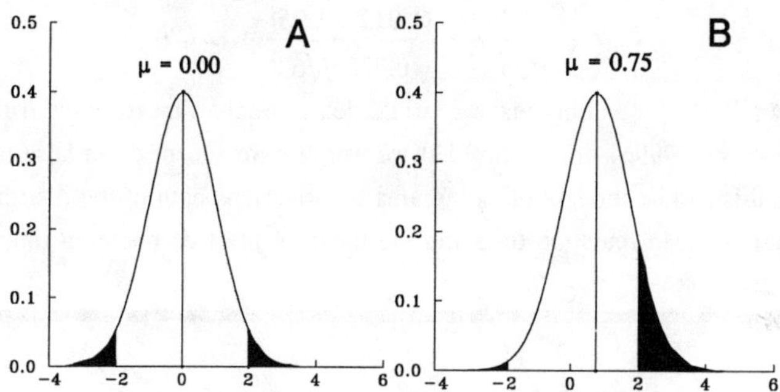

Figure 8-2. Relation between α and β significance levels.

using a chromatographic technique:

1.87, 1.85, 1.89, 1.91, 1.93, 1.86

The average retention time (*rt*) is 1.885 and the standard deviation is 0.0308. The confidence limits for $\alpha = 0.05$, are $1.853 \leq rt_{true} \leq 1.917$, which means that we would expect the true retention time to be within these limits 95% of the time. If the mean retention time at a future date was outside of these limits, the hypothesis that the two sets of retention times are equal at the 5% significance level would be rejected and the material giving rise to the chromatographic peak would not be identified as pyrene. However, unbeknownst to us when we performed the initial analysis, one of the parameters was incorrectly set, and the true retention time should have been 1.95. What is the probability, that the false mean (1.885) for this true retention time (1.95) would have been accepted? First calculate the confidence limits based on the experimental mean retention time (μ_c=1.885) and the confidence limits using the α significance level ($1.853 \leq rt_{true} \leq 1.917$). Use these two limits in the *Z*-transform equation to calculate the *Z*-values based upon a true mean of 1.95.

The two *Z*-values corresponding to the confidence interval limits for the true mean are -2.624 and -7.714. The areas under the normal cumulative distribution function curve associated with these are 0.00435 and 0. β is the sum of the probabilities, 0.00435. Thus

$$Z_L = \frac{(L_L - \mu_0)}{s/\sqrt{n}} = \frac{(1.853 - 1.95)}{0.0308/\sqrt{6}} = -7.714$$

$$Z_U = \frac{(L_U - \mu_0)}{s/\sqrt{n}} = \frac{(1.917 - 1.95)}{0.0308/\sqrt{6}} = -2.624 \quad (24)$$

there is only a 0.4% probability that we would detect a departure from the true retention time or a $\{1\text{-}\beta = 0.9996\}$ probability that we would have accepted the false hypothesis. Graphically, this can be thought of as the area that overlaps both of the distributions for the experimental mean retention time and the mean of the true retention time as shown in Figure 8-2.

Often it is not sufficient to just test one "true" retention time (or mean) but must often be required to test many different means. This gives rise to a set of βs that taken together constitutes the operating characteristic or power curve for a significance test. Figure 8-3 shows the result of calculating the set of $(1-\beta)$s for the range $1.80 \leq$ true retention time ≤ 2.00, in 0.2 increments. Note the shape of the curves. The β curve shows that the further the true mean is from the calculated mean, the greater the probability that a false hypothesis will

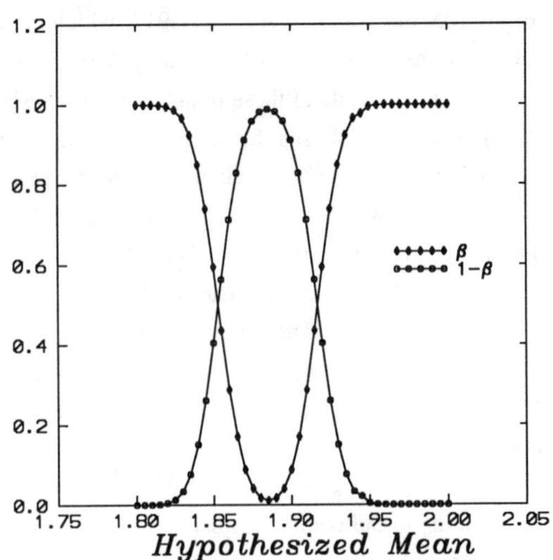

Figure 8-3. Operating characteristic or power curve for the true (hypothesized) mean.

from the calculated mean, the greater the probability that a false hypothesis will be accepted. The closer the true mean is to the calculated mean, the greater the probability that a false hypothesis will be rejected.

In a sense, this testing for both Type I and Type II errors is the basis of risk analysis. One must be able to determine how much risk is acceptable when each decision is made. These are balanced for cost, time, and other resources.

9

TRANSFORMATION OF INSTRUMENT DATA

Smoothing of Instrument Data

General

When data are obtained from an instrument there is noise associated with that signal from the conversion of the analog signal to a digital signal. The noise is partially because of the resolution of the analog to digital converter (ADC) and partially because of the instability of the signal. Figure 9-1 shows a typical baseline that might be acquired from a chromatograph. The figure shows the noise which would appear. Quite often we will want to appear to enhance a signal of interest by averaging the data from the instrument. Realize, of course, that the enhancement is only an appearance since both the signal of interest and

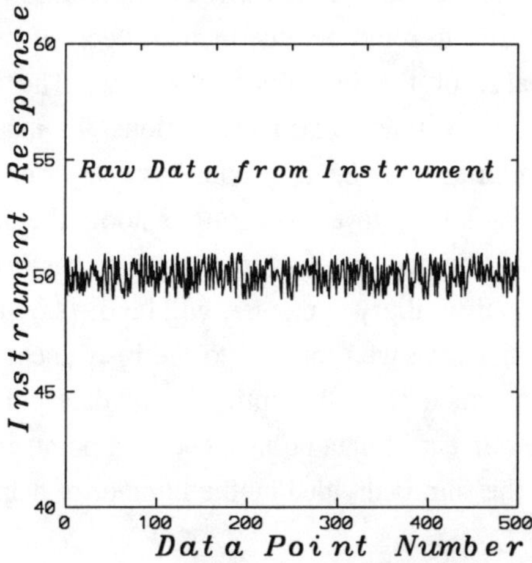

Figure 9-1. Raw data acquired from a chromatograph showing the baseline of the chromatogram.

the noise in the signal are averaged. Both are degraded, but the relative degree of degradation of the noise is significantly larger than the degradation of the signal. Several schemes are available to perform this type of averaging fall under the general rubric of *DIGITAL FILTERING.* Although the general field of digital filtering falls under the aegis of electrical engineering (it is generally concerned with the averaging of frequency signals), the principles have many applications in chemistry. A comparison with which most of us are familiar is the resistor capacitor circuit to filter high-frequency noise from an analog signal. This same type of filter can be applied to a set of digital data and can also be used to filter out high-frequency noise. Consider the set of data digitally acquired from a chromatograph shown in Figure 9-1. The data were acquired at an acquisition rate of 10 Hz*. This data is a portion of a baseline of a chromatogram. Several different schemes of smoothing of this data will be presented.

Digital Signal Processing

Box Car Smooth

When modern instruments are designed, provisions are generally made for automatic acquisition of data in a digital format. In addition, the manufacturer will generally also desire to remove some of the effects of random fluctuations about the central value of the instrument response. The fluctuations occur because of things such as line voltage vacillations. In many cases, the first method of data smoothing will be a simple box car average. Subsequently, additional methods may be employed to further smooth the data. Depending on the type of instrument, the data from which the noise has been minimized will then be transmitted to an auxiliary computer, will be displayed on the instrument panel, or will be further massaged internal to the instrument.

With box car smoothing, the entire set of data points is divided into windows of the same number of data points. The data points in each window are added together, then the sum is divided by the number of data points in the data

* A Hertz (Hz) is one data point/second. Thus 10 Hz would be 10 data points/sec.

window. The result of this division becomes the value for the central data point in that data window. The window is then moved to the right of the last data point in that data set and the calculation repeated. Note that the windows do not overlap. Each raw data point is used only once in the smoothing. By so doing, all of the data points in a window are replaced by a single value. Figure 9-2 shows the result of applying a 7–point box car smooth to the data of Figure 9-1. The inset is an expanded portion of the chromatogram showing several data points and the windows that are used to smooth the data points. This has the disadvantage of reducing the number of data points in the data set, but it is easy to implement averaging in an instrument.

The number of data points that result from the averaging depends on the width of the window and the total number of data points in the data set. If there are n data points and the window is $2m+1$, then the number of data points that are lost is

$$\text{number lost} = n - \frac{n}{2 \cdot m + 1} \cdot 2 \cdot m \tag{1}$$

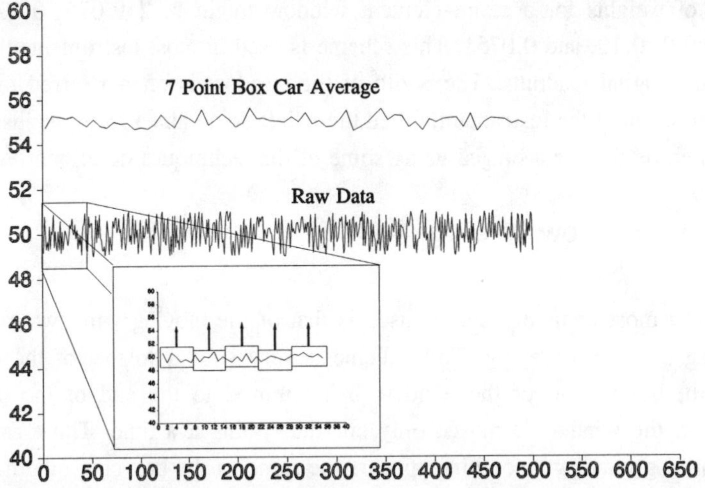

Figure 9-2. Box car smooth; note that the windows do not overlap as shown in the insert.

where the ratio in the equation is the number of windows that can be formed from the number of data points.

The general formula for the box car smooth is

$$Y_j^* = \frac{\sum_{k=-m}^{m} Y_{j+k}}{2m+1} \tag{2}$$

where Y_i^* is the averaged data point, $2m+1$ is the number of data points in the window, and Y_j are the data points used in the average. Equation (2) is for the unweighted case. Each point in the window will receive equal weight in the average. For a box car smooth, this is probably the best, but not the only alternative. For a weighted box car average, the following equation would be used:

$$Y_j^* = \frac{\sum_{k=-m}^{m} C_k \cdot Y_{j+k}}{2m+1} \tag{3}$$

C_k are the weighting constants and can be used to vary the amount of emphasis each point has. If it is felt, for example, that the data points at the edges of the window should receive less weight than the ones in the center of the window, a set of weights for a seven–element window might be { 0.075, 0.10, 0.20, 0.25, 0.2, 0.10, and 0.075}. This scheme is used in most instruments that give periodic digital readouts. The width of the window is often referred to as the time constant of the instrument. Often these data so obtained from the instrument will then be further averaged using some of the techniques described below.

Moving Window Average

The next most common scheme used is that of the moving window smooth or moving box car averaging. This scheme is very similar to that of the box car smooth, but instead of the window being moved to the end of the previous window, the window is moved only one data point at a time. The formula for the moving window smoothing appears the same as the box car smooth.

$$Y_j^* = \frac{\sum_{k=-m}^{m} Y_{j+k}}{2m+1} \tag{4}$$

where j is incremented in steps of 1 rather than in steps of m as in the simple box car averager. The moving window average is depicted in Figure 9-3. Note that in this method that the windows overlap from window to the next. Each data point is used m in the averaging process.

This scheme has three main disadvantages. The first of the disadvantages is that the first transformed data point is always centered on the $(m+1)th$ untransformed data point. And the last transformed data point will be centered on the $[n-(m+1)]th$ data point. The data points before and after the first and last transformed data points are "lost." There are ways to handle this apparent problem of losing the endpoints of the data set. This disadvantage can be overcome by starting data acquisition many data points from the region of interest. The second disadvantage is that the averaging technique always gives a lower amplitude to the signal than the original signal. There is no good way that has been reported in the literature to overcome this disadvantage. One way to minimize the diminution of signal that has been proposed is to use weighting

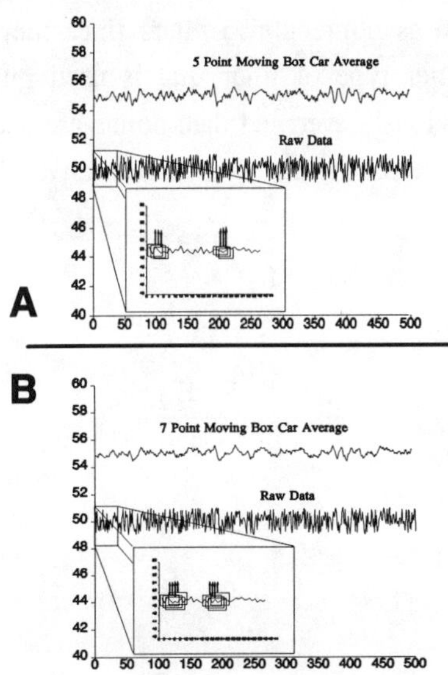

Figure 9–3. 5 and 7 point moving box car. The windows overlap as shown in the insets. A. 5 point smooth B. 7 point smooth

schemes that instead of simple arithmetic averages being obtained, greater weight is given to the data points nearer to the centroid of the window. This is given by the equation

$$Y_j^* = \frac{\sum_{k=-m}^{m} C_k \cdot Y_{j+k}}{2m+1} \tag{5}$$

The last disadvantage for this filter is that a phase shift is induced in the transformed data. The shift is always to the right. This can be a disadvantage when the analysis relies on the location of, for example, a chromatographic peak or an X-ray peak. The phase shift can be overcome by applying the filter a second time from the right to the left instead of the left to the right, i.e., start with the high end of the data series and go to the low end. This induces a phase shift to the left, which has a further disadvantage that the intensity is further diminished.

Recursive Filters

These schemes are both known as nonrecursive filters since they involve only the original data points. Another type of filter that is used quite often is a recursive filter wherein the previously averaged data points are used in the next average. The equation for the general moving box car recursive filter is

$$Y_j^* = \sum_{k=-m}^{0} \frac{Y_{j+k}^*}{m+1} + \sum_{k=1}^{m} \frac{Y_{j+k}}{m} \tag{6}$$

for the unweighted situation and

$$Y_j^* = \sum_{k=-m}^{0} \frac{c_k \cdot Y_{j+k}^*}{m+1} + \sum_{i=1}^{m} \frac{b_k \cdot Y_{i+k}}{m} \tag{7}$$

for the weighted average.

Convolution

As can be seen all of the schemes presented so far have in general used the formula

$$Y_j^* = \frac{\sum_{k=-m}^{m} C_k \cdot Y_{j+k}}{2m+1} \tag{8}$$

This general formula can be thought of as being the basis to formulate other similar smoothing functions. The general scheme is called convolution. The coefficients C_k are called the convoluting integers and N is a normalizing factor that depends on the type of convoluting function being used and will produce different results. For example, the convoluting function for the simple case of a moving window average is a rectangular function. A triangular weighting function would weight the data points at each end of the window as near zero whereas those at the middle of the window would be weighted about 1. The proper choice of convoluting integers can produce derivatives of the original data in addition to data smoothing. This technique, which was popularized by Savitzky and Golay[1] in the mid-1960s, is still very popular and is probably one of the most quoted papers dealing with data transformation in the literature. Unfortunately they made several errors in calculation of the original sets of convoluting integers. The errors were reported by Steinier, et al.[2] in 1972. The convoluting integers which were proposed by Savitzky and modified by Steinier, *et.al.* were calculated using the general scheme of smoothing using a simplified least-square procedure. The sets of convoluting integers for smoothing using quadratic and higher polynomials are given in the two referenced papers by Savitzky-Golay[1] and Steinier et al.[2] The integers for *n*-point smooth, and the first through fourth derivatives are given in the table of Savitzky-Golay convoluting integers in Appendix C.

Example

Acquisition of digital data from a chromatograph or other similar experiment inherently has noise associated with it as previously discussed. A chromatogram showing resolution of three components is shown in Figure 9-4. The data were smoothed with a seven-point quadratic fit. The upper graph in Figure 9-4 is offset from the raw chromatogram and shows the smoothed chromatogram. The convoluting integers (-2, 3, 6, 7, 6, 3 and -2) and the normalizing integer (21) that were used were obtained from the table of convoluting integers found in Appendix C. The first 25 data points in the data stream are shown in Table 9-1. This table shows the data point number, the raw data, and the

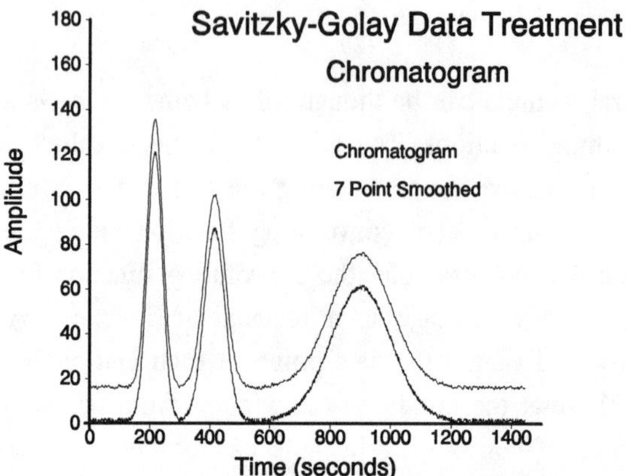

Figure 9-4. **Chromatographic peaks smoothed with the Savitzky–Golay algorithm. The abscissa values of the smoothed chromatogram are offset from the original data.**

smoothed data. Note that the smoothed values are offset from the original data values.

Least–Squares Polynomial Smoothing

All of the techniques that have been presented are special cases of the general technique called least squares polynomial smoothing. In this technique, the subset of the data to be smoothed is fit to a general polynomial (or other function) using least–squares criteria. The subset is usually the same size as one of the windows, which we have discussed above, e.g., 2m+1 data units. In the most rigid form of the technique, the abscissa values do not have to be evenly spaced, but in the more useful of the techniques, the abscissa values are all evenly spaced. This simplifies the application of the smoothing function since for an odd number of values in the window, the abscissa drops out of the final result, and becomes a function of only the ordinate values.

If the general function to be fit to the window is f_i, then the criteria for a least–squares fit is that the following is minimized with respect to each of the fitting parameters by setting the derivative of the sum of the square of the residuals with respect to each parameter equal to zero:

Table 9-1. Chromatographic data for digitally acquired data. 15 has been added to the data smoothed by 7 point Savitzky-Golay quadratic.

Point Number	Raw Data	Smooth Data	Point Number	Raw Data	Smooth Data	Point Number	Raw Data	Smooth Data
1	1.819		10	2.024	16.437	18	0.162	16.302
2	2.146		11	1.614	16.377	19	2.240	16.467
3	1.905		12	0.735	16.740	20	1.916	16.671
4	1.878	16.634	13	2.026	16.595	21	1.046	16.746
5	0.486	16.068	14	2.170	16.434	22	1.783	16.107
6	1.443	15.886	15	0.797	16.482	23	0.476	15.745
7	0.160	15.667	16	0.644	15.786	24	0.331	15.532
8	1.407	16.013	17	1.460	15.762	25	0.314	15.033
9	0.712	16.263						

$$\frac{\partial(x_i - \hat{x}_i)^2}{\partial\beta_j} \qquad (9)$$

This leads to a set of normal equations, which can be solved to obtain the fitting parameters. For example if $R = \beta_0 + \beta_1 * x^j$ is the equation with which the data are to be smoothed, then the normal equations become

$$b_0 \sum_{i=-k}^{+k} 1 + b_1 \sum_{i=-k}^{+k} = \sum_{i=-k}^{+k} R_i$$
$$b_0 \sum_{i=-k}^{+k} x_i + b_1 \sum_{i=-k}^{+k} x_i^2 = \sum_{i=-k}^{+k} R_i \cdot x_i \qquad (10)$$

where b_o and b_1 are estimates of β_o and β_1, respectively. If the xs are assumed to be equally spaced with a spacing of h, then the equations are greatly simplified. By obtaining instrument response values that are evenly spaced with respect to the independent variable (usually time), the effects of the individual variable are eliminated from consideration as will be shown later. The value of x_i is then $i{\bullet}h$. Note that the subscripts refer to the x values in the window and not the subscript for the x values in the total data stream*. For a window of size

*To be totally rigorous, the x values should have a second subscript to indicate the position in the data stream, $x_{i,j}$ where j is the subscript that denotes the position in the data stream. For ease of working with the equations, only the subscript for the position in the window will be used.

three (with instrument response values of R_{-1}, R_o, and R_{+1}), the sums needed for the normal equations become

$$\sum x_i = -h + 0 \cdot h + h = 0$$
$$\sum x_i^2 = h^2 + 0 + h^2 = 2 \cdot h^2$$
$$\sum R_i = R_{-1} + R_0 + R_1 \tag{11}$$
$$\sum R_i \cdot x_i = -h \cdot R_{-1} + 0 \cdot R_0 + h \cdot R_1 = h \cdot (R_1 - R_{-1})$$

which lead to the expressions for the coefficients, b_0 and b_1 of the normal equations:

$$b_0 = \frac{R_{-1} + R_0 + R_{+1}}{3}$$
$$\tag{12}$$
$$b_1 = \frac{-R_{-1} \cdot h + 0 + R_{+1} \cdot h}{2 \cdot h^2} = \frac{-R_{-1} + R_{+1}}{2 \cdot h}$$

Smoothed values for all data points in the window may now be calculated by substituting the values for b_o and b_1 into the model equation above:

$$R_i^* = b_0 + b_1 \cdot x_i$$
$$\tag{13}$$
$$R_i^* = \left[\frac{R_{-1} + R_0 + R{-1}}{3}\right] + \left[\frac{R_{+1} - R_{-1}}{2 \cdot h}\right] \cdot i \cdot h$$

For a window that has three data values, the expressions for the three smoothed values in the window are

$$R_{-1}^* = \left[\frac{5 \cdot R_{-1} + 2 \cdot R_0 - R_{+1}}{6}\right] \tag{14}$$

$$R_0^* = \left[\frac{R_{-1} + R_0 + R_{+1}}{3}\right] \tag{15}$$

$$R_{+1}^* = \left[\frac{-R_{-1} + 2 \cdot R_0 + 5 \cdot R_{+1}}{6}\right] \tag{16}$$

Example

Let us suppose that the first seven data values of a baseline of a chromatogram are {28.93, 29.35, 29.80, 27.35, 28.47, 28.91, and 28.00} and that they were acquired at evenly spaced intervals. Calculate the first five smoothed data values for this data set. Assume that a five-point smooth is desired. Since the data are taken at the beginning of the chromatogram, first assume that a first-order regression will be used to calculate the smoothed data values. The third smoothed data point may be calculated as the sum of the first five data points divided by 5:

$$Y_3^* = R_0^* = \frac{28.93 + 29.35 + 29.80 + 27.35 + 28.47}{5} = 28.78 \tag{17}$$

This is, of course, the same as the R_0^* noted above, but it is actually the third smoothed data point, Y_3^* in the data stream. The first two smoothed data points must now be calculated using the normal equations for a five-point smooth:

$$\begin{aligned} b_0 \sum_{i=-2}^{+2} + b_1 \sum_{i=-2}^{+2} x_i &= \sum_{i=-2}^{+2} R_i \\ b_0 \sum_{i=-2}^{+2} x_i + b_1 \sum_{i=-2}^{+2} x_i^2 &= \sum_{i=-2}^{+2} R_i \cdot x_i \end{aligned} \tag{18}$$

The general equation for the smoothed instrument response will be

$$\begin{aligned} R_i^* &= b_0 + b_1 \cdot x_i \\ &= \left[\frac{R_{-2} + R_{-1} + R_0 + R_{+1} + R_{+2}}{5}\right] + \left[\frac{x_i \cdot (-2 \cdot R_{-2} - R_{-1} + R_{+1} + 2 \cdot R_{+2})}{10 \cdot h}\right] \end{aligned} \tag{19}$$

For any given window, the initial term in Equation (19) will be the same for all smoothed values in that window. For this window, the value will be

$$b_0 = \left[\frac{R_{-2} + R_{-1} + R_0 + R_1 + R_2}{5}\right] = 28.78 \tag{20}$$

from which the values for the first two smoothed data points (Y_1^* and Y_2^* or R_{-2}^* and R_{-1}^*) may now be calculated to be

$$Y_1^* = R_{-2}^* = 28.78 + \left[\frac{-2 \cdot h(-2 \cdot (28.93) - 29.35 + 27.35 + 2 \cdot 28.47)}{10 \cdot h}\right] = 29.36 \tag{21}$$

and

$$Y_2^* = R_{-1}^* = 28.78 + \left[\frac{-2 \cdot h(-1 \cdot (28.93) - 29.35 + 27.35 + 2 \cdot 28.47)}{10 \cdot h} \right] = 29.07 \quad \textbf{(22)}$$

The fourth and fifth smoothed data points are calculated from the sums of the second to the sixth data points and the third to seventh data points, respectively; each being divided by 5. The results are $Y_4^*=28.78$ and $Y_5^*=28.51$.

This method of calculation has the advantage of being able to calculate smoothed values for those end points that were incalculable using just the moving window technique explained earlier. In practice then, the data points in the first window are used to calculate not only the centroid of the first $2m+1$ data values but also to calculate the first $m+1$ smoothed data values. The remaining m values in this window are calculated as the centroids of the next m windows. The same procedure is applied to the last window where the last $m+1$ smoothed values are calculated from the solution to the normal smoothing equations.

It is not necessary to only limit the smoothing equation to the linear first order model; other models such as quadratic or even higher polynomials may also be appropriate. It is conceivable that even such smoothing functions as exponential or logarithmic may be appropriate. The important thing to recall is that the abscissa values must be evenly spaced and that the number of abscissa values in the window should be odd. The requirement of evenly spaced data points is handled quite nicely using most of the analog to digital converter boards on the market today since data acquisition from them is in evenly spaced intervals, if the acquisition software is written correctly.

An additional requirement that has some applicability is to be careful about the sampling frequency, i.e., the interval between sampling points. Too slow a sampling frequency can lead to a condition known as aliasing.[3,4] The rule for a sampling frequency to be valid is that it should be at least twice the frequency of the fastest frequency in the system. This is known as the Nyquist condition.

Smoothing does not introduce any new information into a data set, but only reduces noise. By reduction of noise some signal that was previously obscured by the noise may now be visible. Occasionally, a single outlier in the data stream may introduce false information. The data stream should always be carefully screened to ensure that there are no such outliers in the data set.

References

1. Savitzky, A., and Golay, M.J.E. *Anal. Chem.*, **1964**, **36(8)**, 1627-1639.

2. Steinier, J., Termonia, Y., and Deltour, J. *Anal. Chem.*, **1972**, **44(11)**, 1906-1909.

3. Walraven, R. ***Proceedings of the Digital Equipment Computer User's Society Symposium***, Los Angeles, California, 1982, p. 186.

4. Walraven, R., and Graham, R.C. ***22nd Northeast Regional Meeting of the American Chemical Society***, New Paltz, New York, 1986.

10
DESIGN OF EXPERIMENTS

Introduction

Before design of experiments is discussed, it might do well to discuss experiments first. Natrella[1] has defined experiments as "a considered course of action aimed at answering one or more carefully framed questions". These experiments must be carefully planned before the conduct of the experiment. In this age of shrinking resources for research and development, carefully designed experiments are a must. In Chapter 7 on significance levels and hypothesis testing, allusion was made to comparing more than one set of samples to determine if they were from the same population. Without proper design of experiments, this will not be possible. Often other comparisons will be desired. Any time comparisons are to be made between two sets of data, experimenters must assure themselves and those who will be viewing the results that the experiments have been well designed to avoid spurious effects and to uncover the effects of any lurking variables.

Unfortunately, design of experiments is often a matter of chance with some chemists. Little time is spent examining the system to determine the questions that must be answered. Design of experiments is often nothing more than judicious planning of what questions must be answered. Often, we are able to predict or hypothesize what we expect the results of experiments to be. These predictions then form a basis for generation of a set of experiments that will provide the data from which the hypotheses can confidently be accepted or rejected. Examination of a set of potential answers prior to the conduct of experiments is essential to gaining a fuller understanding of the results. Each question to be answered will provide potential outcomes to experiments. These potential outcomes should be fully explored to optimally design experiments that will provide data to accept or reject a hypothesis. The design of experiments is a field of study to which books[2,3,4,5] have been devoted. Although quite old, the classic book describing scientific experimentation is by E. Bright Wilson.[6] A few of the more common types of experimental designs will be discussed here.

Each of the experimental designs that will be presented emphasizes the possible interdependence of results. In the past, a common method of attempting to uncover the effect of variables was to perform a set of experiments while varying only one variable and holding all others constant. This approach to unraveling the dependence of a system on a variable, such as temperature, ignored the possible interdependence of the variables. The medical profession recognizes that the effect of many drugs is not additive. One drug may have a greater positive or negative effect on a patient if other drugs are present. By proper design of experiments, often this interdependence may be uncovered.

Criteria of Experimental Design

Requisite for a Sound Experimental Design

The most important concept in planning an experimental program is to be able to clearly define the objectives of the program. At the earliest stages of experimentation planning, the objectives must be succinctly stated. Anderson and Bancroft gave some sage advice in their book *Statistical Theory in Research*[7]:

1. The experimenter should clearly set forth his objectives before proceeding with the experiment.
2. The experiment should be described in detail. The treatments should be clearly defined.
3. An outline of the analysis should be drawn up before the experiment is started.

This advice suggests that before any experimentation begins experimenters clearly define all of the variables that they he feel certain will affect the outcome of the experiments. In addition, they should also clearly set out those variables that may be felt of little consequence. Of especial importance here is to decide whether the experiment is to lead to an estimate of a population parameter or to a test of significance of a set of experiments. Experimental conditions must be clearly identified.

Once the experimental objectives are clearly stated, the experimentalist must set out the design in considerable detail. Each of the variables to be evaluated must be stated. A decision must be made whether a set of control treatments is needed. If it is decided that a control is not needed, a decision must be made in advance as to which of the experiments will be the baseline to which results of other experiments will be compared. In this stage of the planning process, one must be extremely careful to determine the full scope of the experimental program to ensure oneself (and one's supervisors) that sufficient resources (time, dollars, and people) are available to accomplish the desired objectives.

A former supervisor of mine made it very clear that it was important to do a job right the first time. If the resources were not there, this supervisor would not do half a job with insufficient resources. Often the situation is that time constraints often lead experimenters into a poorly detailed program with insufficient resources. If the questions are not clearly answered at the end of the program, the experimental regime must be repeated often at much greater expense of time, people, and money. I recently saw a sign hanging in an office that sums up the need to do the job right the first time: "Why is there never enough time to do the job right the first time, but there is always time to do the job a second time?" Of course, people or money could just as easily be substituted for time in this quote.

Before the actual experimentation begins, the method of data analysis should be specified. This will guide the experimenter to ensure that the proper data will be taken to answer the questions of the experimental program. The method of data analysis should include all of the statistical tests that will be applied to the data. Such things as treatment of missing data should be considered. Rarely, if ever, will the experimenter have the entire set of data needed to answer the objectives of the experimental program. Murphy's Law will be expressed at the most inopportune times and in a way that will cause the most damage to the experimental program. This eventuality should be anticipated and protocols designed to account for such situations. Often this step will lead to a relook of the objectives and description of the experimental program. The statistical analysis portion of the experimental program should be concerned with the desired level of significance and confidence. If this stage has been well thought out in advance, the requisite number of samples will be taken. A term used (which will be described in more detail in a later chapter) to describe the level of significance is the precision of the experiment. The experimental program must provide a measure of the precision of the experiment. The precision of the experiment must also be sufficient to ensure that the objectives are clearly met.

Design of experiments is summed up with the following:

1. Have well-defined, clear objectives.
2. Experimental factors should not be obscured by other variables.
3. Experiments should be free of bias.
4. Provide a measure of the precision of the experiments.
5. Ensure that the precision of the experiments is sufficient to meet the objectives of the experimental program.
6. Ensure that sufficient resources are available to complete the designed project.

Factors

A goal of experimentation is to obtain data from which conclusions may be drawn about the effect of the random variables on the outcome of the experiment. The variables are often called *factors*. Factors of an experiment that might be easily controlled are ones such as *temperature, pressure, concentration, wavelength*, or *time.* The factors also might be ones over which the experimenter has little control such as *different analysts, different instruments, or instrumental noise.* The experimenter should define those factors that are felt to be most important. The factors are evaluated using groupings of experiments.

Experimental System

Experiments are conducted to provide data to validate hypotheses. Any experiment can be considered to involve a set of variables that may be measured or controlled. The goal of the experiment is to ascertain which of the input variables are of the most influence on the outcomes of the experiments. The ones that influence the outcome of the experiment are termed factors and are determined from the experiments that are performed. The scientific method is designed around the concept that a hypothesis is formulated, experiments are performed to provide data with which to validate the hypothesis, and conclusions are drawn. From the conclusions, new hypotheses are formulated and the process repeated. A goal of the scientific method and the use of the experimental system for chemical research should be to determine the variables that have a direct effect on the outcome of the experiment and could thus be termed factors.

An experimental system may be described as a black box in which transformations are made to the variables that are input to the experiment and the output variables are measured, as illustrated in Figure 10-1. As the output variables are measured a set of data may be obtained from which the influence of each of the input variables may be ascertained. There is a possibility that all factors may not be recognized before experimentation begins and are not controlled during the experiment. These uncontrolled factors lead to randomness in the data. As better control is gained over factors, better control is gained over the experimental system. There is also the possibility that a factor may be recognized but control of that factor may not be possible or feasible. Output variables that are influenced by input factors are called responses. The magnitude of the response to a factor may be small compared to the magnitude of the change in the factor. The experimenter should be able to determine whether the response being exhibited is a small response to a large change in the factor or whether the factor has no influence over the response. The small response to a large change in a factor is called *robustness*.

Variables that appear to influence an experiment yet do not are called *latent* or *lurking* variables. They may be thought of as masquerading or disguising another variable. These lurking variables hide the true identity of the responsible factor. An example of a lurking variable is the intensity of the sunlight, which was seen to cause an increase in the NO_2 concentration in the atmosphere in a smog-prone area during the morning hours. The NO_2

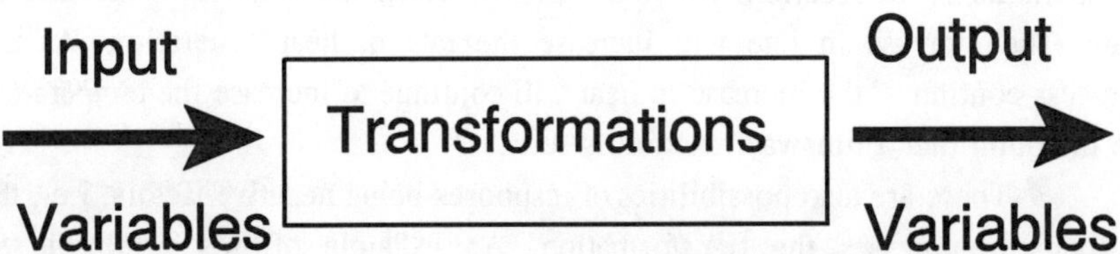

Figure 10-1. The generalized system black box showing general input and output variables.

concentration, however, should have decreased during the same period if reacting photochemically. As a result of knowing what the proper behavior of the concentration of the NO_2 should be, experimenters were able to recognize that the increase was the result of the presence of oxidizing materials, primarily ozone, which oxidized the nitrogen monoxide to nitrogen dioxide. As the nitrogen dioxide was formed, it would photochemically decompose to the nitrogen monoxide and atomic oxygen. The atomic oxygen would react with molecular oxygen to form more ozone. The kinetics of the reactions are such that the oxidation of nitrogen monoxide occurs much more rapidly than does the photochemical dissociation of the nitrogen dioxide. Consequently, the nitrogen dioxide concentration was seen to increase with time of day and intensity of sunlight. The reactions are summarized in the following equations:

$$\begin{aligned} NO + O_3 &\rightarrow NO_2 + O_2 \\ NO_2 &\rightarrow NO + O \\ O + O_2 &\rightarrow O_3 \end{aligned} \tag{1}$$

Another classic example of a lurking variable that is often given is the relation between time of year that grapes are picked and the sweetness of the wine that is produced. The lurking variable is the time of year. In actuality the controlling factor is the sugar content of the grapes. As the year progressed, the grapes became more mature and had a higher sugar content. The higher sugar content in the grapes resulted in a sweeter wine.

As the experimental system is being described, there may be a response that will also be recognized as a factor for the block. Consider the case of a chemical reaction that generates heat. The experimental system may be considered as a chemical reaction (transformation) that has an input variable, temperature, and an output variable, heat. If the heat that is generated in the reaction is not adequately dissipated, the heat will serve to increase the temperature of the reaction, which in turn will cause the reaction to increase its rate. The increase in rate will increase the rate of heat generation. As the process continues, the increase in heat will continue to increase the temperature to the point that a runaway reaction is seen.

There are also possibilities of responses being negative factors, i.e., the response suppresses the transformation. An example of this would be an endothermic reaction that pulls heat from the surroundings. If there is inadequate heat exchange with the surroundings to supply the necessary heat, the temperature in the reaction will decrease. As the temperature decreases, the rate of reaction decreases, which causes a lower amount of heat to be needed. Eventually, the temperature will be such that the reaction may be totally suppressed.

The last two examples are illustrations of feedback in experimental systems. In feedback, an output variable (a response) also acts as a factor. Feedback mechanisms can be used as controls in experimental systems. Feedbacks are examples of LeChatelier's principle in practice. When a system at or near equilibrium is stressed, the system will shift the position of equilibrium to relieve the stress.

Generally more than one transformation block will be required to completely describe an experimental system. As more blocks are added to the experimental system, many of the responses or output variables will become input variables to another block. As the blocks are nested or linked, potential interactions between factors may become evident. The experimenter will want to test each of the input variables for each block to determine which are factors.

An example of an experimental analytical system is a chromatography system, such as shown in Figure 10-2. The figure shows just the major components of a general chromatograph and not all the portions of an analytical system. An analytical system would need to contain all of the elements discussed in Chapter 1. Additional blocks could also be added to this figure to make a more complete chromatographic system such as sample preparation or data recording. A set of input variables that may be constructed as influencing the outcome of a chromatographic experiment includes such things as pressure, temperature, flow rate, the phase of the moon, the number of coincident cosmic rays, the column packing, the liquid phase on the solid support, the length of the column, and the sample being analyzed. The transformations that occur are the equilibria that are established within the chromatographic system. And the output variables that may be measured are the response of detector(s) that may be part of the chromatograph. A simplified description of a chromatograph may be thought of as being three interconnected processes, injection, separation, and detection, as shown in Figure 10-2.

This is a very simplified schematic for a chromatograph, but the concept of generating a block diagram for an experiment can be used to define some of the variables that may influence the final output variables of an experiment. Figure 10-2 above contains only variables that we know affect the

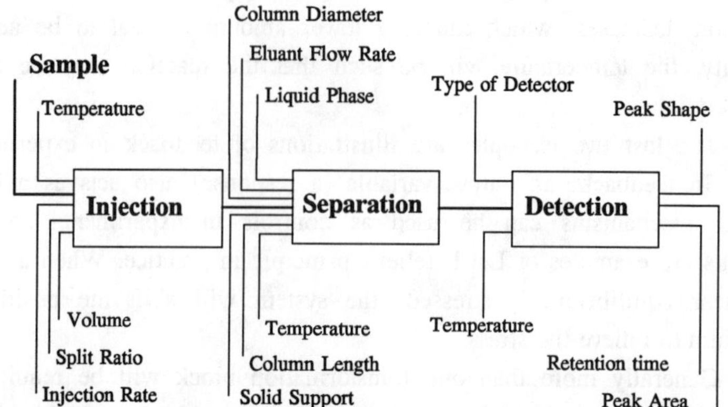

Figure 10-2. A general chromatography system expressed as an experimental system. This shows just three components of a chromatograph.

output. As we begin experimentation, we may not know all of the variables that may affect the output, but as the experiments are performed, those input variables that affect the output variables will become evident.

Experimental Designs

After the experimental system has been defined using the above principles, the variables that have been identified as being potential factors of the experimental system must be classified as being of little or large significance in the outcome of experiments. To make the determination of the significance of the variables, experiments must be designed that will unravel the significance and the potential interaction of the variables. The following sections will introduce several common experimental designs that will aid in the determination of significance and interaction.

Randomized Block

Data may often be separated into smaller, more homogeneous subsets in which the data more closely resemble one another. These smaller subsets are called blocks. The blocks may contain more than one data point. An example of a randomized block is found in acid rain monitoring. A set of samplers is placed in a network. The position of each sampler is fixed in time and position, i.e., the samplers do not move. The amount of rain collected in each sampler is measured each week and chemical analyses are performed to determine the chemical composition. The data for the concentration of a particular analyte or amount collected each week for all of the samplers would become a block. The entire set of blocks for a given system would constitute the data set. When the data are analyzed, the experimenter would like to have a predictor for the concentration of a variable that depends on the relationship of the analyte concentration as it varies from week to week and the relationship of the analyte concentration between samplers. The predictor is given the symbol, $\hat{x}$, and denotes the predicted value of the random variable. The statistical model for a given analyte in a block would be

$$\hat{x}_{mk} = \mu + \alpha_m + \beta_k + \epsilon_{mk} \tag{2}$$

where μ is the overall mean of the data,

$$\mu = \frac{\sum_m \sum_k x_{mk}}{m \cdot k} \tag{3}$$

α_m is the mean of the mth block (the week),

$$\alpha_m = \frac{\sum_k x_{mk}}{m} \tag{4}$$

β_k is the effect of the kth factor (the sampler),

$$\beta_k = \frac{\sum_m x_{mk}}{k} \tag{5}$$

and ϵ_{mk} is the residual left from the prediction and represents the difference between the predicted value and the actual data value:

$$\epsilon_{mk} = x_{mk} - \hat{x}_{mk} = x_{mk} - (\mu + \alpha_m + \beta_k) \tag{6}$$

The data for a randomized block would be schematically represented as shown in Table 10-1. The data for such a block design would be analyzed by either a one-way or a two-way analysis of variance as discussed in Chapter 11. The set of data for a randomized block shown in Table 10-2 represents weekly concentrations of sulfate ion in rainfall for several sites in Illinois. The locations of the sites are shown in Chapter 12. The results of the analysis of variance (ANOVA) are

SOURCE	*df*	*SS*	*MS*	*F*
weeks	51	5513.5	108.1	9.83
stations	5	40.3	8.1	0.736
ERROR	255	2796.3	11.0	
TOTAL	311	8350.1		

I will not discuss the interpretation of the analysis of variance table that is shown in this chapter, but will leave that until a later chapter. There are two other plots that are generally constructed from the ANOVA results. These are shown in Figure 10-3. The two plots are the normal probability plot and the plot of residuals vs predicted value. These two plots test the assumptions for the use of the ANOVA technique.

Factorial Design

The previous example indicated only one level for each factor by assuming that the only variable was the sampler that was used in the experiment. However, if there were several manufacturers of samplers, each of which had several variables that could affect the outcome of the experiment, then each of the variables for each of the samplers would need to be investigated. Some of the variables that might affect the outcome of collection of rainfall would be the sampler manufacturer, the resistance that is required to expose the collection

Table 10-1. Schematic Representation of Data for a Randomized Block Design. The rows represent different treatments.

Factor

x_{11}	x_{12}	x_{13}	...	...	...	x_{1j}
x_{21}	x_{22}	x_{23}	...	...	...	x_{2j}
x_{31}	x_{32}	x_{33}	...	...	...	x_{3j}
...	...	...	...	...	...	...
...	...	...	...	x_{km}	...	...
...	...	...	...	...	...	...
x_{i1}	x_{i2}	x_{i3}	...	...	...	x_{ij}

Table 10-2. Sulfate Ion Concentration for Six Sites in Illinois

IL18	IL11	IL19	IL35	IL47	IL63
2.09	2.57	4.44	1.83	2.45	1.90
2.30	4.10	1.12	0.60	0.79	0.82
3.60	4.20	3.60	1.80	3.68	2.77
3.18	2.42	2.82	1.61	2.32	4.05
3.10	2.40	0.69	2.01	0.78	1.47
4.60	4.09	4.60	3.25	4.13	8.43
6.25	3.72	8.05	2.94	2.85	3.61
4.46	5.00	6.57	5.00	5.00	3.72
5.54	5.47	1.22	7.27	5.59	5.84
3.30	2.62	3.47	3.21	2.25	3.19
2.46	4.93	3.97	3.70	4.74	4.63
4.59	1.91	3.40	7.80	4.18	16.25
2.48	2.50	2.32	3.46	3.63	1.98
4.20	3.65	0.91	6.20	5.74	8.09
3.10	2.43	2.30	2.47	2.46	2.29
7.52	5.10	4.64	4.95	4.76	5.24
2.30	2.30	5.10	4.52	2.30	6.76
3.40	1.72	8.54	2.68	2.24	1.65
4.06	3.09	3.16	2.68	2.30	3.64
5.35	4.63	3.63	6.43	6.05	6.39
3.33	2.27	4.30	2.48	2.38	3.45
19.39	9.86	5.12	6.00	10.69	4.99
2.49	2.30	4.18	6.80	5.43	7.78
4.73	3.94	3.39	1.19	2.58	2.12
5.11	4.02	2.30	2.26	4.04	10.05
1.52	1.59	3.43	2.22	2.94	1.49
1.56	5.82	3.53	3.61	1.56	2.15
1.39	4.46	3.54	2.30	4.47	2.30
3.08	6.08	3.51	4.09	1.97	2.77
5.93	8.16	4.16	2.30	2.30	2.30
1.39	5.55	5.08	5.77	4.99	4.54
2.30	2.30	2.30	5.21	4.72	4.62
6.25	6.37	6.65	7.86	4.59	3.25
3.36	3.74	8.18	3.98	3.46	1.77
3.32	2.04	2.78	1.65	2.39	2.58
3.63	2.66	4.12	1.64	4.26	3.15
3.24	4.29	3.11	3.22	2.06	2.52
2.30	7.78	8.89	2.53	7.85	1.88
2.30	7.17	2.30	5.41	2.30	2.30
2.01	1.74	3.02	0.52	1.40	1.40
2.65	2.30	7.08	2.30	2.30	2.30
2.51	2.03	2.65	2.98	2.06	4.59
1.74	2.17	2.27	2.17	2.20	3.90
5.08	2.66	6.45	2.38	1.54	2.30

Table 10-2 (continued). Sulfate Ion Concentration for Six Sites in Illinois

IL18	IL11	IL19	IL35	IL47	IL63
1.53	1.52	1.87	1.06	2.30	1.87
2.64	1.84	2.27	1.90	2.02	1.76
0.79	1.62	1.40	1.57	2.27	1.36
0.83	1.36	1.48	0.51	0.51	0.83
2.30	1.72	2.30	1.78	2.12	4.77
2.89	2.64	3.57	5.48	4.44	2.37
0.94	1.14	2.30	0.80	0.87	1.13

vessel to the precipitation, the temperature to which the sensor is heated to dry the sensor after the cessation of precipitation, the magnetic direction to which the sensor is directed, the presence of potentially interfering objects in a cone about the sensor, or the number of times that a sampler opens or closes during a precipitation event. The list could go on for many more variables. To obtain information that will allow the investigator to determine which of the variables are important will require that all combinations of variables be investigated. If the major variable is the sampler manufacturer then a subset of the factors that could affect the outcome of the collection of precipitation would be shown in Table 10-3.

The number of experiments required for a three level experiment on three factors with 8 treatments is 3x3x8 or 72 experiments. In general, the

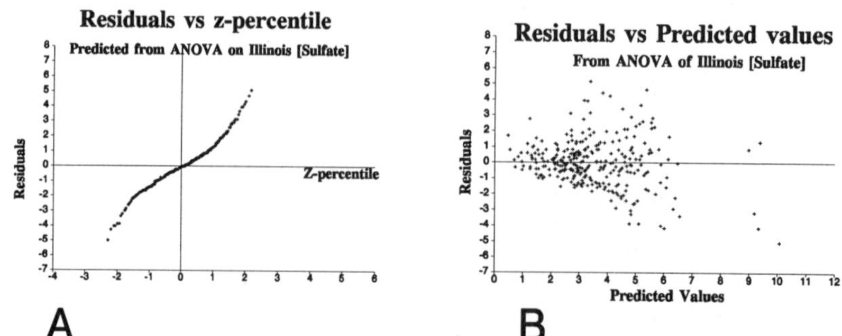

Figure 10-3. Plots made from ANOVA results. A. Normal Probability Plot. B. Residuals vs the predicted values.

number of experiments required for a full factorial experimental design is the number of factors multiplied by the number of levels for each factor multiplied by the number of treatments for the design. This type of design may also be used to determine interactions between factors.

As can be seen as the number of factors and the number of levels increases, the number of experiments required to delineate the interactions and influence of each factor increases very dramatically. Ways to reduce the number of experiments are required.

Latin Square

The randomized block design will enable the experimenter to determine the nonrandom effects of a particular treatment. But it will not necessarily enable the experimenter to determine if there are systematic variations because of the way the treatments are effected or implemented. For example, with the example of the acid rain monitoring network, it may be desirable to determine if there is an interaction between sampler location and type of sampler. To test this, one would need to collect samples each week from all of the samplers and the samplers would need to be rotated from sampling location to sampling location each week. This would enable the experimenter to determine if there is a systematic problem with a given sampler. The model for this design would be

$$\hat{x}_{ijk} = \mu + \alpha_i + \beta_j + \gamma_k + \epsilon_{ijk} \tag{7}$$

where μ is the overall mean,

$$\mu = \frac{\sum_i \sum_j \sum_k x_{ijk}}{(i+j) \cdot k} \tag{8}$$

α_i is the average effect of the ith column (the location),

$$\alpha_i = \frac{\sum_j \sum_k x_{ijk}}{j \cdot k} \tag{9}$$

β_j is the average effect of the jth column (the sampler),

$$\beta_j = \frac{\sum_i \sum_k x_{ijk}}{i \cdot k} \tag{10}$$

γ_k is the average effect of the kth treatment (the week)

Table 10-3. Factorial Experiment Design: Collection of Precipitation

	Sampler								
	Manufacturer A			**Manufacturer B**			**Manufacturer C**		
Week	Temperature	Sensor Resistance	Sensor Orientation	Temperature	Sensor Resistance	Sensor Orientation	Temperature	Sensor Resistance	Sensor Orientation
1	amount	amount	amount	amount	amount	amount	amount	amount	amount
2	amount	amount	amount	amount	amount	amount	amount	amount	amount
3	amount	amount	amount	amount	amount	amount	amount	amount	amount
4	amount	amount	amount	amount	amount	amount	amount	amount	amount
5	amount	amount	amount	amount	amount	amount	amount	amount	amount
6	amount	amount	amount	amount	amount	amount	amount	amount	amount
7	amount	amount	amount	amount	amount	amount	amount	amount	amount
8	amount	amount	amount	amount	amount	amount	amount	amount	amount
9	amount	amount	amount	amount	amount	amount	amount	amount	amount
10	amount	amount	amount	amount	amount	amount	amount	amount	amount
11	amount	amount	amount	amount	amount	amount	amount	amount	amount
12	amount	amount	amount	amount	amount	amount	amount	amount	amount

$$\gamma_k = \frac{\sum_i \sum_j x_{ijk}}{i + j} \tag{11}$$

ϵ_{ijk} is the random error or residual associated with each measurement and is calculated from

$$\epsilon_{ijk} = x_{ijk} - \hat{x}_{ijk} = x_{ijk} - (\alpha_i + \beta_j + \gamma_k) \tag{12}$$

The residuals must be normally distributed with a mean of zero and a constant standard deviation.

The Latin Square design is shown schematically in Table 10-4. The letters in the blocks indicate location. The Latin Square experimental design is analyzed using analysis of variance. As seen in Table 10-4, for each treatment the levels are scrambled in an orderly fashion. An example of the Latin Square is to evaluate the performance of different samplers to ensure no interaction of the sampler with sampling location. In this example, the treatment would be week and the samplers would be indicated by the letters *A, B, C,* and *D*.

Youden Blocks

As was shown, in a randomized block of a multifactor experiment with many levels of each factor being tested, the number of experiments that must be performed rapidly increased. This is particularly true if there is suspected interaction between any of the factors. The number of experiments may be reduced by only allowing two levels for each factor. A disadvantage of this type of design is that the total effect of differing levels of a factor may not be discerned. Plackett and Burman[8] described a method for designing experiments

Table 10-4. Latin Square Experimental Design

Treatment	FACTOR Levels			
1	*A*	*B*	*C*	*D*
2	*B*	*C*	*D*	*A*
3	*C*	*D*	*A*	*B*
4	*D*	*A*	*B*	*C*

for 7, 11, and 15 factors. Youden[9] used this concept for the ruggedness testing of analytical methods. Let us examine a factorial scheme for seven factors with two levels in each factor. Let *A, B, C, D, E, F*, and *G* be the main levels for each of the factors, and *a, b, c, d, e, f,* and *g* be the secondary levels for each factor. Each of the factors represents a different step in an analytical method (e.g., digestion with HCl or $HClO_4$), or an experimental condition (temperature=55°C or 75°C). A set of experimental parameters will then be completely specified by selecting either a lower or upper case letter for each of the factors. Unfortunately, there are a large number of combinations of these levels and factors that can be obtained (2^7 or 128). It is possible to obtain subsets of these that give a balance between the levels and from which information is relatively easy to extract. Such a set is shown in Table 10-5. Whenever eight determinations are split into two groups of four on the basis of one of the letters, all other factors cancel out within each group. Therefore, each of the seven factors is evaluated by the eight determinations. For example, to determine the effect of the variability in the experimental system caused by factor (*C* or *c*) examine the two averages

$$\bar{C} = \frac{s + u + w + y}{4}$$
$$\bar{c} = \frac{t + v + x + z}{4} \tag{13}$$

Compute and rank each of the differences (*A-a*, *B-b*, etc). It can then be quickly determined by inspection if one or two factors affect the results of the experiment.

The determinations that are to be averaged are found by examining Table 10-5. The conditions for the particular experiment are also found from Table 10-5 and Table 10-6. For example, the conditions for extraction for determination number 1 are extraction time of 1 hour, 0.1 N hydrochloric acid, glass fiber thimble, hexane, 50°C in light.

An estimate of the error associated with an analytical method may be found from such a design by estimating the standard deviation. The sum of the square of the differences (*A-a*, *B-b*, etc.) [calculated as above] is multiplied by 2/7 and the square root is then calculated. If the standard deviation is abnormally large, the reason for that deviation must be sought in the experimental system. Alternatively, the standard deviation may be calculated from the set of determined values (*s, t, u, v, w, x, y*, and *z*).

Table 10-5. Youden Table Showing Eight Combinations of Seven Factors

Factor Level	Combination or Determination Number							
	1	2	3	4	5	6	7	8
A or a	A	A	A	A	a	a	a	a
B or b	B	B	b	b	B	B	b	b
C or c	C	c	C	c	C	c	C	c
D or d	D	D	d	d	d	d	D	D
E or e	E	e	E	e	e	E	e	E
F or f	F	f	f	F	F	f	f	F
G or g	G	g	g	G	g	G	G	g
Observed result	s	t	u	v	w	x	y	z

Example

A laboratory manager is concerned with a new analytical method being developed to determine the amount of a pesticide in a soil sample. The method involves the extraction of the pesticide from the soil using hexane in a Soxhlet extractor. Several variables are thought to be possible factors of the experiment. Before a full factorial design is applied, the manager wants to limit the number of experiments as much as possible to determine those variables that will likely be factors of the extraction. He has decided to use a Youden block design for the initial investigation. The variables thought to be of concern are shown in Table 10-6. The results of eight different determinations with these factors varied are shown in Table 10-7. The averages for the various variables are formed using

Table 10-6. Potential Factors for Extraction

Factor Designator	Variable	Level	
		A	B
A	Extraction time	1 hour	3 hour
B	Acid concentration	0.100 *N*	0.05 *N*
C	Thimble material	Glass fiber	Cellulose
D	Type of acid	HCl	$HClO_4$
E	Solvent	Hexane	Toluene
F	Temperature	50°C	75°C
G	Amount of light	Light	Dark

Table 10-7. Results of Extraction Determinations

Determination number	1	2	3	4	5	6	7	8
Letter designator	s	t	u	v	w	x	y	z
Percent recovery	89	91	96	95	60	85	75	81

the letters from Table 10-5. (e.g., the average for the variables A and a are

$$\bar{A} = \frac{s + t + u + v}{4}$$

$$\bar{a} = \frac{w + x + y + z}{4} \quad (14)$$

Similar averages are formed for each of the other pairs. The differences and squares of the differences are then computed from the averages. The results of calculating the averages, difference of averages, and square of differences are shown in Table 10-8. The largest difference is the difference between variable A (extraction time of 1 hr) and variable a (extraction time of 3 hr). The type of acid has no effect and should not be considered as a factor. Of moderate effect are the influence of light, the concentration of acid, and the temperature. The other effects that will bear further investigation are the thimble material and the solvent. This set of experiments has eliminated from further consideration the type of acid to be used in the extraction (HCl and $HClO_4$ have an equal effect). Ones that may bear no further investigation (or at least limited further experimentation) are the influence of light and the concentration of acid.

Analysis of Variance Model

A factorial design may also be analyzed by an analysis of variance model. The model used to describe the factorial design is

$$\hat{x}_{ijk} = \mu + \alpha_i + \beta_j + \gamma_{ij} + \epsilon_{ijk} \quad (15)$$

where x_{ijk} is the kth observation of factors i and j. μ is the overall mean

$$\mu = \frac{\sum_i \sum_j \sum_k x_{ijk}}{i \cdot j \cdot k} \tag{16}$$

α_i is the effect of the *i*th level of factor 1

$$\alpha_i = \frac{\sum_i x_{ijk}}{i} \tag{17}$$

β_j is the effect of the *j*th level of factor 2

$$\beta_j = \frac{\sum_j x_{ijk}}{j} \tag{18}$$

γ_{ij} is the effect of interaction between the *i*th level of factor 1 and the *j*th level of factor 2

$$\gamma_{ij} = \frac{\sum_i \sum_j x_{ijk}}{i \cdot j} \tag{19}$$

and ϵ_{ijk} is the random error of the *ijk*th observation and is the residual from the predicted and actual data value

$$\epsilon_{ijk} = x_{ijk} - \hat{x}_{ijk} = x_{ijk} - (\alpha_i + \beta_j + \gamma_{ij}) \tag{20}$$

The residuals should be normally distributed with a mean of zero and constant standard deviation.

Separation of Means

Table 10-8. Calculation of the Effect of the Potential Factors

Average	A	B	C	D	E	F	G
	92.75	81.25	88.00	84.00	87.75	81.25	86.00
Average	a	b	c	d	e	f	g
	75.25	86.75	88.00	84.00	80.25	86.50	82.00
Difference of Averages	(A-a) 17.50	(B-b) -5.50	(C-c) -8.00	(D-d) 0.00	(E-e) 7.50	(F-f) -5.25	(G-g) 4.00
Difference²	(A-a)² 306.25	(B-b)² 30.25	(C-c)² 64.00	(D-d)² 0.00	(E-e)² 56.25	(F-f)² 27.56	(G-g)² 16.00

When the hypothesis of equality of means has been rejected, it would be desirous to know which of the means are different from the others. A rather simple test known as Duncan's *separation of means* test may be used to indicate which of the means is different. To use this test, each of the means and each of the variances associated with each mean must first be calculated. To use Duncan's separation of means test requires that the variances are not statistically different. A test such as Bartlett's or Cochran's can be used to determine if the variances of the set are equal (or, more correctly, not significantly different). Let us first examine each of these tests. Each of these tests uses a test which is analogous to the F-test to determine if the variances are statistically different.

Cochran's Test for Homogeneity of Variance

Let us consider a case where we have k means with k associated variances. To use Cochran's test to determine homogeneity of variance, the number of elements in each sample must be the same. Each of the means and variances was calculated with n variables. It is desired to determine if the variances are statistically different. The null hypothesis for this test is

$$H_0: \; s_1^2 = s_2^2 = \cdots = s_k^2 \tag{21}$$

and the alternative is that at least one of the variances is not equal to the others. The test statistic is calculated as

$$g = \frac{\text{largest } s^2}{\sum_k s_k^2} \tag{22}$$

The resulting test statistic from the calculation is compared to the F-Table with k degrees of freedom in the denominator and n degrees in the numerator.

Bartlett's Test for Homogeneity of Variance

Bartlett's test is slightly more involved since it does not require all samples to be of equal size as does Cochran's test. The null hypothesis is the same as for Cochran's test. The pooled variance must be calculated to use Bartlett's test: The pooled variance is an average of all of the variances weighted by the number of degrees of freedom used to calculate each of the variances. n_i is the

$$s_p^2 = \frac{\sum_i (n_i - 1) \cdot s_i^2}{\sum_i (n_i) - k} \tag{23}$$

number of observations for the *i*th variance and *k* is the number of variances that have been calculated. The remaining variables are given in the following equations:

$$M = (N - k) \cdot \ln (s_p^2) - \sum_i (n_i - 1) \cdot \ln (s_i^2) \tag{24}$$

$$df_1 = k - 1$$
$$df_2 = \frac{k+1}{A^2} \tag{25}$$

$$A = \frac{1}{3 \cdot (k-1)} \left[\sum \left(\frac{1}{n_i - 1} \right) - \frac{1}{(N - k)} \right] \tag{26}$$

$$b = \frac{df_2}{1 - A + 2 / df_2} \tag{27}$$

$$F = \frac{df_2 \cdot M}{df_1 \cdot (b - M)} \tag{28}$$

In these equations, df_1 and df_2 are the degrees of freedom in the numerator and denominator, respectively. Bartlett's test is quite sensitive to the distribution that describes the data, working best for data that are normally distributed. Generally, the test should be used only for data that are normally distributed. If there is a question about the distribution of the data, it should not be used.

If it is determined that the variances are not significantly different after using either Bartlett's Test or Cochran's test, a pooled variance may be calculated using equations given earlier. Duncan's test for separation of the means may be accomplished only if it is shown that the variances of the individual means are not significantly different.

To use Duncan's test, each of the *k* means and the pooled standard deviation are calculated. The means are then sorted in ascending sequence. Differences between all of the means are then calculated by subtracting the smallest mean from the largest, then the next smallest mean from the largest, and so forth up to the next largest mean being subtracted from the largest mean. This calculation is then repeated with the smallest mean up to the third largest mean, each being subtracted from the next largest mean. A triangular matrix of differences is then formed. If A is the largest mean, B the next largest mean, C the third largest mean, D the next to the smallest mean, and E the smallest mean, then the matrix that would be formed is shown in Table 10-9.

Duncan's values (D_R) are obtained for each of the differences by looking at the table of Duncan's values in Appendix C for the number of means included in the range of the difference. For example, in Table 10-9 above the smallest mean subtracted from the largest mean has five values included in the range. A significance parameter is obtained from the table for each of the integers less than or equal to 5 but greater than 1. In this example, there would be Duncan's values obtained for 5, 4, 3, and 2. Duncan's values for the next row of the matrix will be obtained for 4, 3, and 2. The number of degrees of freedom that will be used to enter the table is the number of differences that can be formed from the means:

$$\textit{Degrees of freedom} = \sum_{i=1}^{k} (i-1) \tag{29}$$

Each of the values obtained from the table is then multiplied by the pooled standard deviation

$$\textit{Critical value} = D_R \cdot s_p \tag{30}$$

to obtain the critical values for the test. The critical values are termed the *least significant ranges.* The test statistics are the differences of the means. If the difference exceeds the least significant range, the difference is considered significant. One exception to the significance rule is that if a difference is deemed not significant then any difference in the subset is also considered not significant.

Example

Suppose there are five analysts who determine the amount of cholesterol in blood samples. The variances have been shown to be statistically the same and have a pooled standard deviation of 4.420. The averages for the analysts are

A	B	C	D	E
101.0	105.6	112.8	99.6	120.0

Table 10-9. Number of Means in the Range for Duncan's Separation of Means Test

Differences between the means			
5	4	3	2
A-E	A-D	A-C	A-B
	B-E	B-D	B-C
		C-E	C-D
			D-E

We desire to determine if there is a difference between the mean of the analysts. First sort the means in ascending sequence ($D<A<B<C<E$) and then form the differences of the means as shown in Table 10-10.The values obtained from the table for the Duncan's values are ($p=5, 4, 3, 2$) D_R = 3.43, 3.37, 3.3 and 3.07; the least significant ranges are 15.2, 14.9, 14.6, and 13.9. For only E-D and E-A are the differences significant. This indicates that for the range of four means from 99.6 to 112.8 and the range of three means from 105.6 to 120.0, we cannot distinguish between the analyst performance. If we want to improve our ability to distinguish between the means, we must improve the precision (decrease the standard deviation, s_p) or increase the number of analyses performed by each analyst.

Table 10-10. Number of Means in the Range for Duncan's Test

Difference of the Means			
5	4	3	2
(E-D) 20.4	(E-A) 19.0	(E-B) 14.4	(E-C) 7.2
	(C-D) 13.2	(C-A) 11.8	(C-B) 7.2
		(B-D) 6.0	(B-A) 4.6
			(A-D) 1.4

References

1. Natrella, M. **Experimental Statistics**. National Bureau of Standards Handbook 91, 1966, p. 11-1.

2. Box, G.E.P, Hunter, W.G., and Hunter, J.S. **Statistics for Experimenters**. Wiley, New York, 1978.

3. Cochran, W.G., and Cox, G.M. **Experimental Design**. 2nd ed. Wiley, New York, 1957.

4. Fisher, R.A. **The Design of Experiments**. 9th ed. Hafner Press, New York, 1971.

5. Myers, J.L. **Fundamentals of Experimental Design**. 3rd ed. Allyn and Bacon, Boston, 1979.

6. Wilson, E.B. ***An Introduction to Scientific Research***. McGraw-Hill, New York, 1952.

7. Anderson, R.L., and Bancroft, T.A. **Statistical Theory in Research**. McGraw-Hill, New York, 1952, p. 223.

8. Plackett, R.L. and Burman, J.P. The design of optimum multifactorial experiments, **Biometrika**, 1946, **33**, 305.

9. Youden, W.J. Statistical techniques for collaborative tests in **Statistical Manual of the Association of Official Analytical Chemists**. Association of Official Analytical Chemists, Arlington, Va, 1975, pp. 34–35.

11

ANALYSIS OF VARIANCE (ANOVA)

General

There are many instances where the data that have been obtained must be compared to other data sets or to known values. The comparison of the data may be accomplished in two ways; a determination of whether the (1) means or (2) standard deviations (or variances) are statistically indistinguishable. To this point, we have discussed only techniques to determine if two means or two variances are equal. Much of the data acquired in the chemical laboratory are of a much more complex nature and there are multiple (more than two) means and/or multiple (more than two) variances to compare. A first inclination of the novice user of statistical techniques is to make the comparisons one pair at a time. Unfortunately, when this approach is taken we must recognize that the experiments that produced the means and variances are not isolated events. The probability of making the correct decision on a single pair is calculable. But when many nearly simultaneous experiments have been conducted, the probabilities of making many such decisions are not isolated but may be connected. If the probability of an event cannot be separated from the probability of another event, we must consider the joint probability of the events. Equations for this were given in Chapter 2. To use these equations requires knowledge of the dependence of the probability of the two events, and the probability that a second event will occur given that the first has already occurred.

For simplicity in the following, it will be assumed that the events are not connected but are independent. Fortunately, relations are known that allow the calculation of joint probabilities when events are independent. If α is the probability of event A and β is the probability of event B, the probability that both event A and event B will occur is $\alpha \bullet \beta$. The general equation for joint probability of many events is

$$\textit{Probability of } A, B, C, \cdots = P_A \cdot P_B \cdot P_C \cdot \cdots = \prod_i P_i \quad (1)$$

This equation can be used to determine what the probability is of making five correct decisions each at the 95% probability level. The 95% probability level is chosen since that is the usual confidence level used in statistical tests. The probability of making one correct decision is 95%, but the probability of making two correct decisions at 95% for each is $(0.95)^2$ or only 90.25%. For five, the probability has dropped to 77.4%. The probability of making five correct decisions in a row at the 95% confidence level is only 77.4%. From this, the reader should understand that the technique of making multiple *t*-tests should not be used.

Fortunately, there are statistical tests that can be used to determine whether differences exist between means or variances. We have already seen examples of these with Duncan's test for separation of means and Bartlett's test and Cochran's test for homoscedasticity (homogeneity of variance). We have also indicated several times that the variance of an analytical method is not comprised of just the variance of the signal from the analytical instrument, but is comprised of all sources of variation in the method from the sampling to the reporting of the analytical results. We shall present a technique that will allow us to separate the contribution to the total variance of the several components of the variance. An example of an experimental system that has several components of variance to be separated is the chromatograph that was described in Chapter 10. In a chromatographic system, there are many potential sources of variability in the peaks that are observed. Some of these input variables will have greater effect than others. For example, changing the column length from 2 m to 2.25 m will likely have a much smaller effect on the separation of two compounds than changing the column liquid phase from OV-101 to a Carbowax. The analysis of variance (ANOVA) will allow us to determine the contribution of variability that each input variable makes to the total variability. We are then in a position to better determine experimental conditions for a new situation.

ANOVA implies that we are able to determine whether a significant difference in means exists by examining the variances in a set of data. One might ask how it is possible to use variances to determine whether means are equal or not. Remember that a variance of the mean of an analytical method is the sum of the variances of the operations that make up the method. Therefore, if the variances of the components of a method can be separated, it can be determined if the means are equal. Recall from Chapter 7 that the best estimator of the population mean was the arithmetic mean because it yielded the lowest estimate of the population variance. The population mean if estimated by a different combination of random variables always yielded a larger variance for the mean. If the separation of the components of the mean yields equal variance for each combination, then all components have the same mean. If the variances are not equal, it can be concluded that the individual means are not equal.

If two or more samples of data come from the same population (i.e., they have means and variances that are equal), then combining the samples will affect neither the population mean nor the population standard deviation. In more mathematical terms, if the mean of sample a equals the mean of sample b ($\mu_a = \mu_b$); and the variance of sample a equals the variance of sample b ($\sigma_a^2 = \sigma_b^2$), then the mean of the combined sample will be $\mu_{a+b} = \mu_a = \mu_b$ and the variance of the combined sample will be $\sigma_{a+b}^2 = \sigma_a^2 = \sigma_b^2$. However, consider that two samples are combined whose population means are not equal, but whose variances are equal. If the number of elements in each sample are equal ($n_a = n_b$) then $\mu_{a+b} = (\mu_a + \mu_b)/2$. But the variance will be much larger. It is because of this increase in a variance and being able to partition the total variance that we can use this method to detect differences in means.

One Way ANOVA

Comparison of Multiple Means of One Variable

A common situation in analytical chemistry is to have a set of samples that is used as comparisons for determining whether the performance of an analytical method has changed with time. A set of analytical results is obtained for the samples over many weeks. For simplicity, we will assume that there is only one sample that is to be analyzed. A portion of the sample is analyzed n times in each of the weeks. To limit the number of potential sources of variability, we will also assume that the same instrument and the same analyst have been used for all analyses. It should be recognized that many different concentrations, many analysts, many instruments, and many laboratories should be used to gain the maximum amount of information about the performance of the method. Such a set of data is given in Table 11-1. This data represent the analysis of the same

Table 11-1. A Single Material Analyzed 10 Times Each Week for 5 Weeks

	Week				
Determination	**1**	**2**	**3**	**4**	**5**
1	5.095	5.284	5.040	5.401	5.139
2	5.235	5.339	5.172	5.102	5.155
3	5.332	5.456	5.020	5.188	5.431
4	5.437	5.448	5.117	5.405	5.298
5	5.398	5.067	5.356	5.479	5.073
6	5.121	5.209	5.258	5.204	5.323
7	5.304	5.337	5.142	5.090	5.104
8	5.137	5.119	5.190	5.114	5.382
9	5.150	5.177	5.432	5.284	5.403
10	5.087	5.333	5.102	5.241	5.180

sample 10 times each week for 5 weeks. The data are arranged in tabular form with the weeks being represented by columns and the rows representing the determination number. We wish to determine if the mean from week to week has changed. The null and alternative hypotheses are

$$H_0\colon \mu_1 = \mu_2 = \mu_3 = \mu_4 = \mu_5$$
$$H_a\colon \mu_1 \neq \mu_2 \neq \mu_3 \neq \mu_4 \neq \mu_5 \tag{2}$$

The variance must be partitioned into two components: between columns (weeks) and within columns, i.e., we want to determine whether the major source of variation in the data is within a week or between weeks. The total variance of the data set may be calculated from the familiar equation for the variance:

$$\sigma^2 = \frac{\sum_i \sum_j x_{ij}^2 - \left[\left(\sum_i \sum_j x_{ij}\right)^2 / N\right]}{N - 1} \tag{3}$$

The numerator of this equation is the important portion with which we are concerned. This portion is known as the total sum of squares. For any data set, the total sum of squares is the average of the square of the sum of the data values [$(\Sigma x_{ij})^2/n$] subtracted from the sum of the squares of the individual data values (Σx_{ij}^2):

For the data set in Table 11-1, the total sum of squares is 41.5079. If this is

$$\sum_i \sum_j x_{ij}^2 - \frac{\left(\sum_i \sum_j x_{ij}\right)^2}{N} \tag{4}$$

divided by the total number of degrees of freedom (n-1=49), we obtain the total variance, 0.8471.

The within–weeks (columns) sum of squares is found similarly, except that we are concerned about the variance within a column as opposed to the total variance, therefore we sum the variances of the individual weeks (columns). First calculate the sum of squares for each week using

$$\sum_i x_{ij}^2 - \frac{\left(\sum_i x_{ij}\right)^2}{n_i} \tag{5}$$

where the subscript i is the designator for the rows and j is the designator for the columns. These sum of squares are then summed over the number of columns

$$Sum\ of\ squares_{within} = \sum_j \left[\sum_i x_{ij}^2 - \frac{\left(\sum_i x_{ij}\right)^2}{n_i} \right] \tag{6}$$

to obtain 0.7981. When divided by the number of degrees of freedom for within columns (ν = N-number of columns = 50 - 5 = 45), the result is the within columns variance (0.0177).

Next, the between– or among–weeks variance must be calculated. This is accomplished by calculating the sum of squares for each row and summing over the number of rows

$$\sum_i \left[\left(\sum_j x_{ij}^2 \right) - \frac{\left(\sum_j x_{ij}\right)^2}{n_j} \right] \tag{7}$$

and then summing the results over each of the rows

$$Sum\ of\ Squares_{between\ weeks} = \sum_i \left(\sum_j x_{ij}^2 - \frac{\left(\sum_j x_{ij}\right)^2}{n_j} \right) \tag{8}$$

For this data set, the sum of squares for between weeks is 0.0490. If this is divided by the degrees of freedom for between weeks (number of groups-1=4), the variance between weeks is found to be 0.0123. Now that we have two variances (between weeks and within weeks, we can then calculate the required F-statistic by dividing the among week variance by the within week variance

Number of groups	5
Number of observations	50
Total Mean	5.2378
Total Variance	0.8471

Anova Table

Source of Variation	DF	SS	MS	F-Statistic
Among Groups	4	0.0490	0.0123	0.6907
Within Groups	45	0.7981	0.0177	
Total	49	0.8471		

Group Statistics

Group	N	Sum	U-SSQ	Mean	CV	SD	SE(CV)
[1]	10	52.30	273.64	5.23	2.49	0.13	0.557
[2]	10	52.77	278.61	5.28	2.49	0.13	0.56
[3]	10	51.83	268.78	5.18	2.55	0.13	0.57
[4]	10	52.51	275.88	5.25	2.64	0.14	0.59
[5]	10	52.49	275.66	5.25	2.54	0.13	0.57

Figure 11-1. Analysis of variance table for the data in Table 11-1.

(0.0123/0.0177 = 0.69). These results are summarized in an ANOVA Table such as in Figure 11-1. Examine the *F*-statistic and compare it to the critical value of *F* obtained from the table of values of the cumulative distribution function for the F-Distribution in Appendix C (for $\alpha=0.05$, $\nu_{\text{numerator}}$ of 4, $\nu_{\text{denominator}}$ of 45) of 2.59. The *P*-value is 0.61. This tells us that the between-weeks variation is not significant. The most meaningful reason for the nonsignificance of the between-weeks variation is the magnitude of the within-weeks variation. The within-weeks variation is swamping out the between-weeks variation. To improve the capability of the ANOVA method to determine between-weeks variation, the standard deviation of within weeks must first be made smaller by either making the analytical method more precise or by increasing the number of analyses in a week.

Partitioning of Variance

As has been explained many times before, the total variance of any method or system is the sum of the variances of the individual operations or components that can affect the variability of the method. Using ANOVA techniques, the total variance of a method can be partitioned into the individual components of variability so that the contribution to the total variance may be estimated. The total variance for a one-way ANOVA is the sum of the between columns variance and the within column variance:

$$\sigma^2_{total} = \sigma^2_{between} + \sigma^2_{within} \tag{9}$$

To be able to separate the variability according to the contribution of between and within sources of variation, we must be able to assign an expected mean square to each of the sources of variability. The easiest of these to assign is the variance of the within variability. This variance is the expected mean square for within column variation.

$$\text{Mean square}_{\text{within}} = MS_w = \frac{\text{sum of squares}_{\text{within}}}{\text{degrees of freedom}_{\text{within}}} = \sigma_{\text{within}}{}^2 = \sigma^2 \tag{10}$$

The expected mean square for between columns is the summation of the variance of within columns (σ^2) plus the variance between columns adjusted for the number of elements in each column:

$$\text{Mean square}_{\text{between}} = \frac{\text{sum of squares}_{\text{between}}}{\text{degrees of freedom}_{\text{between}}} = \sigma^2 + n \cdot \sigma_{t}{}^2 \tag{11}$$

where n is the number of elements in a column. This presupposes that the number of elements in each column is the same. If the number of elements is not the same, then n is the average number of elements per column. From these last two equations, the variance of each source of variability (within and between) can be estimated. The total variance is calculated using Equation (9). After the total variance of the method has been calculated, we can then calculate the percentage contribution of each of the sources to the total variability. In other words, we are able to estimate how much of the variance is explained by the variability of each of the sources. The percentage contribution of each of the sources may be found from

$$Percent = \frac{\sigma_i^2}{\sum_i \sigma_i^2} \cdot 100 \tag{12}$$

Example

Examine a set of data where four laboratories made five replicate determinations of purity on samples of 1-bromopropane (also known as NEMAGON, a pesticide to treat nematodes). Care was taken to ensure homogeneity in sampling. Identical samples were also sent to each laboratory. The results of the analytical determinations are given in Table 11-2. The between column sum of squares is

$$SS_c = \frac{369^2}{5} + \frac{371^2}{5} + \frac{345^2}{5} + \frac{359^2}{5} - \frac{1444^2}{20} = 84.80 \tag{13}$$

The total sum of squares is

$$SS_T = 73^2 + 75^2 + 73^2 + \dots - \frac{1444^2}{20} = 95.20 \tag{14}$$

and the within column sum of squares is

$$SS_W = 95.20 - 84.80 = 10.40 \tag{15}$$

The ANOVA Table is given in Table 11-3. As can be seen, the variance between laboratories is much more significant than within laboratories. The standard deviation for within laboratories is given as the square root of 0.650 or 0.806. The variance for variability between laboratories is given as

Table 11-2. Determination of 1-Bromopropane by Four Different Laboratories

	Lab A	Lab B	Lab C	Lab D	
	73	74	68	71	
	75	74	69	72	
	73	75	69	72	
	75	74	70	71	
	73	74	69	73	
mean	73.8	74.2	69.0	71.8	
Σx_{ic}	369	371	345	359	$\Sigma\Sigma x_{ij}=1444$

Table 11-3. Anova Table for Analysis of 1-Bromopropane for Data in Table 11-2

Source	SS	DF	MS	Expected MS
Between labs	84.80	3	28.27	$\sigma^2 + 5\,\sigma_b^2$
Within labs	10.40	16	0.650	σ^2
Total	95.20	19		

$F = 28.27/0.650 = 43.49$
$F_{\alpha=0.05,3,16} = 3.24$

$$
\begin{aligned}
s^2 &= 0.650 \\
s^2 + 5\,s_c^2 &= 28.270 \\
s_c^2 &= 5.524 \\
s_c &= 2.35
\end{aligned}
\qquad (16)
$$

These numbers give a measure of the precision for between–laboratory variability (2.35) and also within–laboratory variability (0.806). As might be expected the between–laboratory variability is much greater than within–laboratory variability. The total variance of the method would be 0.650 + 5.524 = 6.174. The variability in the results explained by the within–laboratory variance is 10.53%, whereas the variability in the results explained by the between–laboratory variance is 89.47%.

Unequal Number of Observations Per Column

Although the calculations are somewhat easier if the number of observations per column are equal, that is not a prerequisite to performing a one-way ANOVA. ANOVA may be performed even if the columns have different numbers of observations. Table 11-4 shows a data set that has an unequal number of elements in each column. The ANOVA Table for this data set is given in Table 11-5. The only difference about this ANOVA and the previous example is the number of elements in each column and the coefficient of the variance of between variance in the expected mean square column (5.5 rather than 5.0); 5.5 is the average number of observations per column.

Table 11-4. Data Table for Unequal Numbers of Observations per Column

	Lab A	Lab B	Lab C	Lab D	
	73	74	68	71	
	75	74	69	72	
	73	75	69	72	
	75	74	70	71	
		75	68	70	
		74	69		
		74			
Mean	**74.00**	**74.29**	**68.83**	**71.2**	
Σx_{ic}	**296.**	**520.**	**413.**	**356.**	$\Sigma\Sigma x_{ij}$ = **1585**

Two Way ANOVA

If we have two independent variables, e.g., catalyst and temperature, interaction may be possible between the variables. We would not be able to analyze such a data set by the one-way ANOVA, but must use the two-way ANOVA. In the two-way ANOVA, we are concerned about the effect of variable 1, the effect of variable 2, and the effect of the interaction between the two variables. To test for the interaction will require that replication be conducted at each of the combinations of variable 1 and variable 2. Without replication no estimate of within-cell variability can be made. The determination of interaction is

Table 11-5. ANOVA Table for the Data in Table 11-4

Source	SS	DF	MS	Expected MS
Between	115.89	3	38.63	$\sigma^2 + 5.5\sigma_c^2$
Within	11.06	18	0.61	σ^2
Total	**126.95**	**21**		

$F_{calculated}$ = **62.87**
$F_{\alpha=0.05,3,18}$ = **3.16**

dependent on knowing whether the variability between cells is because of variability between factors or because the variability within cells is larger than the variability between cells. The total variance of a set of experiments for a two–way ANOVA is attributable to between factor 1, between factor 2, within cells, and between cells. To determine if there is interaction, another component of the total variance must be determined. Only with replication of analysis in a cell (that is the intersection of factor 1 and factor 2) can the within–cell variance be determined. Without the within–cell variance, this aspect of the total variance cannot be determined. If no within–cell variance is obtained, the total variance cannot be partitioned. If no replication of analyses is conducted, the effect of interaction cannot be determined.

To determine the effect of two independent variables requires one additional set of calculations beyond that performed for the one-way ANOVA and that is to determine the sum of squares for the rows in addition to the sum of squares for columns. The representative data matrix for the two-way analysis of variance is given in Table 11-6. The calculational formulas required for the two–way analysis of variance without replication in cells are given in Table 11-7.

Without Replication in Cells

The expected mean squares for two–way ANOVA without replication in cells

Table 11-6. Representative Data Matrix and Calculational Formulas for Row and Column Totals with no Replication in Cells[a]

	Variable A					
Variable *B*	*W*	*X*	*Y*	*Z*	*j*	Row Totals
w	x_{11}	x_{12}	x_{13}	x_{14}	x_{1j}	Σx_{1c}
x	x_{21}	x_{22}	x_{23}	x_{24}	x_{2j}	Σx_{2c}
y	x_{31}	x_{32}	x_{33}	x_{34}	x_{3j}	Σx_{3c}
z	x_{41}	x_{42}	x_{43}	x_{44}	x_{4j}	Σx_{4c}
i	x_{i1}	x_{i2}	x_{i3}	x_{i4}	x_{ij}	Σx_{ic}
Column Totals	Σx_{r1}	Σx_{r2}	Σx_{r3}	Σx_{r4}	Σx_{rj}	$\Sigma\Sigma x_{ij}$

[a] *r*, number of rows; *c*, number of columns

Table 11-7. Calculational Formulas for Two–Way Analyis of Variance Without Replication

Source	SS	DF	MS	Expected MS
Columns	$SS_C = \frac{\sum_c \left(\sum_r x_{rc}\right)^2}{r} - \frac{\left(\sum_c \sum_r x_{rc}\right)^2}{r \cdot c}$	c-1	$\frac{ss_c}{DF_c}$	$\sigma^2 + r \bullet \sigma_c^2$
Rows	$ss_r = \frac{\sum_r \left(\sum_c x_{rc}\right)^2}{c} - \frac{\left(\sum_r \sum_c x_{rc}\right)^2}{r \cdot c}$	r-1	$\frac{ss_r}{DF_r}$	$\sigma^2 + c \bullet \sigma_r^2$
Residual	$ss_{residual} = ss_t - ss_c - ss_r$	(r-1)•(c-1)	$\frac{ss_{residual}}{DF_{residual}}$	σ^2
Total	$ss_t = \sum_r \sum_c x_{rc}^2 - \frac{\left(\sum_r \sum_c x_{rc}\right)^2}{r \cdot c}$	(r•c)-1		

are calculated slightly differently than the expected mean squares for the one way ANOVA. The principle is the same, however. The general approach for determining the mean square for any source of variability is to first define the model that tries to explain the data. In the case of a two–way ANOVA without replication in a cell, the model is

$$x_{rc} = \mu + \alpha_r + \beta_c + \epsilon_{rc} \tag{17}$$

We have previously learned that the total variance is the sum of the individual variances of the operations comprising the system. The variance of the two–way ANOVA with no replication will be the sum of the between–column, the between–row, and the residual variances. The mean square for residual is then the variance of the residuals.

$$MS_{residuals} = \frac{SS_{residuals}}{DF_{residuals}} = \sigma^2_{residuals} = \sigma^2 \tag{18}$$

The mean square of the between–column is the sum of the residual variance and the between–column variance multiplied by the number of rows:

$$MS_{columns} = \frac{SS_{columns}}{DF_{columns}} = \sigma^2 + n_{rows} \cdot \sigma^2_{columns} \tag{19}$$

and the mean square for between–rows will be the sum of the variance for the residual and the variance of the between–rows multiplied by the number of columns:

$$MS_{rows} = \frac{SS_{rows}}{DF_{rows}} = \sigma^2 + n_{columns} \cdot \sigma^2_{rows} \tag{20}$$

Example

Let us suppose that three analysts have determined the percentage of calcium oxide in four batches of Portland Cement. The analysis data and the ANOVA Table are shown in Table 11-8. Table 11-8 shows the calculational formulas necessary for the calculation of the sum of squares for rows, columns, residuals, and totals. The two ratios that must be calculated to determine whether batch or analyst or both are significant are the ratio of $MS_{batch}/MS_{residual}$ and $MS_{analyst}/MS_{residual}$. These are $F_{batch}=200.97/8.56=23.49$ ($F_{0.05,3,6}=4.36$) and $F_{analyst} = 160.33/8.56 = 18.74$ ($F_{0.05,2,6} = 5.14$). Thus we can conclude that both effects are significant. We can use the expected mean square to

Table 11-8. Data matrix and ANOVA table for the determination of CaO in batches of Portland Cement.

Batch of Cement	Analyst a	b	c	Row Total
1	20	27	25	**72**
2	1	17	15	**23**
3	16	28	30	**74**
4	14	27	29	**70**
Column Total	**51**	**89**	**99**	**239**

SOURCE	SS	DF	MS	Expected MS
Between Analysts	320.67	2	160.33	$\sigma^2 + 4\,\sigma_c^2$
Between Batches	602.92	3	200.97	$\sigma^2 + 3\,\sigma_r^2$
Residual	51.33	6	8.56	σ^2
Total	974.92	11		

determine the contribution to the variance for each of the independent variables. The standard deviation for the residuals is

$$\sigma = \sqrt{8.56} = 2.93 \tag{21}$$

The standard deviation for the batch is

$$s_{batch} = \sqrt{\frac{200.97 - 8.56}{3}} = \sqrt{64.14} = 8.01 \tag{22}$$

and the standard deviation for the analyst is

$$s_{analyst} = \sqrt{\frac{160.33 - 8.56}{4}} = \sqrt{37.94} = 6.16 \tag{23}$$

The total variance is $(2.93)^2 + (8.01)^2 + (6.16)^2 = 110.6$. The standard deviation of the residual may be considered as a measure of the precision of the overall method regardless of the batch and analyst if interaction is assumed unimportant. The amount of variability that is explained by the variance of the batch is about 58% [(64.14/110.6)•100], by the variance of the analyst is about 34% [(37.9/110.6)•100], and by neither batch nor analyst (residual) is about 8% [(8.56/110.6)•100)].

With Replication in Cells

The model for analysis of a sample design that has replication in the cells is

$$x_{rc} = \mu + \alpha_r + \beta_c + \gamma_{rc} + \epsilon_{rc} \quad (24)$$

where γ_{rc} is the term in the model that accounts for the interaction. As stated earlier, replicate analyses must be performed at each combination of variables to make this determination. The factorial block sampling design is used for the data collection. Each of the cells at which a row and a column intersect must contain multiple observations. The general form of the data set for a two-way analysis of variance is shown in Table 11-9. The equations necessary to analyze a set of data for a factorial design with multiple observations per cell are shown in Table 11-10.

Table 11-9. Typical data set for ANOVA with replication in cells r rows, c columns, n per cell

	Variable *A*						
Variable *B*	1	2	j	Replicate Total			Row Total
A	x_{111}	x_{121}	x_{1j1}	Σx_{11k}	Σx_{12k}	Σx_{1jk}	$\Sigma\ (\Sigma x_{1jk})$
	x_{112}	x_{122}	x_{1j2}				
	x_{113}	x_{123}	x_{1j3}				
	x_{11k}	x_{12k}	x_{1jk}				
B	x_{211}	x_{221}	x_{2j1}	Σx_{21k}	Σx_{22k}	Σx_{2jk}	$\Sigma\ (\Sigma x_{2jk})$
	x_{212}	x_{222}	x_{2j2}				
	x_{213}	x_{223}	x_{2j3}				
	x_{11k}	x_{12k}	x_{1jk}				
i	x_{i11}	x_{i21}	x_{ij1}	Σx_{i1k}	Σx_{i2k}	Σx_{ijk}	$\Sigma\ (\Sigma x_{ijk})$
	x_{i12}	x_{i22}	x_{ij2}				
	x_{i13}	x_{i23}	x_{ij3}				
	x_{i1k}	x_{i2k}	x_{ijk}				
Column Totals	$\Sigma\ (\Sigma x_{i1k})$			$\Sigma\ (\Sigma x_{i2k})$			$\Sigma\ (\Sigma x_{ijk})$
				Grand Total			$\Sigma\Sigma\Sigma x_{ijk}$

Table 11-10. Equations to analyze for Interaction with Multiple Observations per Cell with a Factorial Design

Source	DF	MS	Expected MS
Columns	c-1	SS_c/DF_c	$\sigma^2 + r\,n\,\sigma_c^2 + n\,\sigma_i^2$
Rows	r-1	SS_r/DF_r	$\sigma^2 + c\,n\,\sigma_r^2 + n\,\sigma_i^2$
Interaction	(c-1)(r-1)	SS_i/DF_i	$\sigma^2 + n\,\sigma_i^2$
Subtotal	$r\,c$-1	SS_S/DF_S	
Within combination			
(residual)	$r\,c\,(n$-1)	SS_w/DF_w	σ^2
Total	$r\,c\,(n$-1)		

Example

A factorial experiment was designed to test the amount of monomer remaining in a polymerization reaction. The two variables tested were time of reaction and temperature of reaction. For the time of reaction, three levels were chosen (1, 2, and 3 hr), and for the temperature, two levels were chosen (100° and 110°C). The data and the ANOVA table results are shown in Table 11-11. It should be noted that the expected mean squares shown in Table 11-11 are valid only for random effects, i.e., levels for each factor have been chosen at random from all available levels.

In testing for significance, the interaction mean square is first tested versus the within combination mean square to determine if there is significant interaction. If the interaction mean square is not significantly different than the within combination, then significant interaction does not occur. The ratio is 0.37/0.06=6.02, the critical value for $F_{0.05,2,6}$=5.14 (P=0.96). Therefore, we conclude that significant interaction does occur at the α=0.05 significance level. The significance of the temperature is then tested. Since the interaction term was shown to be significant, the expected mean square for the effect of temperature is the sum of the interaction variance and twice the variance of the time. The ratio that is needed is the ratio of the $MS_{temperature}/MS_i$ = 7.52/0.37 = 20.56. This

Table 11-11. Data for percent monomer remaining in a polymerization reaction. ANOVA table for the data.

Temperature	Time (hr) 1	2	3	Replicate Totals			Row Totals
100°C	5.4	2.7	2.1	**10.2**	**5.0**	**4.0**	**19.2**
	4.8	2.3	1.9				
110°C	4.2	0.4	0.4	**8.4**	**0.9**	**0.4**	**9.7**
	4.2	0.5	0.0				
Row Totals	**18.6**	**5.9**	**4.4**				**28.9**

Source	SS	DF	MS
Time	30.43	2	15.22
Temperature	7.52	1	7.52
Interaction	0.73	2	0.37
Subtotal	38.68	5	
Within combination	0.37	6	0.061
Total	39.05	11	

is compared to a critical value of $F_{0.05,1,2}=18.5$ ($P=0.953$). Therefore, the temperature effect is significant. The next comparison is for the time. The MS_{time} is divided by MS_i to obtain 15.22/0.37 = 41.60. The critical value is 19.0 ($P=0.953$). Therefore, we conclude that this effect is also significant at the 5% significance level.

The variance of each of the sources of variability is next calculated. The standard deviation of within combinations is 0.06. The interaction standard deviation is $\sigma^2+2\sigma_i^2=0.73$, from which a standard deviation of interaction of 0.58 is calculated. The standard deviation for the effect of time is calculated from $\sigma^2+2\sigma_i^2+2\sigma_r^2=MS_r$ to be 1.90 $[15.22=0.06+2(0.37)+4\sigma_c^2]$. And the standard deviation for the effect of temperature is calculated from $\sigma^2+2\sigma_i^2+4\sigma_c^2=MS_c$ to be 1.83 $[7.52=0.06+2(.37)+2\sigma_r^2]$. The per cent amount of variability that is explained by each variable is within combination (0.8%); interaction (4.5%); temperature (45.7%); and time (49.0%).

The calculations would have had to be handled slightly differently if the interaction and within combination had not proven to be significantly different. In that case the sum of squares for each would have been added and divided by the number of degrees of freedom for the sum. This would then have been the MS_{within} that would have been used in calculations.

Conclusion

The ANOVA is a powerful method of data analysis that allows the chemist to explain the portion of the variability of the experiment that can be assigned to each variable. The method requires a significant amount of data to conduct a meaningful analysis. The method may be further extended to more than two factors.

12
REVIEW OF MATRIX MATHEMATICS

Introduction

Much of what has been presented has been easily represented without the benefit of a compact notation. The concepts to be presented in the rest of the book will rely quite heavily on and will benefit greatly from a compact notation for sets of data. The compact notation chosen is the data matrix. All of the concepts presented so far do not demand a method to represent multidimensionality of the data set. The data shown for the sulfate concentration in atmospheric precipitation samples belied the fact that each week that the concentration of sulfate ion was found, concentrations of a multitude of other analytes were also determined; nor did the presentation intimate that there were also multiple sampling locations each week; nor was it intimated that a single sample may have been analyzed more than once. This complete sampling description yields a data set with at least four possible dimensions: *time*, *replication of analysis*, *sampling location*, and *multiple analytes*. For each of these data elements, the dimension may ran from 1 to n.

To present the remaining statistical techniques, matrix techniques will be required. To ensure that the reader is aware of the operations that can be conducted more easily with matrices, a review of matrix operations is presented.

Data Vectors

Grouping of n common data values (e.g., [Cl^-] may be accomplished to give a one-dimensional array called a data vector. Vectors of dimensions $1 \; x \; n$ or $n \; x \; 1$ and are known as row and column vectors, respectively (Figure 12-1). Vectors describe a common property of an object. Objects that are described by data vectors are extremely varied. Concentration of sulfate ion in rain samples,

$$\begin{bmatrix} x_1 \\ x_2 \\ x_3 \\ \cdots \\ x_n \end{bmatrix} \qquad \begin{bmatrix} x_1 \; x_2 \; x_3 \cdots x_n \end{bmatrix}$$

Figure 12–1. Row and Column vectors of a data set for values of the random variable, *X*.

chromatographic output, amount of water pumped from a well, gasoline usage of cars, and temperature profile with depth in a lake are all examples of data vectors. Each of the data vectors consists of the set of values of a random variable. There is no limitation on the number of data elements in a data vector.

Data vectors may be configured depending on the goal of the statistical analysis to be performed. An example of a data vector was suggested to be the concentration of sulfate ion in rain samples. The data vector to describe this may be configured to calculate the average value of sulfate at a single location with samples taken over time or to determine the average sulfate ion concentration in a single week over an area. A sample data vector for the first case is shown in Table 12-1*. Dates are shown for convenience only, the important portion of the table is the column of sulfate ion concentration. The sampling locations in the state of Illinois for the second case are graphically portrayed in Figure 12-2 and Table 12-2 shows the data vector for the concentration of sulfate ion at seven sampling locations. The data vector for the first case has 52 data elements; but the data vector for the second case has but 7 data elements.

Vectors may be graphically represented as points on a line as shown in Figure 12-3. The graphic depiction of points on a line indicates that the elements in a data vector are not constrained to be numerically in sequence, but may be in any order and random in magnitude. It can be shown that one should be suspicious if there are many data points in sequence that are monotonically decreasing or increasing.

Data vectors are described and characterized by a mean and a standard deviation. Any data set may be considered as a data vector if the values in the

---

* From the National Atmospheric Deposition Program/National Trends Network for sulfate ion concentration at Hubbard Brook, New Hampshire, 1985.

Table 12–1. Sulfate Concentrations (mg/L) for 1985 for Weekly Atmospheric Deposition Samples obtained at Hubbard Brook, New Hampshire

DATE	$[SO_4^{2-}]$	DATE	$[SO_4^{2-}]$	DATE	$[SO_4^{2-}]$
850108	2.90	850514	3.71	850910	0.22
850115	-- [a]	850521	1.49	850917	0.91
850122	6.02	850528	3.91	850924	0.00
850129	6.43	850604	3.31	851001	0.00
850205	1.84	850611	0.89	851008	--
850212	1.44	850618	1.78	851015	1.56
850219	0.39	850625	2.52	851022	3.29
850226	1.84	850702	0.44	851029	4.41
850305	--	850709	2.30	851105	0.00
850312	1.54	850716	1.59	851112	2.06
850319	1.44	850723	5.17	851119	0.87
850326	4.20	850730	0.63	851126	2.76
850402	1.97	850806	0.24	851203	0.70
850409	3.21	850813	3.95	851210	4.78
850416	10.2	850820	1.92	851217	1.377
850423	--	850827	1.00	851224	4.26
850430	--	850903	1.87	851231	--
850507	1.64				

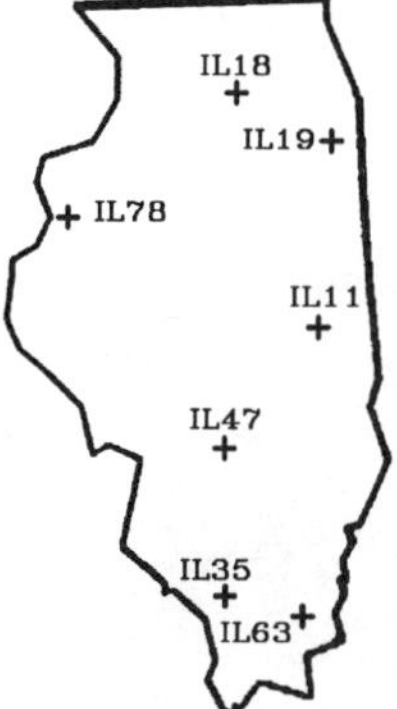

Figure 12–2. Sampling sites from the National Atmospheric Deposition/National Trends Network in the state of Illinois.

Table 12–2. Sulfate Ion Concentrations at Seven Sampling Locations in Illinois

IL11	IL18	IL19	IL35	IL47	IL63	IL78
2.57	2.09	4.44	1.83	2.45	1.9	2.35

data set represent multiple values of a single random variable.

Data Matrices

Several vectors may be joined together to form a data *matrix*. To do this, however, generally requires that each vector be the same length. If missing data values are the cause for unequal lengths, the missing values should be estimated using a technique such as described by Hicks.[1] A general matrix is depicted in Figure 12-4. Matrices are designated by the values of the random variable being enclosed in brackets ([ ]) or braces ({ }), or by placing a tilde (~) over the uppercase letter representing the matrix, $\tilde{A}$. A matrix may also be identified by using the lowercase letter followed by the dimensions of the matrix as subscripts and enclosed in parentheses ($(a_{n,m,1,\ldots})$). The convention that will be followed is that the first subscript will always designate the row and the second subscript will designate the column. Thus for the matrix (a_{nm}), there are n rows and m columns. The general matrix element for this matrix would be a_{ij} . A specific element, such as a_{23}, would designate the second row and the third column. An alternative way to designate an element is to separate the subscripts

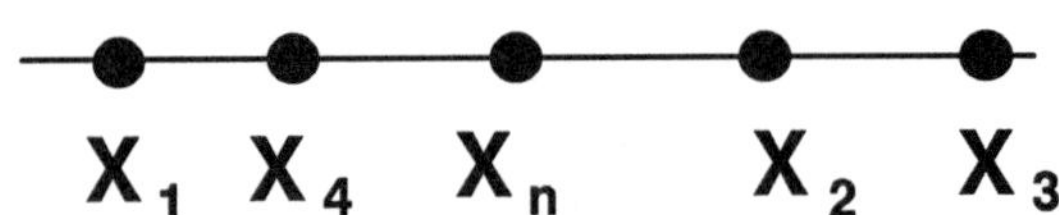

Figure 12–3. Data points represented on a number line.

$$\tilde{A} = (a_{n,m}) = \begin{bmatrix} a_{1,1} & a_{1,2} & a_{1,3} & \cdots & a_{1,m} \\ a_{21} & a_{2,2} & a_{2,3} & \cdots & a_{2,m} \\ \cdots & \cdots & \cdots & a_{i,j} & \cdots \\ a_{n,1} & a_{n,2} & a_{n,3} & \cdots & a_{n,m} \end{bmatrix}$$

Figure 12–4. **Representation of a general matrix ($a_{n,m}$).**

by a comma, ($a_{i,j}$) to eliminate any possible ambiguity. A matrix is square only if the number of rows equals the number of columns.

Addition and Subtraction of Matrices

Two matrices are said to be compatible if the number of rows in both matrices is the same and the number of columns in both matrices is the same. Two matrices are identified as being equal if they are compatible and every element in the first matrix equals the corresponding element in the second matrix:

$$\tilde{A} = \begin{bmatrix} 1 & 4 \\ 2 & 3 \\ 4 & 8 \end{bmatrix} \quad \tilde{B} = \begin{bmatrix} 1 & 4 \\ 2 & 3 \\ 4 & 8 \end{bmatrix} \quad \tilde{A} = \tilde{B} \tag{1}$$

Matrices may be added or subtracted (Figure 12-5), but only if they are compatible. Addition or subtraction is accomplished by adding or subtracting corresponding elements. Note that the addition operation is commutative and associative.

$$\tilde{D} = \tilde{C} + \tilde{A} = \begin{bmatrix} 3 & 5 \\ 7 & 1 \\ 1 & 9 \end{bmatrix} + \begin{bmatrix} 1 & 2 \\ 2 & 3 \\ 4 & 8 \end{bmatrix} = \begin{bmatrix} 4 & 7 \\ 9 & 4 \\ 5 & 17 \end{bmatrix}$$

$$\tilde{E} = \tilde{C} - \tilde{A} = \begin{bmatrix} 3 & 5 \\ 7 & 1 \\ 1 & 9 \end{bmatrix} - \begin{bmatrix} 1 & 2 \\ 2 & 3 \\ 4 & 8 \end{bmatrix} = \begin{bmatrix} 2 & 3 \\ -2 & -2 \\ -4 & 1 \end{bmatrix}$$

Figure 12–5. **Addition and subtraction of Matrices.**

$$\tilde{A} + \tilde{B} = \tilde{B} + \tilde{A} \tag{2}$$

One operation of interest and importance in matrix mathematics is the transpose of a matrix (Figure 12-6). The transpose of a matrix is obtained by interchanging the rows and columns of the matrix and is designated either by a prime (') or a T after the matrix name, e.g. $(a_{ij})^T$ or $(a_{ij})'$. The importance of this operation will be shown shortly.

A matrix may be multiplied by a scalar by multiplying each element of the matrix by the scalar (Figure 12-7). The matrix that results has the same number of rows and columns as the original matrix.

Two matrices may be multiplied together only if the number of elements in each row of the first matrix is equal to the number of elements in each column of the second matrix. Multiplication is given by the equation

$$\tilde{A} \cdot \tilde{B} = \tilde{C} \qquad c_{ij} = \sum_{k=1}^{n} a_{ik} \cdot b_{kj} \tag{3}$$

and can be seen to be the sum of the elements of the ith row multiplied by the elements of the jth column. An example is shown in Figure 12-8 for the multiplication of two matrices, $\tilde{A}$ and C^T. Notice that matrices need not be square to be multiplied. Also be aware that multiplication is not commutative. Consider the question whether $A \times C^T = C^T \times A$. In the first instance, right multiplication of A by C^T yields a 3X3 matrix (Figure 12-8) and the second instance (left multiplication by C^T) results in a 2X2 matrix (Figure 12-9). As can be seen, multiplying together two matrices of different dimensions results in a new matrix having the number of rows of the first matrix and the number of columns of the second matrix. It is left to the reader to determine whether multiplication of matrices is associative, i.e., consider the question whether $A \times (B \times C) = (A \times B) \times C$.

Zero and Identity Matrices

$$\tilde{B} = \begin{bmatrix} 1 & 4 \\ 2 & 3 \\ 4 & 8 \end{bmatrix} \qquad \tilde{B}^T = \begin{bmatrix} 1 & 2 & 4 \\ 4 & 3 & 8 \end{bmatrix}$$

Figure 12–6. Transpose of a matrix; note that the columns and rows are interchanged.

$$a \cdot \tilde{B} = \begin{bmatrix} a \cdot 1 & a \cdot 2 \\ a \cdot 2 & a \cdot 3 \\ a \cdot 4 & a \cdot 8 \end{bmatrix}$$

Figure 12–7. Multiplication of a matrix by a scalar.

$$A \cdot C^T = \begin{bmatrix} 1 & 2 \\ 2 & 3 \\ 4 & 8 \end{bmatrix} \cdot \begin{bmatrix} 3 & 7 & 1 \\ 5 & 1 & 9 \end{bmatrix} = \begin{bmatrix} 13 & 9 & 10 \\ 21 & 17 & 29 \\ 52 & 36 & 76 \end{bmatrix}$$

Figure 12–8. Multiplication of two matrices, $\tilde{A} \bullet C^T$.

$$C^T \cdot A = \begin{bmatrix} 3 & 7 & 1 \\ 5 & 1 & 9 \end{bmatrix} \cdot \begin{bmatrix} 1 & 2 \\ 2 & 3 \\ 4 & 8 \end{bmatrix} = \begin{bmatrix} 21 & 35 \\ 43 & 85 \end{bmatrix}$$

Figure 12–9. Multiplication of two matrices, $C^T \bullet A$.

Two matrices that are of importance are the unity or identity matrix and the zero matrix (Figure 12-10). The unity matrix is a square matrix that has a value of 1 for each diagonal element; all of the off-diagonal elements are zero. In the zero matrix, all of the elements of the matrix are equal to zero. A square matrix when multiplied by a second matrix that results in the identity matrix is called the inverse of the second matrix and is designated A^{-1} . The inverse of a matrix is defined only for square matrices. Multiplication by the inverse of a matrix and

$$\tilde{I} = \begin{bmatrix} 1 & 0 & 0 & \cdots & 0 \\ 0 & 1 & 0 & \cdots & 0 \\ 0 & 0 & 1 & \cdots & 0 \\ \cdots & \cdots & \cdots & \cdots & \cdots \\ 0 & 0 & 0 & \cdots & 1 \end{bmatrix} \qquad \tilde{0} = \begin{bmatrix} 0 & 0 & 0 & \cdots & 0 \\ 0 & 0 & 0 & \cdots & 0 \\ 0 & 0 & 0 & \cdots & 0 \\ \cdots & \cdots & \cdots & \cdots & \cdots \\ 0 & 0 & 0 & \cdots & 0 \end{bmatrix}$$

Identity *Zero*

Figure 12–10. The identity Matrix consists of all zeroes except ones on the diagonal. All elements of the zero matrix are equal to zero.

multiplication by the identity matrix are the only multiplication operations that are both associative and commutative.

Determinants

Before the calculation of the inverse of a matrix can be explained, a definition of a determinant must be given. Determinants map the space of square matrices to real number space, which means that the determinant of a matrix is a single number, a scalar. Determinants thus are scalar representations of square matrices. The determinant of a matrix is denoted by $|A|$. The determinant of a 2x2 matrix (a_{ij}) is given as

$$|A| = \begin{vmatrix} a_{1,1} & a_{1,2} \\ a_{2,1} & a_{2,2} \end{vmatrix} = a_{1,1} \cdot a_{2,2} - a_{2,1} \cdot a_{1,2} \tag{4}$$

The definition for a determinant of higher order is not as easily depicted. A determinant may be evaluated by expanding the determinant along a row or down a column. To expand a determinant, each element of the row (or column) is multiplied by the subdeterminant formed by eliminating the row and column of that matrix element. The subdeterminant is called the *cofactor*. For example, the cofactor of a_{11} for the 2x2 determinant above is a_{22}. Consider the 4 x 4 matrix shown in Figure 12-11, the cofactor of a_{22} is also shown in Figure 12-11. The product of the element and its cofactor must also be multiplied by $(-1)^{i+j}$. We now have a complete definition for a determinant:

$$\det A = \sum_{i=1}^{k} (-1)^{i+j} \cdot a_{ij} \cdot \text{cofactor}(a_{ij}) \tag{5}$$

Of course, it should be recognized that for any determinant of dimension greater than 2 that the cofactor must also be expanded.

Inverse of a Matrix

With the definition of a determinant and a cofactor the definition for an adjoint of a matrix, A^{A} can be given. To form an adjoint, one must first transpose the matrix. After the matrix is transposed, each element of a matrix is replaced by the cofactor of the matrix element multiplied by $(-1)^{i+j}$. If each element of the

$$\begin{bmatrix} a_{11} & a_{12} & a_{13} & a_{14} \\ a_{21} & a_{22} & a_{23} & a_{24} \\ a_{31} & a_{32} & a_{33} & a_{34} \\ a_{41} & a_{42} & a_{43} & a_{44} \end{bmatrix}$$

General 4x4 Matrix

$$\begin{bmatrix} a_{11} & a_{13} & a_{14} \\ a_{31} & a_{33} & a_{34} \\ a_{41} & a_{43} & a_{44} \end{bmatrix}$$

cofactor

Figure 12–11. **A generalized** 4 x 4 **matrix and the cofactor of element** a_{22}.

resulting adjoint matrix is then divided by the determinant of the original matrix, the resulting matrix is the inverse of the original matrix. Realize, of course, that division by a scalar is synonymous with multiplication of the matrix by the reciprocal of the scalar. It should be recognized that not all matrices have an inverse, only those whose determinant is non-zero.

The inverse of a 2x2 matrix is thus given by

$$A^{-1} = \frac{1}{\det A} \begin{bmatrix} a_{22} & -a_{21} \\ -a_{12} & a_{22} \end{bmatrix} \tag{6}$$

The inverse of a general square matrix is

$$A^{-1} = \frac{1}{\det A} \begin{bmatrix} cof(a_{11}) & -cof(a_{21}) & \cdots & (-1)^{1+m} \cdot cof(a_{ml}) \\ -cof(a_{12}) & cof(a_{22}) & \cdots & (-1)^{2+m'} \cdot cof(a_{m2}) \\ \cdots & \cdots & \cdots & \cdots \\ (-1)^{m+1} \cdot cof(a_{im}) & (-1)^{m+2} \cdot cof(a_2 m) & \cdots & (-1)^{m+m} \cdot cof(a_{mm}) \end{bmatrix} \tag{7}$$

The last of the concepts concerning matrices and determinants which will be needed is the trace. The trace of a determinant is the sum of the diagonal elements of the matrix.

$$\text{Trace}\, A = \sum_{i=1}^{m} \sum_{j=1}^{m} a_{ij} \tag{8}$$

where $i{=}j$. The trace of the general 2x2 matrix is $a_{11}+a_{22}$. The trace of any square matrix is then:

$$\text{Trace} = \sum_{i=j} a_{ij} = a_{11} + a_{22} + a_{33} + \cdots \tag{9}$$

Example

Let us find the inverse of the 4x4 matrix, $\tilde{A}$:

$$\begin{bmatrix} 1 & 3 & 8 & 1 \\ 4 & 1 & 6 & 2 \\ 3 & 4 & 5 & 7 \\ 7 & 1 & 2 & 9 \end{bmatrix}$$

The determinant of the matrix is found by expanding along the top row. The cofactor matrices of the matrix elements of the first row are

$$\begin{bmatrix} 1 & 6 & 2 \\ 4 & 5 & 7 \\ 1 & 2 & 9 \end{bmatrix} \quad \begin{bmatrix} 4 & 6 & 2 \\ 3 & 5 & 7 \\ 7 & 2 & 9 \end{bmatrix} \quad \begin{bmatrix} 4 & 1 & 2 \\ 3 & 4 & 7 \\ 7 & 1 & 9 \end{bmatrix} \quad \begin{bmatrix} 4 & 1 & 6 \\ 3 & 4 & 5 \\ 7 & 1 & 2 \end{bmatrix}$$

Each of the cofactor matrices must be expanded using cofactors also. As an example the full expansion of the first matrix element leads to the following as the cofactor of a_{11}:

$$\begin{vmatrix} 1 & 6 & 2 \\ 4 & 5 & 7 \\ 1 & 2 & 9 \end{vmatrix} = \left(1 \cdot \begin{vmatrix} 5 & 7 \\ 2 & 9 \end{vmatrix} - 6 \cdot \begin{vmatrix} 4 & 7 \\ 1 & 9 \end{vmatrix} + 2 \cdot \begin{vmatrix} 4 & 5 \\ 1 & 2 \end{vmatrix} \right) = (31 - 6 \cdot 29 + 2 \cdot 3) = -137$$

The adjoint of $\tilde{A}$ is

$$\begin{bmatrix} -137 & 136 & 90 & -85 \\ -98 & 178 & 160 & -142 \\ 88 & -70 & -62 & 54 \\ 109 & -110 & -74 & 79 \end{bmatrix}$$

and the determinant of $\tilde{A}$ is 82. The inverse of $\tilde{A}$ is

$$A^{-1} = \begin{bmatrix} -1.671 & 1.659 & 1.098 & -1.037 \\ -2.415 & 2.171 & 1.951 & -1.732 \\ 1.073 & -0.854 & -0.756 & 0.659 \\ 1.329 & -1.341 & -0.902 & 0.963 \end{bmatrix}$$

The trace of $\tilde{A}$ is $1+1+5+9 = 16$.

Reference

Hicks, C. R. ***Fundamental Concepts in the Design of Experiments***. Holt, Rinehart &Winston, New York, 1973.

13
LINEAR MODELS

General

Whenever we have data, we will fit the data to a general equation. The fitting equation is called a model, which consists of a mathematical relationship between the variables. The parameters of the model are to be obtained to be able to predict future values of the dependent variable when future values of the independent variable are known. The other goal of a model is to reduce the size of the data set. By describing the data set with a model equation, the size of the data set has been reduced from n to the number of parameters in the model. The equation

$$R_i = b_0 + b_1 \cdot x_i \tag{1}$$

that we used in the smoothing of analytical data was an example of a model.

Models may be divided into two classes, linear and nonlinear. The distinction between the two is often confused in the minds of many individuals and some of the linear models in common use are often termed nonlinear models. In this text, the term linear model will indicate any model for which the coefficients or parameters of the model are first order (i.e., have an exponent of 1) and no two coefficients are multiplied or divided by each other. A nonlinear model will then be one that has coefficients whose exponents are not equal to 1 or that have coefficients multiplied or divided by another coefficient. A linear model may contain almost any transformed variable of the independent variable, such as x^2, $SIN(x)$, e^x, and so forth. Inclusion of such functions in a model in no way impacts on the model being considered linear. Calculation of fitting parameters for linear models is much simpler than for nonlinear models. Many different schemes have been used to fit nonlinear models including Fein-Marquardt, simplex optimization, and Levinson-Marquardt.

Matrix Formulation of Models

An example of a linear model was previously encountered when the smoothing of analytical data using a linear least–squares approach was considered using the function

$$R_i = b_0 + b_1 \cdot x_i \tag{2}$$

and obtained the normal equations

$$\begin{aligned} b_0 \cdot \sum 1 + b_1 \cdot \sum x_i &= \sum R_i \\ b_0 \sum x_i + b_1 \cdot \sum x_i^2 &= \sum R_i \cdot x_i \end{aligned} \tag{3}$$

to estimate the fitting parameters, b_o and b_1. These equations may be solved by casting them in matrix form. If the following represent the matrices of the coefficients and the various sums

$$\tilde{B} = \begin{bmatrix} b_0 \\ b_1 \end{bmatrix} \tag{4}$$

$$\tilde{X} = \begin{bmatrix} \sum 1 & \sum x_i \\ \sum x_i & \sum x_i^2 \end{bmatrix} \tag{5}$$

$$\tilde{Y} = \begin{bmatrix} \sum R_i \\ \sum x_i \cdot R_i \end{bmatrix} \tag{6}$$

a matrix equation can be formulated as

$$\tilde{Y} = \tilde{X} \cdot \tilde{B} \tag{7}$$

The inverse of X is given by

$$\tilde{X}^{-1} = \frac{1}{Det\,\tilde{X}} \begin{bmatrix} \sum x_i^2 & -\sum x_i \\ -\sum x_i & N \end{bmatrix} \tag{8}$$

If both sides of Equation (7) are left multiplied by the inverse of X, it is evident that the resulting matrix will be the matrix of coefficients

$$\tilde{B} = \tilde{X}^{-1} \cdot \tilde{Y} = \frac{1}{Det\,\tilde{X}} \begin{bmatrix} \sum x_i^2 \cdot \sum R_i - \sum x_i \cdot \sum x_i \cdot R_i \\ N \cdot \sum x_i \cdot R_i - \sum x_i \cdot \sum R_i \end{bmatrix} \tag{9}$$

or after expanding the determinant of X,

$$\tilde{B}=\begin{bmatrix}b_o\\ b_1\end{bmatrix}=\begin{bmatrix}\dfrac{\sum x_i^2\cdot\sum R_i-\sum x_i\cdot\sum x_i\cdot R_i}{N\cdot\sum x_i^2-(\sum x_i)^2}\\ \\ \dfrac{N\cdot\sum x_i\cdot R_i-\sum x_i\cdot\sum x_i\cdot\sum R_i}{N\cdot\sum x_i^2-(\sum x_i)^2}\end{bmatrix} \tag{10}$$

General Fitting

Let us now consider a general fitting equation:

$$Y=\sum b_i\cdot f(x_i) \tag{11}$$

If $f(x_i)$ is x_i^i, a polynomial would result:

$$Y=\sum b_i\cdot x_i^i \tag{12}$$

From this equation, many different models could be formulated to fit a set of analytical data. Among the models would be three rather trivial models that typify the method: a model that contains only b_o, a model that contains only b_1, and a model that contains both b_o and b_1. Let us consider ways to solve each of these, then a way to find solutions to a general model.

The Y_is calculated from a model will, in general, seldom be exactly equal to the actual ordinate values obtained by experimentation, but will have some portion that is not accounted for by the model. This unexplained portion is called the residual of the experimental value that cannot be explained by the model. If the residual amount is given the symbol, ϵ_i, then the experimental Y_is can be thought of as

$$Y_i^{experimental}=Y_i^{estimated}+\epsilon_i \tag{13}$$

The estimated Y_i is often represented by

$$Y_i^{estimated}=\hat{Y} \tag{14}$$

where now we must also estimate the ϵ_i's in addition to the b_i and we can formulate equations for three situations. Let us first consider a situation where we wish to estimate Y_i from 2 experiments with no dependence upon the x_i (i.e., the fitting parameter b_1 is equal to zero), then

$$Y_i = \beta_0 + \epsilon_i \tag{15}$$

and an attempt could be made to formulate equations

$$\begin{aligned} y_1 &= b_0 + \epsilon_1 \\ y_2 &= b_0 + \epsilon_2 \end{aligned} \tag{16}$$

Unfortunately, we would then be trying to estimate three unknowns (b_o, ϵ_1, and ϵ_2) from two equations. This is an impossible situation from which to obtain an analytic solution. An additional constraint that the derivative of the sum of squares of the residuals with respect to b_o be zero must be invoked:

$$\frac{\partial \Sigma \epsilon_i^2}{\partial b_0} = 0 \tag{17}$$

In matrix notation*, the equation becomes

$$X \cdot B + \mathrm{E} = Y \tag{18}$$

where X is a matrix of x values, B is the matrix of b coefficients, E is the matrix of residuals, and Y are the experimental data values. The equation can be rearranged to find a value for the residual matrix to be

$$\mathrm{E} = Y - X \cdot B \tag{19}$$

To solve this matrix equation for the matrix of fitting parameters, the other three matrices, X, B, and Y, must be known or at least be able to be estimated. These three matrices are given by

$$X = \begin{bmatrix} 1 \\ \\ 1 \end{bmatrix} \qquad \mathrm{E} = \begin{bmatrix} \epsilon_1 \\ \\ \epsilon_2 \end{bmatrix} \qquad Y = \begin{bmatrix} y_1 \\ \\ y_2 \end{bmatrix} \tag{20}$$

Left multiplication of the residual matrix by its transpose, E^T, obtains the sum of squares of the residuals:

$$\mathrm{E}^T \cdot \mathrm{E} = \begin{bmatrix} \epsilon_1 & \epsilon_2 \end{bmatrix} \cdot \begin{bmatrix} \epsilon_1 \\ \\ \epsilon_2 \end{bmatrix} = \begin{bmatrix} \epsilon_1^2 + \epsilon_2^2 \end{bmatrix} \tag{21}$$

but E is also equal to $Y - X \bullet B$, therefore, $\mathrm{E}^T = (Y - X \bullet B)^T$. Thus, the derivative of $\mathrm{E}^T\mathrm{E}$ with respect to b equals

---

*The tilde above the upper case letter will be dropped in the rest of the discussions. Recognize that an upper case letter will be used to designate a matrix.

$$X^T(Y - X \cdot B) = 0 \tag{22}$$

Expanding the product and setting equal to each other

$$X^T \cdot X \cdot B = X^T \cdot Y \tag{23}$$

If both sides are multiplied by the inverse

$$(X^T \cdot X)^{-1} \tag{24}$$

the following is obtained:

$$(X^T \cdot X)^{-1} \cdot (X^T \cdot X) \cdot B = (X^T \cdot X)^{-1} \cdot X^T \cdot Y$$

$$B = (X^T \cdot X)^{-1} \cdot X^T \cdot Y \tag{25}$$

where B is the matrix of b coefficients according to the above fitting equation. This is a general equation for the solution of a linear model. As will be seen in the following, this equation can be used to evaluate any set of data. The matrices that are formed, the $\boldsymbol{X}$, $\boldsymbol{X}^{\mathrm{T}}$, and $\boldsymbol{Y}$ matrices, to obtain the coefficient matrix, $\boldsymbol{B}$, have the dimensions

$\boldsymbol{X}$ (n x p)
$\boldsymbol{Y}$ (n x 1)
$\boldsymbol{B}$ (p x 1)

where n is the number of data pairs (x_i,y_i) and p is the number of coefficients being obtained.

Let us evaluate b_0 for the two experiment design according to the matrices given in Equation (20).

$$X^T = [1 \quad 1] \qquad X^T \cdot X = 2 \qquad [X^T \cdot X]^{-1} = \left[\frac{1}{2}\right]$$

$$X^T \cdot Y = [1 \quad 1] \cdot \begin{bmatrix} y_1 \\ y_2 \end{bmatrix} = [y_1 + y_2] \tag{26}$$

Thus b_o is the estimate of the arithmetic average of the Y_is:

$$B = [b_0] = \left[\frac{1}{2}\right] \cdot [y_1 + y_2] = \left[\frac{y_1 + y_2}{2}\right] \tag{27}$$

Consider a different model for fitting data- a straight line through the origin. The fit equation will be

$$y_i = b_1 x_i + \epsilon_i \tag{28}$$

since b_0 is constrained to be equal to zero. Therefore the B matrix will contain only the estimated value of the slope, b_1. The other matrices that will be needed are

$$X = \begin{bmatrix} x_{11} \\ x_{21} \end{bmatrix} \qquad X^T = \begin{bmatrix} x_{11} & x_{21} \end{bmatrix}$$

$$Y = \begin{bmatrix} y_{11} \\ y_{21} \end{bmatrix} \qquad (X^T X)^{-1} = \begin{bmatrix} \dfrac{1}{x_{11}^2 + x_{21}^2} \end{bmatrix} \tag{29}$$

$$X^T \cdot Y = \begin{bmatrix} x_{11} & x_{21} \end{bmatrix} \cdot \begin{bmatrix} y_{11} \\ y_{21} \end{bmatrix} = \begin{bmatrix} x_{11} \cdot y_{11} + x_{21} \cdot y_{21} \end{bmatrix}$$

from which the coefficient matrix, B, is calculated to be

$$B = \begin{bmatrix} b_1 \end{bmatrix} = \begin{bmatrix} \dfrac{1}{x_{11}^2 + x_{21}^2} \end{bmatrix} \cdot \begin{bmatrix} x_{11} \cdot y_{11} + x_{21} \cdot y_{21} \end{bmatrix} = \begin{bmatrix} \dfrac{x_{11} \cdot y_{11} + x_{21} \cdot y_{21}}{x_{11}^2 + x_{21}^2} \end{bmatrix} \tag{30}$$

The third model contains both of the fitting parameters:

$$y_i = b_0 + b_1 x_i + \epsilon_i \tag{31}$$

The matrices required to solve for both b_o and b_1 are the same as for the other two cases:

$$B = \begin{bmatrix} b_0 \\ b_1 \end{bmatrix} \qquad X = \begin{bmatrix} 1 & x_{21} \\ 1 & x_{22} \end{bmatrix} \qquad X^T = \begin{bmatrix} 1 & 1 \\ x_{21} & x_{22} \end{bmatrix}$$

$$X^T \cdot X = \begin{bmatrix} 2 & x_{21} + x_{22} \\ x_{21} + x_{22} & x_{22}^2 + x_{22}^2 \end{bmatrix} \qquad Y = \begin{bmatrix} y_{11} \\ y_{12} \end{bmatrix} \tag{32}$$

$$(X^T \cdot X)^{-1} = \frac{1}{Det\, X^T \cdot X} \begin{bmatrix} x_{21}^2 + x_{22}^2 & -(x_{21} + x_{22}) \\ -(x_{21} + x_{22}) & 2 \end{bmatrix}$$

from which we obtain

$$B = \begin{bmatrix} b_0 \\ b_1 \end{bmatrix} = \frac{1}{Det\ X^T \cdot X} \begin{bmatrix} (x_{21}^2 + x_{22}^2) \cdot (y_{11} + y_{21}) - (x_{21} + x_{22}) \cdot (y_{11} \cdot x_{21} + y_{21} \cdot x_{22}) \\ - (x_{21} + x_{22}) \cdot (y_{11} + y_{21}) + 2 \cdot (x_{21} \cdot y_{11} + x_{22} \cdot y_{21}) \end{bmatrix} \quad (33)$$

The general solution to the matrix formulation of linear regression for the two parameter model

$$y = b_0 + b_1 \cdot x \quad (34)$$

is

$$B = \begin{bmatrix} b_0 \\ b_1 \end{bmatrix} = (X^T \cdot X)^{-1} = \frac{1}{Det\ X^T \cdot X} \begin{bmatrix} \sum_{i=1}^{n} x_{2i}^2 \sum_{i=1}^{n} y_{il} - \sum_{i=1}^{n} x_{2i} \sum_{i=1}^{n} (x_{2i} \cdot y_{il}) \\ - \sum_{i=1}^{n} x_{2i} \sum_{i=1}^{n} y_{1i} + n \cdot \sum_{i=1}^{n} x_{2i} \cdot y_{1i} \end{bmatrix} \quad (35)$$

where

$$Det\ X^T \cdot X = \left[n \cdot \sum_{i=1}^{n} x_{2i}^2 - \left(\sum_{i=1}^{n} x_{2i} \right)^2 \right] \quad (36)$$

The beauty of using this formulation is that it is quite simple to fit any general function of x to a linear regression [linear referring to the fact that all of the coefficients are first order and not that the $f(x)_i$s are necessarily first order] to any summation of functions of x.

$$y = \sum b_i \cdot f(x_i) \quad (37)$$

For example, the model

$$y = b_0 + b_1 \cdot \mathrm{SIN}(x_i) + b_2 \cdot x_i + b_3 \cdot x_i^2 + b_4 \cdot \mathrm{COS}(x_i) \quad (38)$$

may be fit very easily by correctly forming the required matrices and the values of the functions of x. The required matrices are:

$$B = \begin{bmatrix} b_0 \\ b_1 \\ b_2 \\ b_3 \\ b_4 \end{bmatrix} \quad Y = \begin{bmatrix} y_1 \\ y_2 \\ y_3 \\ y_1 \\ \ldots \\ y_n \end{bmatrix} \quad X = \begin{bmatrix} 1 & \mathrm{SIN}(x_1) & x_1 & x_1^2 & \mathrm{COS}(x_2) \\ 1 & \mathrm{SIN}(x_2) & x_2 & x_2^2 & \mathrm{COS}(x_2) \\ 1 & \mathrm{SIN}(x_3) & x_3 & x_3^2 & \mathrm{COS}(x_3) \\ \ldots & \ldots & \ldots & \ldots & \ldots \\ 1 & \mathrm{SIN}(x_n) & x_n & x_n^2 & \mathrm{COS}(x_n) \end{bmatrix} \quad (39)$$

Interpretation of a Model Equation

The equation just developed for the two-parameter model, Equation (34), can be interpreted in another fashion besides being used as the means to which data are fit. The quantity, Y, may also be thought of as the expected value of the y_is and the X can be thought of as the expected value of the x_is. Then the equation is interpreted to mean that the expected value of Y can be predicted from the expected value of X. This leads us to formulate two measures of the deviations of the y_is from expected values—the deviation from the expected value of Y

$$y_i - \bar{Y} \tag{40}$$

and the deviation of the predicted value of a single y_i from the experimental value of y_i

$$y_i - \hat{Y}_i \tag{41}$$

If the deviations in Equation (40) are squared and summed, the sum of squares about the mean is obtained:

$$SS_{mean} = \sum (y_i - \bar{Y})^2 \tag{42}$$

If the deviations in Equation (41) are squared and summed, the sum of squares of the residuals is obtained:

$$SS_{residual} = \sum (y_i - \hat{y})^2 \tag{43}$$

A third deviation may also be obtained as the deviation of the predicted value from the expected value. If these deviations are squared and summed, the sum of squares about the regression is obtained:

$$SS_{regression} = \sum (\hat{y} - \bar{y})^2 \tag{44}$$

It should be obvious that the SS_{mean} is the sum of $SS_{regression}$ and $SS_{residual}$. From such an analysis, the ideal model will be one that has a very large $SS_{regression}$ and a very small $SS_{residual}$.

Each of the sum of squares may be obtained as products of matrices:

$$SS_{mean} = Y^T \cdot Y \tag{45}$$

$$SS_{regression} = B^T \cdot X^T \cdot Y \tag{46}$$

$$SS_{residual} = Y^T \cdot Y - B^T \cdot X^T Y \tag{47}$$

where the vector Y is the vector of values of the dependent variable, X is the matrix of values of the independent variable, and B is the vector of coefficients of the model. The X matrix may be composed as earlier described.

Model Fitting with Analysis of Variance

A set of data consisting of pairs (x_i, y_i) may also be analyzed with an analysis of variance (ANOVA) model. In the ANOVA model, two sources of variability will be considered, variability due to the regression and variability due to the errors or residuals in the fitted terms. The ANOVA table thus constructed will have the following form:

Source of Variability	Degrees of Freedom	Sum of Squares	Mean Square
Regression	p	B^TX^TY	$MS_{regression}$
Residual	$n - p$	$Y^TY - B^TX^TY$	$MS_{residual}$
Total	n	Y^TY	

Thus it should be seen that the ANOVA table can be used to determine if the regression is significant or not.

Correlation Coefficient

As with other statistical operations, a determination of how well the model fits the data is desired. One of the means to determine the degree to which the values of the dependent variable are explained by the independent variable is to calculate the statistic called r, the correlation coefficient. A summary of some important aspects of r and r^2 is given in Table 13-1. r will range from -1 to +1. ± 1 are perfect correlations and 0 is no correlation between the variables. r is given by

Table 13-1. Aspects of *r*, the Correlation Coefficient

1. r will always lie in the interval, $-1 \leq r \leq 1$
2. The value of r will be independent of the measurement units of x or y
3. The value of r will be the same regardless of the regression of y *on* x or x *on* y
4. The square of r is the coefficient of determination and gives a measure of the amount of the variance in the dependent variable explained by the variation in the independent variable

$$r = \frac{\sum_i (x_i - \bar{x}) \cdot (y_i - \bar{y})}{(n-1)\sqrt{s_x^2 \cdot s_y^2}} \qquad \textbf{(48)}$$

Examination of this equation leads one to understand that the important governing portions of the equation are the two differences, $(x_i - \bar{x})$ and $(y_i - \bar{y})$, which indicate that if a relationship is strongly positive that any of the x values which are greater than the mean of x will be paired with a value of y, which is greater than the mean of y. The overall consequence of this is that the product will be greater than 0. Similarly, if a value of x is less than the mean of x, then the x–value will be paired with a y–value that is less than the mean of y, then the product will still be greater than zero. However, if the value of x that is greater than the mean of x is paired with a y–value that is less than the mean of y, the product will be less than 0. A similar argument holds for a value of x that is less than the mean of x paired with a value of y that is greater than the mean of y.

The product of the differences should not be used as the indicator of the strength of the relation between x and y since the product is strongly dependent on the units in which x and y are expressed. For example, we could express the relation between x and y in meters, then the product might be 64; but if the relation were expressed in millimeters, the product would be 64,000, and if the relation were expressed in kilometers, the product would be 0.064. To eliminate this dependence on the units, the product is normalized using the standard deviations of x and y.

As with any statistic which we calculate, there is a null hypothesis associated with the statistic. The null hypothesis is that $r=0$ and the alternative hypothesis is that $r \neq 0$. Elements of the ANOVA table can be used to obtain the significance of the correlation. The quotient

$$\frac{\mathrm{MS}_{\mathrm{regression}}}{\mathrm{MS}_{\mathrm{residual}}} \qquad \mathbf{(49)}$$

can be compared to the F–table to obtain the critical value. The null hypothesis is that $r=0$. If the quotient is larger than the critical value of F, the null hypothesis is rejected.

The square of the correlation coefficient, r^2, is often called the coefficient of determination and is akin to variance. r^2 is related to how much of the variation in the dependent variable is explained by the variation in the independent variable. A correlation coefficient of 0.9 would give a coefficient of determination of 0.81 and the variation of the independent variable would account for 81% of the variation in the dependent variable. Some care must be taken when comparing the results of the model (i.e., the predicted values) to the original data values. Suppose there are several values of the independent variable for which multiple values of the dependent variable are present. The regression model, since it is only in actuality fitting coefficients to means of the dependent variable, will only be fitting to the number of pairs of repeat values. For example, five values of the independent variable and 20 observations of the dependent variable for each of the values of the independent variable would appear to have 100 pairs of data. In actuality, there are only five pairs of data—one for each pair $(x_i, \bar{y}_i)$. Thus if a four- or five-coefficient model is used to predict the correlation, it should be no surprise that the model does such a creditable job of the prediction.

As an alternative to using the table of values for the F distribution for determination of significance of r, one may calculate a t-value from the equation

$$t_{test} = \frac{r \cdot \sqrt{n-2}}{\sqrt{1-r^2}} \qquad \mathbf{(50)}$$

where the t –value will be compared to a critical value of t with an α level of significance and n-2 degrees of freedom.

One very important assumption is made when we perform a linear regression using either of the two above methods (matrix or normal equations):

all of the variation is assumed to be in the ordinate values and not in the abscissa values. This may be a valid assumption if the magnitude of the x-variations is very small compared to the magnitude of the variations in the y-values. For example, in preparation of a calibration curve, the variation in the y-values are often of the magnitude of about 1–2% or more. Depending on how the standards were prepared, the variation of the concentration values is usually on the order of 0.1% or less. In this case, we are probably justified in assuming that the variation in the x-values (concentrations) may be ignored when compared to the variation of the y-values (instrument responses). As instruments progress in complexity and in reproducibility, the y-variations may start to be on the same order as the x-values. Different techniques of preparation of calibration curves will then need to be accomplished.

Another very important assumption made in this vein is that the distribution of both the x– and the y–values is normal. The correlation coefficient then is a measure of how well a bivariate normal distribution would describe the two variables. This, unfortunately, is beyond the scope of this book to determine if the joint distribution is bivariate normal.

Analysis of Residuals

As regression models are obtained, one may calculate an amount that is not explained by the model for each data point. This amount "left over" is termed the residual and is calculated from the equation

$$\epsilon_i = y_i - \hat{y}_i \tag{51}$$

where y_i is the data value and $\hat{y}_i$ is the predicted value from the model. An examination of these residuals can help us to determine several things about the appropriateness of the model. Let us assume for sake of explanation that we are strictly dealing with the linear regression model:

$$\hat{y}_i = b_0 + b_1 \cdot x_i \tag{52}$$

in which case, we can determine from the residuals:

1. Whether the best model is the linear regression given in Equation (52).
2. Whether the mean of the residuals is zero.

3. Whether the residuals are homoscedastic.
4. Whether the error terms are independent.
5. Whether the residuals are normally distributed.
6. Whether it is possible that more than one independent variable should be used to explain the variation in the dependent variable.
7. Whether all of the values of the dependent variable are fit by the regression equation or whether there a few that are radically aberrant.

We shall now examine how we may use the residuals to determine each of the foregoing seven conditions. First the condition of normality of the residuals. The easiest way to determine this is through the normality plot given in Chapter 6. To use the normality plot, one must calculate each of the residual values and sort these in ascending sequence. Then calculate the percentiles represented by the number of data points in the data set using the equation:

$$i\text{th } \textit{percentile} = \left[\frac{100 \cdot (i - 0.5)}{n}\right] \qquad \textbf{(53)}$$

where n is the number of pairs of data used in the regression (if using a first-order polynomial as the fit model).

One should examine these residuals to determine if there is a systematic variation of the residuals with the values of the independent variable. It is not sufficient to use just the correlation coefficient to determine how well the data are explained by the model. The residuals that are obtained should be plotted and examined graphically. An informal way to determine whether a linear model is appropriate for the data is to graph the data points with a plot of the model function superimposed on it. Examination of such a plot can often indicate whether the model is appropriate or not. As you examine such a plot, the number of data points "above" and "below" the model line should appear to be about the same as shown in Figure 13-1, for example.

An alternative method to superimposing the model function on a scatter plot would be to calculate the set of residuals and then plot them against the values of the independent variable from which they were obtained. This method of determining the validity of the model is generally more appropriate if there is a very steep dependence of the dependent variable on the independent variable. A very small change in x results in a very large change in y. Let's examine an example to see the method of application of these principles.

Example

If we have a set of data such as shown in Table 13-2, we would first calculate a linear model to the data. The data in the table represents pairs of nitrate and sulfate ion concentrations in rainfall. The linear model which is chosen for this data is the linear first order polynomial:

$$\left[SO_4^{2-}\right] = b_0 + b_1 \cdot \left[NO_3^-\right] \tag{54}$$

The sulfate data is plotted as a function of the nitrate concentration and is shown in Figure 13-1. The predicted concentrations of sulfate ion in Table 13-2 were obtained from the known nitrate concentration and equation (54).

One should also determine that the residuals are normally distributed with a mean of zero and constant variance over the range of the independent variable. To do so, two approaches are available. The first is to plot the residuals as a function of the value of sulfate ion predicted from the model. This plot is shown in Figure 13-2. One can also construct the normal probability plot introduced in chapter 6. The normal probability plot is shown in Figure 13-3.

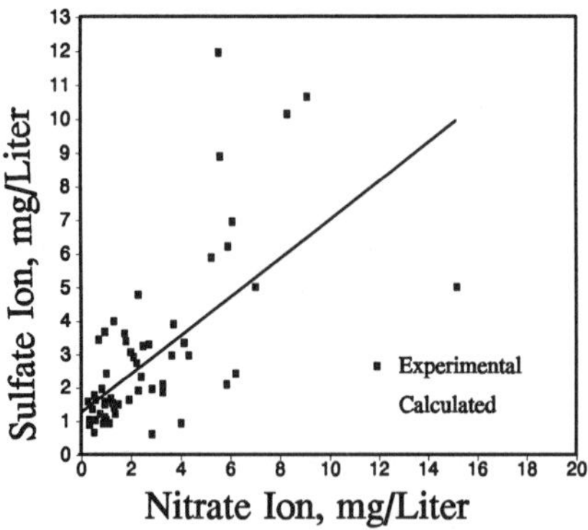

Figure 13-1. Sulfate ion concentration as a function of nitrate ion concentration. The concentrations of the ions were determined in samples of rainwater.

Table 13-2. Data for the Relation Between Concentration of Nitrate and Sulfate Ions in Rainwater; Also Shown is the $[SO_4^{2-}]_{pred}$

$[NO_3^-]$	$[SO_4^{2-}]$	$[SO_4^{2-}]_{pred}$	$[NO_3^-]$	$[SO_4^{2-}]$	$[SO_4^{2-}]_{pred}$
0.17	1.05	1.377	1.76	3.33	2.292
0.29	0.94	1.446	1.79	1.65	2.310
0.34	0.66	1.475	1.87	2.36	2.356
0.43	1.03	1.526	1.93	3.05	2.390
0.48	1.57	1.555	2.04	2.92	2.454
0.49	0.89	1.561	2.22	2.78	2.557
0.52	1.30	1.578	2.23	4.79	2.563
0.57	0.79	1.607	2.24	1.91	2.569
0.58	1.42	1.613	2.32	2.30	2.615
0.69	1.34	1.676	2.38	3.23	2.650
0.70	0.97	1.682	2.67	3.27	2.817
0.71	3.41	1.688	2.78	2.99	2.880
0.78	1.18	1.728	2.79	2.04	2.886
0.79	1.33	1.734	2.8	0.55	2.892
0.82	1.30	1.751	3.13	2.12	3.082
0.87	3.69	1.780	3.23	1.91	3.139
0.93	1.09	1.814	3.63	2.85	3.370
0.94	1.36	1.820	3.68	3.87	3.398
0.99	2.36	1.849	3.72	3.29	3.421
1.11	1.88	1.918	3.92	0.94	3.537
1.11	1.65	1.918	4.13	3.39	3.658
1.18	1.52	1.958	4.30	3.05	3.756
1.18	1.45	1.958	5.20	5.89	4.274
1.21	1.17	1.976	5.53	8.86	4.464
1.21	1.87	1.976	5.77	2.12	4.602
1.25	2.03	1.999	5.84	6.23	4.643
1.26	4.03	2.004	6.07	7.01	4.775
1.27	1.51	2.010	6.16	2.39	4.827
1.33	1.93	2.045	7.03	5.08	5.328
1.62	2.13	2.212	8.24	10.16	6.025
1.66	3.59	2.235	9.10	10.76	6.520

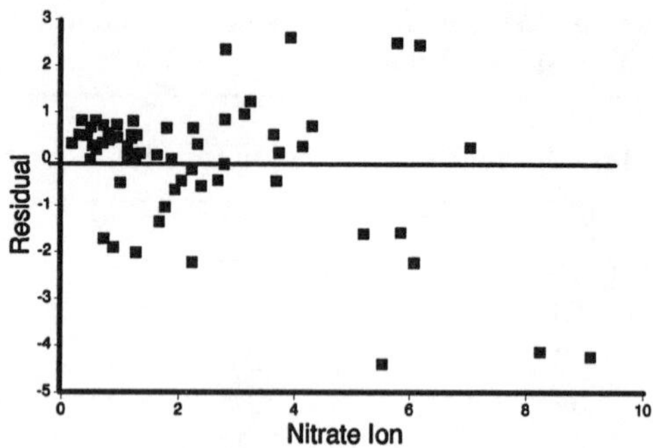

Figure 13-2. Residuals of the model of sulfate as a function of nitrate concentration.

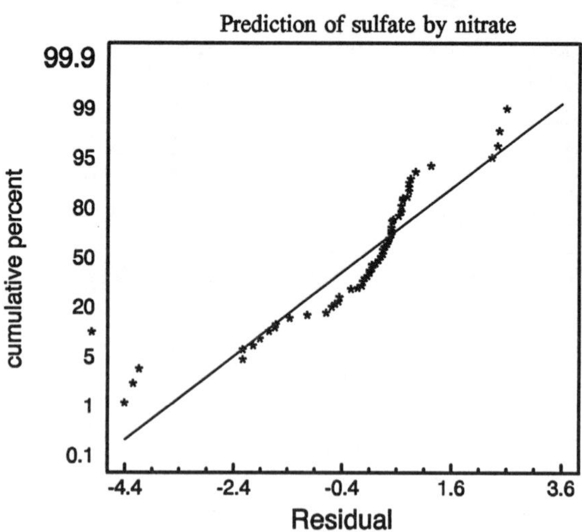

Figure 13-3. Normal probability of residuals from the regression of $[SO_4^{2-}]$ predicted by $[NO_3^-]$.

14
QUANTITATION OF ANALYTES

General

Of what importance is quantitation? How is quantitation accomplished? What are some pitfalls to be careful of when quantitating analytes in a sample? Eisenhart has said "until a measurement operation has ... attained a state of statistical control, it cannot be regarded in any logical sense as measuring anything at all."[1] A major aspect of statistical control is knowing the relationship between the response of an instrument and the concentration of the analyte to which the instrument is responding. To do so will require that the relationship be mathematically expressed. Several different techniques of quantitation will be presented. The most common of the methods is the use of the linear first-order regression called a calibration curve. Other methods for single analytes that can be used are single standard, internal standard, and standard additions.* Several methods will be presented that also may be used for simultaneous calculation of concentration of multiple analytes.

Consideration of Standard Analytical Reference Materials

A major factor in good quantitation procedures is to ensure that we have standards that adequately represent the material(s) being studied. There are several things to consider when selecting a standard analytical reference material (SARM), such as stability, matrix, traceability to The National Institute of Standards and Technology (NIST) (formerly National Bureau of Standards), homogeneity, representation of material being analyzed, availability, cost, and

* The method of standard additions is also known as spiking. It is an extremely useful method when only a few samples are to be analyzed or when the presence of interferences is suspected.

ease of application to desired analytical instrument. Without an adequate SARM, we cannot be sure of any analytical measurement that is made. For the rest of the discussion about quantitation, it will be assumed that adequate SARMs have been obtained and are being used.

Calibration Pitfalls

The major pitfall that individuals fall into when using calibration curves is that they try to extend the range of calibration beyond the range of the standards that were used to prepare the calibration curve. The calibration curve is valid only for the range of the standards that have been used for its preparation.

There is no calibration curve that may be extended indefinitely to high or low concentration. At some point, because of inherent characteristics of the instrumentation, the true response will depart from the model. Response from many instruments will flatten at high and low concentrations, the low end because no signal is being produced and the high end because of detector saturation.

If it is necessary to analyze samples whose concentrations are beyond the range of the calibration curve, one of two courses of action is necessary. One of the actions that is possible is to prepare, from the SARMs, calibration standards with higher or lower concentrations until the concentration of the unknown sample has been bracketed by at least two standards.

An alternative method is to find ways to treat the sample being analyzed to bring its concentration within the range of the calibration curve. If the solution is of higher concentration than the highest standard, then dilution of the sample with an appropriate matrix has been a satisfactory solution. Alternatively, if the concentration of the sample is too low for the calibration curve a means to concentrate the sample must be attempted. The literature is replete with ways to accomplish this including solid–liquid extraction, distillation, liquid–liquid extraction, evaporation, adsorption on material such as TENAX or XAD resins, etc.

Calibration Curve

The most prevalent method of quantitation is to analyze a set of calibration standards, the concentrations of which are known quite accurately. The calibration standards are prepared from standard analytical reference materials that adhere to the criteria proposed for reference materials. The SARMs from which the calibration standards are prepared may be either the primary standard or a secondary standard traceable to the primary standard. In calibration of an instrument, data on some instrument response such as peak height, peak area, absorbance at a given wavelength, and diffusion current are recorded as a function of the concentration of the analyte in the calibration standard. A scatter plot may be drawn that shows a general tendency for the instrument to regularly vary with the concentration of the analyte. A curve is then fit to this data to obtain a functional dependence of the instrument response with the concentration. Whichever model is chosen should have some theoretical basis. Not all instruments respond in a linear fashion, e.g., the electrical potential of a solution is related to the logarithm of the concentration, and the response of the flame photometric detector on a gas chromatograph to sulfur compounds is quadratic. Thus the model should reflect this known functional dependence, if possible. It may be possible to mathematically transform the data to be able to use the models we have been discussing. The most generally used model for the description of the instrumental response as a function of the concentration of analyte is the linear first-order regression of the form

$$\text{Response} = b_o + b_1 \cdot [\text{Analyte}] \tag{1}$$

and has the general appearance as shown in Figure 14-1. The regression coefficients are obtained using the techniques given in Chapter 13.

How do we make use of the regression equations? One of the most common uses of the regression equations is the calibration of analytical instruments so as to be able to determine concentration of analytes in samples.

Formulation of the calibration curve

The general formulation of the calibration curve is given by the relationship in Equation (1) where [Analyte] is the concentration of the analyte expressed in appropriate units. This is the model that has been used for several different situations. We obtain estimates of the two parameters using either least squares procedure used above, namely the two normal equations
or the matrix approach

$$B = (X^T \cdot X)^{-1} \cdot (X^T \cdot Y) \tag{3}$$

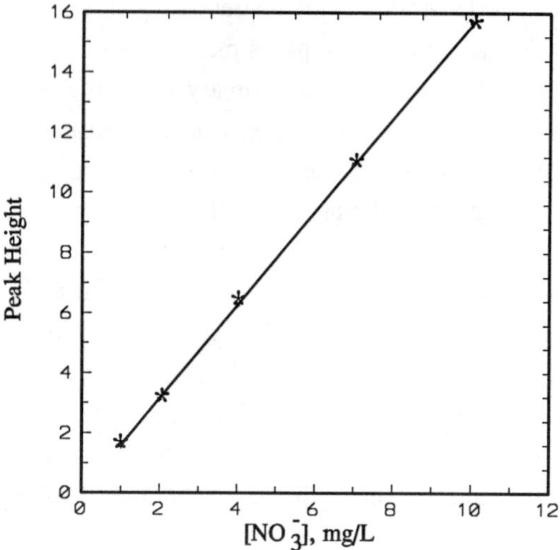

Figure 14-1. Relationship between concentration of nitrate ion and the peak area from injection into an ion chromatograph. This graph is commonly called a calibration curve.

$$b_0 \cdot N + b_1 \cdot \sum x_i = \sum R_i$$
$$b_0 \cdot \sum x_i + b_1 \cdot \sum x_i^2 = \sum x_i \cdot R_i \qquad (2)$$

Realize, of course, that these approaches are mathematically equivalent, but procedurally and algorithmically different. The matrix approach involves the summation of products of elements of matrices, which result in the equivalent sums shown in the normal equations.

The method of expression of the response of the instrument is a difficult one to define as to a general form that it should take. The instrumental response is very dependent on the type of analytical instrument from which the response was recorded. In chromatographic analyses, the form of the response will generally be the peak area or peak height. In polarographic measurements, the form of the response will be the limiting current. The form of the response will also depend greatly on the method of detection of the response. The output of the detector will be measured by some means. In chromatographic methods, the signal is generally an analog voltage, which must be either recorded directly on a strip chart recorder (archaic and out-dated) or converted to a digital voltage for subsequent treatment by computer software. In methods that detect radiation (e.g., nuclear activation analysis, X-ray diffraction, and X-ray fluorescence), the response is often recorded as the number of photons that are striking the detector per unit time. The nature of the instrumental response is rather immaterial to the formulation of the calibration relation.

Example

Ion chromatography is a technique often used to quantitate anions and/or cations in solution. The technique involves the separation of the ions with a separator column followed by suppression of the background conductance of the eluant by a suppressor column. Quantitation is performed by injection of samples of known concentration and measuring the height or area of the resulting peak. A range for quantitation of nitrate in rain samples is 1.0 to 10 mg/L of nitrate. A series of standards were prepared by dissolving reagent grade sodium nitrate ($NaNO_3$) in deionized water. The peak areas that resulted from the injection are shown in Table 14-1. Obtain the calibration curve for the quantitation of nitrate. The calibration curve may be expressed in either of two ways, graphically or as a mathematical model. The graphic solution is given as the scatter plot of peak area as a function of [NO_3^-]. This is shown in Figure 14-2.

The regression parameters for the mathematical model are found using the matrix approach and depend on the matrices for X, X^T and Y:

$$\text{Peak Area} = 0.0037019 + 1.5668 \cdot \left[NO_3^-\right] \tag{4}$$

Table 14-1. Calibration Data for Quantitation of Nitrate Ion in Aqueous Samples

[NO_3^-]	Peak Area	[NO_3^-]	Peak Area	[NO_3^-]	Peak Area
1.000	1.575	4.000	6.264	10.000	15.675
1.000	1.575	4.000	6.274	10.000	15.677
1.000	1.568	4.000	6.278	10.000	15.662
1.000	1.563	4.000	6.276	10.000	15.665
2.000	3.131	7.000	10.979		
2.000	3.134	7.000	10.972		
2.000	3.141	7.000	10.965		
2.000	3.137	7.000	10.975		

where s_y and s_x are the standard deviations of the y– and x–values; n is the

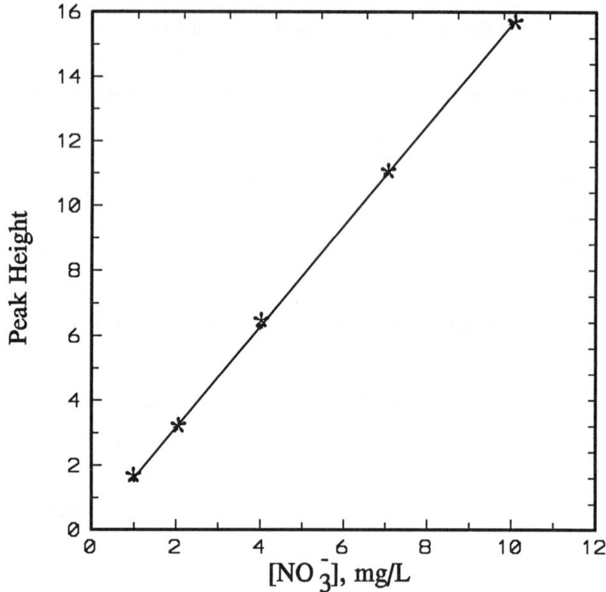

Figure 14-2. Graphic calibration for the quantitation of nitrate ion using ion chromatography.

After the estimates of the parameters are obtained, we would also like to be able to determine confidence limits about the parameters and also about the calibration curve itself.

Confidence Limits

Slope and Intercept

Before we can calculate the confidence limits on the slope and intercept we must calculate the standard deviation of y on x ($s_{y/x}$):

$$s_{y/x} = \sqrt{\frac{(n-1)}{(n-2)} \cdot \left(s_y^2 - b_1^2 \cdot s_x^2\right)} \tag{5}$$

number of pairs of data from which the regression parameters are calculated. This standard deviation is also called the standard error of the estimate. It is also equal to the standard deviation of the residuals.

With this value for $s_{y/x}$, we can now calculate the standard deviations for the slope

$$s_{b_1} = \frac{s_{y/x}}{\sqrt{\sum_i (x_i - \bar{x})^2}} \tag{6}$$

and intercept

$$s_{b_0} = s_{y/x} \cdot \sqrt{\frac{\sum_i x_i^2}{n \cdot \sum_i (x_i - \bar{x})^2}} \tag{7}$$

These values of s_{b_0} and s_{b_1} are used with the Student's t-value having n-2 degrees of freedom (where n is the number of data points that were used to obtain the parameters) to calculate the confidence limits on the slope and intercept as

$$\begin{aligned} &b_1 \pm t_{\alpha,n-2} \cdot s_{b_1} \\ &b_0 \pm t_{\alpha,n-2} \cdot s_{b_0} \end{aligned} \tag{8}$$

A good argument can be made that the intercept of nearly all (if not, all) calibration curves should be at coordinates (0,0). The response of an instrument should be zero if the concentration of the analyte is zero. In reality, rarely will be the intercept be numerically equal to zero, but will have some value indicated by b_0. Thus, an obvious use of these equations is to determine if the intercept of the calibration curve is statistically indistinguishable from 0. Ideally, all calibrations will include 0 in the confidence interval. If the value of 0 is not in the interval, there is justification to perform further investigation to see if there is a systematic error that needs to be corrected. If the intercept is not zero, a bias of some type might exist. That potential bias should be investigated. A systematic error could go undetected if the chemist forced the intercept to be zero by using the equation

$$y = b_1 \cdot x \tag{9}$$

since the regression equation ignores any possible deviation from a zero intercept. It is good practice to first investigate the actual intercept to determine if it is statistically indistinguishable from zero. It if is not different than zero, then the regression Equation (9) may be used for construction of the actual calibration curve. A few instruments will have a slight offset from the zero intercept because of electronics or instrument construction. In this case, no bias or systematic error is present, but the two–parameter regression equation should be used.

Y-Values

We also desire to calculate confidence limits about different y-values. This is the same as calculating confidence limits about the calibration curve. Two different situations are presented: confidence limits about the true mean *y*-values and confidence limits about individual *y*-values. To calculate confidence limits about the true mean *y*-value, we make use of a *t*-equation:

$$(b_0 + b_1 \cdot x) \pm t_{\alpha/2,\nu} \cdot s_{y/x} \cdot \sqrt{\frac{1}{n} + \frac{(x-\bar{x})^2}{s_x^2}} \qquad \textbf{(10)}$$

To calculate the confidence limits, it is assumed that the relative variance about calibration points is constant throughout the regression regime and be considered as the average variance of *y* on *x*. Different values of *x* are assumed from which *y*–values are obtained; these *y*-values are then substituted in the above equation to determine the overall shape of the confidence bands about the calibration curve.

A slightly different situation is called for if it is desired to calculate a confidence limit about a *SIngle* or *Individual* *y*–value. In this case, the term under the radical must have $1/m$ (where *m* is the number of repeat analyses performed on each calibration standard or on each sample) added to obtain the equation

$$(b_0 + b_1 \cdot x) \pm t_{\alpha/2,\nu} \cdot s_{y/x} \cdot \sqrt{\frac{1}{m} + \frac{1}{n} + \frac{(x-\bar{x})^2}{s_x^2}} \qquad \textbf{(11)}$$

n is the number of standards with different concentrations. The term with *m* in the denominator is added to account for the repetitions of each standard. The number of repetitions affects the degrees of freedom of the calculation. The $1/m$ term accounts for this change in the number of degrees of freedom. The regression equation effectively examines the mean of responses for each concentration. The confidence limit in Equation (11) recognizes that there is more than one determination per standard.

It must be realized that this equation is applicable only to calculate a confidence limit about a single predicted *y*-value. As with the true mean value, confidence bands may be obtained by substituting individual *x* –values into the calibration curve equation to calculate individual *y*–values and calculating confidence limit values for each of these values. The band will be slightly wider. A representative graph showing exaggerated confidence limits about a regression for true values and for individual values is shown in Figure 14-3.

X-Values

The last situation for which we would like to calculate confidence limits is to calculate confidence limits about the *x*-values obtained from the calibration curve. The regression curve may be solved for *x* from which the value for an individual concentration may be obtained by substituting the instrument response (*y*) into the regression equation

$$x = \frac{y - b_0}{b_1} \tag{12}$$

However, we can make use of the variance of *y* on *x* to obtain an estimate of the confidence limits for the calculated concentration using the following equation:

The number of analytical standards necessary to obtain an adequate calibration curve must be determined. This will largely depend on the degree of confidence that one desires in the analytical results. The breadth of the confidence limits about the regression equation is approximately inversely proportional to the square root of the number of analytical standards that are used; the more standards that are used, the tighter will be the confidence bands about the calibration curve. However, it will also be shown that the confidence limits are roughly inversely proportional to the square root of the number of repetitions of each standard that are analyzed; the more repetitions of a standard, also, the tighter will be the confidence bands. The degree to which the analyzed

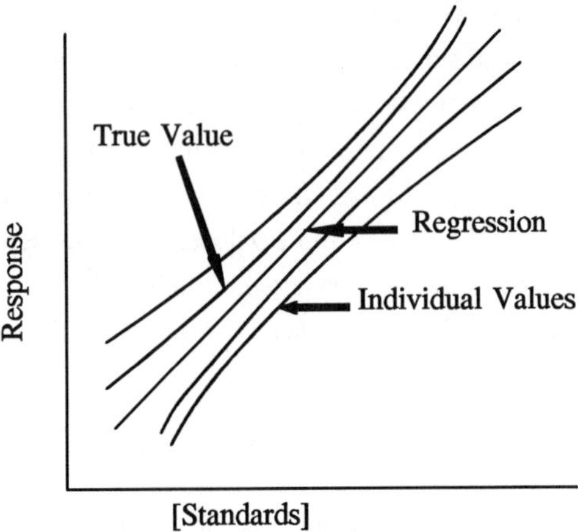

Figure 14-3. Typical calibration curve showing regression, confidence limits about the true values of the regression, and confidence limits about individual values of x.

concentration must be known will in large measure determine how many analytical standards must be used to obtain the calibration curve. One must also realize that analyzing two or three standards multiple times does not introduce greater confidence in the regression. Only the magnitude of the confidence limits is changed. One should analyze as many standards as many times as feasible to obtain the best degree of confidence in the regression. The standards should also be spread over the range of concentrations one expects in the samples to be analyzed. The more standards, Greater will be the confidence in the regression if many diverse concentrations of standards are analyzed multiple times.

Example

Using the data in Table 14-2 (data for the calibration of chloride using ion chromatography), construct the calibration curve for chloride, and also the confidence limits about the calibration curve. A peak height of 2.80 was obtained for the analysis of a sample containing an unknown amount of chloride; calculate the concentration and the confidence limit on the concentration. The matrices needed to calculate the calibration curve and confidence limits are

Table 14-2. Calibration data for Chloride using Ion Chromatography.

Concentration	Response	Concentration	Response
5.0	1.20	15.0	3.60
5.0	1.21	15.0	3.70
5.0	1.19	15.0	3.65
5.0	1.20	15.0	3.63
5.0	1.21	15.0	3.60
10.0	2.40	20.0	4.80
10.0	2.42	20.0	4.77
10.0	2.44	20.0	4.84
10.0	2.39	20.0	4.80
10.0	2.40	20.0	4.90

$$X^T \cdot X = \begin{bmatrix} 20 & 250 \\ 250 & 370 \end{bmatrix}$$

$$(X^T \cdot X)^{-1} = \begin{bmatrix} 0.3 & -0.02 \\ -0.02 & 0.0016 \end{bmatrix}$$

$$X^T \cdot Y = \begin{bmatrix} 60.35 \\ 906.45 \end{bmatrix} \tag{14}$$

$$B = (X^T \cdot X)^{-1} \cdot (X^T \cdot Y) = \begin{bmatrix} -.004 \\ 0.2417 \end{bmatrix}$$

The B matrix gives the values for the slope and intercept. The confidence limits calculated for the calibration line and for individual values are given in Table 14-3. The chloride concentration of the sample whose [Cl$^-$] is unknown is calculated from

$$[\mathrm{Cl}^-] = x = \frac{(y - b_0)}{b_1} = \frac{2.80 - (-0.004)}{0.24172} = 11.601 \tag{15}$$

Before the confidence limits can be calculated, we must calculate the values of several standard deviations, s_x (5.735), s_y (1.387), and $s_{y/x}$:

Table 14-3. Calibration data with confidence limits. XCL (line) are the confidence limits for the Calibration line; XCL (value) are the confidence limits for the values.

Concentration mg/L	Peak Height	Calculated Peak Height	UCL line	LCL line	UCL value	LCL value
5	1.20	1.2046	1.2982	1.1110	1.3207	1.1231
5	1.21	1.2046	1.2982	1.1110	1.3207	1.1231
5	1.19	1.2046	1.2982	1.1110	1.3207	1.1231
5	1.20	1.2046	1.2982	1.1110	1.3207	1.1231
5	1.21	1.2046	1.2982	1.1110	1.3207	1.1231
10	2.40	2.4132	2.4476	2.3788	2.4901	2.3891
10	2.42	2.4132	2.4476	2.3788	2.4901	2.3891
10	2.44	2.4132	2.4476	2.3788	2.4901	2.3891
10	2.39	2.4132	2.4476	2.3788	2.4901	2.3891
10	2.40	2.4132	2.4476	2.3788	2.4901	2.3891
15	3.60	3.6218	3.6562	3.5874	3.6987	3.5760
15	3.70	3.6218	3.6562	3.5874	3.6987	3.5760
15	3.65	3.6218	3.6562	3.5874	3.6987	3.5760
15	3.63	3.6218	3.6562	3.5874	3.6987	3.5760
15	3.60	3.6218	3.6562	3.5874	3.6987	3.5760
20	4.80	4.8304	4.9240	4.7368	4.9465	4.7246
20	4.77	4.8304	4.9240	4.7368	4.9465	4.7246
20	4.84	4.8304	4.9240	4.7368	4.9465	4.7246
20	4.80	4.8304	4.9240	4.7368	4.9465	4.7246
20	4.90	4.8304	4.9240	4.7368	4.9465	4.7246

$$s_{y/x} = \sqrt{\frac{n-1}{n-2} \cdot \left(s_y^2 - b_1^2 \cdot s_x^2\right)} \qquad (16)$$

$$s_{y/x} = 0.03741$$

The confidence limits on the concentration are calculated using Equation (13):

$$\frac{2.80+0.004}{0.24172} \pm \frac{2.10 \cdot 0.03741}{0.24172} \cdot \sqrt{\frac{1}{5} + \frac{1}{20} + \frac{\left(\frac{2.80+0.004}{0.24172} - 12.5\right)^2}{32.8902}} \qquad (17)$$

$$11.600 \pm 0.170$$

The confidence limits can be improved by increasing the number of standards at each concentration, which will allow a greater estimation of the precision of the analysis.

Example

Determine if the intercept is significantly different than zero. The hypotheses are

$$H_0 : b_0 = 0$$
$$H_a : b_0 \neq 0 \tag{18}$$

After calculating the standard deviation of the intercept from Equation (8),

$$s_{b_0} = s_{y/x} \cdot \sqrt{\frac{\sum_i x_i^2}{n \cdot \sum_i (x_i - \bar{x})^2}}$$
$$s_{b_0} = 0.03741 \cdot \sqrt{\frac{3750}{20 \cdot 625}} \tag{19}$$
$$s_{b_0} = 0.0205$$

the confidence limits on the intercept are calculated

$$b_0 \pm t_{\alpha/2,n-2} \cdot s_{b_0}$$
$$-0.004 \pm 2.10 \cdot 0.0205 \tag{20}$$
$$-0.004 \pm 0.0430$$

which encompasses zero. Therefore, the hypothesis that the intercept equals zero cannot be rejected at the 95% confidence level.

Alternative Methods for Calculation of Concentration

General

We have discussed the most popular method (and also the most studied method) for calculation of the concentration of an analyte, namely the calibration curve. Yet the calibration curve, with all of its utility, is limited in its applications. We indicated that the most important criteria for construction of the calibration curve is that standard analytical reference materials must be available. What are we to do if such materials are not available? Another situation may be that although SARMs are available, the time and expense to construct a calibration curve for a single sample do not justify that expenditure of time and expense. Often the samples we are analyzing are very diverse in their matrix and do not lend themselves to analysis using a calibration curve because of the difficulty of duplicating the matrix. How do we handle situations such as these? An additional problem is that many instruments that are in use in the analytical laboratory today determine mass not concentration, therefore the amount of sample placed in the instrument becomes critical. The reproducibility and repeatability of manual injection of a sample into a gas chromatograph are major sources of error. How can we compensate for this source of error and minimize its impact? Several alternative methods are available for calculation of concentration. However, it should be recognized that these alternative methods do not afford the degree of confidence in the final result that using a method with a well-established calibration curve generates.

Earlier, several different types of calibrations including the calibration curve were discussed. Let us now discuss the alternative methods when some of the problems in the previous paragraph are encountered.

Comparison to a Single Standard

Occasionally, we may have need to quantitate an analyte in a one of a kind sample. In this situation, we would probably not want to spend the time and effort to prepare a full calibration chart with all of its ramifications. Neither do we want to rely totally on comparison to a material that is very foreign in nature to the analyte. We can quantitate with a single standard providing a couple of criteria are met. The primary criterion that should be met is that the concentration of the analyte in the standard should be very close to the concentration of the analyte in the sample. Another criterion is that the matrix in which the standard is prepared is very similar to the matrix of the sample. The major assumption in this method is that the intercept of a calibration curve, if it was determined, would be 0. An additional criteria that should be adhered to is that the method should be one that yields concentration rather than one that is a mass detection method. If the detection method is mass, the analyst should be especially careful to reproduce injection volumes or sample sizes of the reference and the sample.

Several standards may need to be prepared before one is obtained that has the concentration of analyte near the concentration of analyte in the sample. The calculations are very straightforward. The concentration of the analyte in either the standard or in the sample will be the response divided by the slope of the calibration curve. The ratio of these two concentrations is obtained:

$$\frac{[A]_{\text{sample}}}{[A]_{\text{standard}}} = \frac{\text{Response}_{\text{sample}} / b_1^{\text{sample}}}{\text{Response}_{\text{standard}} / b_1^{\text{standard}}} \tag{21}$$

but presumably the slope of analyte in the standard and the slope of the analyte in the sample will be equal, therefore the concentration of the analyte in the sample will be found from

$$[A]_{\text{sample}} = \frac{[A]_{\text{standard}} \cdot \text{Response}_{\text{sample}}}{\text{Response}_{\text{standard}}} \tag{22}$$

This method should be used only in situations where the confidence interval of the obtained concentration is not important. If the confidence interval is needed, it may be obtained by several analyses of the sample and the mean and standard deviation from these multiple determinations together with a t-value may be used to calculate an interval using the equations given previously.

Internal Standard

Instruments that determine the amount of a material (as opposed to the concentration) placed in the instrument are prone to errors in constructing a calibration curve because of the relatively poor precision in inserting a constant amount of sample in the instrument. One of the ways proposed to overcome this dilemma is to use a standard that is internal to the sample, thus the name internal standard. The internal standard can be formulated in several ways. One option is to use the analyte of interest as the standard. This presents a complication to determine and separate the effect of the amount of the analyte in the sample and the amount of analyte that was added. The calculations for this particular technique will be discussed in the section on standards addition rather than here. The other technique that uses the internal standard is to use a material that mimics the analyte. Internal standards are most often used for chromatographic techniques of analysis.

When calibration curves were discussed, the model used to represent the calibration curve was the first–order polynomial

$$\text{Response} = b_0 + b_1 \cdot [\text{Analyte}] \tag{23}$$

For the instrument that determines mass rather than concentration, the following model should be used:

$$\text{Response} = b_0 + b_1 \cdot \text{Mass}_{\text{Analyte}} \tag{24}$$

and can be considered as the response function for the analyte. The slope, b_1, is the response factor for the analyte. Each analyte will have its own response function and response factor.

Criteria that should be considered for the selection of an internal standard that mimics the analyte are given. First, the internal standard should have a response function that is similar to the analyte. Second, if the technique is a chromatographic technique, the internal standard should have a retention time similar, but resolved from the analyte of interest. (With some multivariate modeling, this may not be a total requirement.) Third, the internal standard should be stable under the conditions of the analysis. Fourth, the internal standard should be in some measure traceable to the National Institute of Standards and Technology (NIST). Fifth, the response functions of both the analyte and the internal standard should be characterized. Sixth, the last criterion for an internal standard is that the selected internal standard should not be one of the analytes for which analysis is being performed.

The mathematics are much simplified if the intercept of the response function is not significantly different than zero. This last requirement presupposes that calibration curves for the analyte and the internal standard have been prepared with great care. Why should we go to the trouble to prepare a calibration curve for both the analyte and the internal standard? The main reason will be that although this will initially take more time than just establishing one calibration curve, we shall see that the calculations with the internal standard are much simplified over the calibration curve calculations and that the major source of variation in the y-values from the instrument (namely that the variation in the mass inserted in the instrument) is eliminated. We shall also see that this satisfies the major assumption given in the section on linear regression that the variation in the independent variable (mass) is small compared to the variation in the dependent variable (instrument response).

The technique of quantitation by internal standard is accomplished by adding a small amount of a solution of the internal standard to a known amount of solution of the sample. This spiked solution is then analyzed. If we were quantitating using a calibration curve, the amount of analyte would be found by substituting the instrument response into the calibration equation and solving for the concentration.

$$[\text{Analtye}] = \frac{\text{Response}_{\text{analyte}} - b_0^{\text{analtye}}}{b_1^{\text{analyte}}} \tag{25}$$

Likewise, the concentration of the internal standard (*IS*) would be found in the similar manner:

$$[\text{IS}] = \frac{\text{Response}_{\text{IS}} - b_0^{IS}}{b_1^{\text{IS}}} \tag{26}$$

However, since we are concerned with instruments having mass detection, the variance of the concentration could be quite large. Thus if we take the ratio of the last two equations, we obtain

$$\frac{[\text{Analyte}]}{[\text{IS}]} = \frac{\left(\text{R}_{\text{analyte}} - b_0^{\text{analyte}}\right) / b_1^{\text{analyte}}}{\left(\text{R}_{\text{IS}} - b_0^{IS}\right) / b_1^{\text{IS}}} \tag{27}$$

from which we can now solve for the concentration of the analyte directly to be

$$[Analyte] = \frac{[IS] \cdot \left(R_{Analyte} - b_0^{Analyte}\right) b_1^{IS}}{\left(R_{IS} - b_0^{IS}\right) \cdot b_1^{Analyte}} \tag{28}$$

Since we have obtained the ratio of the two concentrations, and the concentration of the internal standard is known, the amount injected into the instrument will be immaterial since that volume injected will be the same for both the analyte and for the internal standard. (Remember that concentration times volume gives a function of the mass, e.g. grams, moles, etc.)

Simple Standard Addition

Several types of analytical techniques are very prone to be matrix dependent. One of the major problems with analytical techniques is that of interferences. Among the more well known are atomic absorption and atomic emission spectrophotometric,[2,3] polarographic,[4,5] and X-ray techniques.[6] X-ray fluorescence, for example is known to have both positive and negative interferences, i.e., the presence of one element may absorb part of the secondary X-rays produced by another element or the secondary X-rays

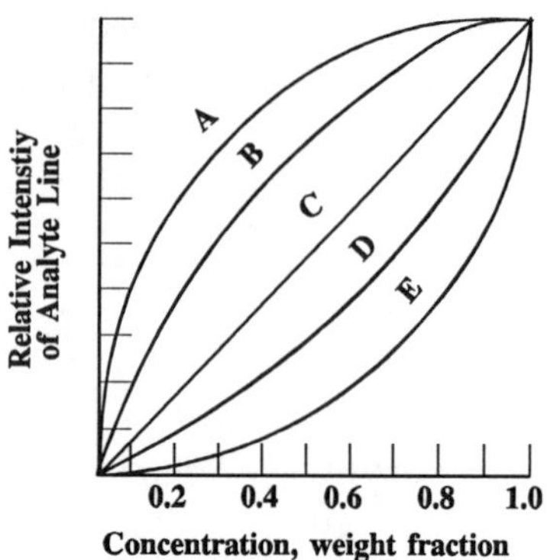

Figure 14-4. Potential calibration curves in X-ray spectrometric analysis showing the effect of positive and negative enhancements of X-ray fluorescence.

produced may be sufficiently energetic to generate additional fluorescence by elements in the sample.[7] Figure 14-4 shows the potential calibration curves when an element may exhibit positive enhancement (curves A and B), no effect (curve C), and negative interference effects (curves D and E) in X–ray fluorescence analysis.

An additional condition when sample matrix is quite important is when a sample has been obtained that is unique and is unlikely to be required to be analyzed in a similar sample in the future. Standards are particularly hard to prepare for these types of samples since one of the criterion for a good SARM is that the SARM should have the same matrix as the samples. A technique that has been developed to overcome this problem is that of standards addition or spiking.[8,9] In this technique, the sample serves as its own standard. Samples may be either solid or liquid and the standards are prepared using that sample. It should be noted that if the sample is standard, the problem of the number of particles becomes quite important as discussed earlier. However, if the sample is liquid, the problem is greatly simplified. The following discussion will focus on the technique of differential pulsed polarography (DPP). The matrix greatly influences the two major quantities determined by DPP, namely the half-wave potential (from which the identity of the analyte can be determined or confirmed) and the derivative of the diffusion current (from which the concentration of the analyte may be determined). To analyze a sample by DPP, the diffusion current on a hanging mercury drop electrode is measured just before and just after the application of a step in the potential. The two currents are subtracted and recorded. The trace that is recorded is a plot of the change of the diffusion current as a function of the potential. The trace is somewhat similar to a chromatographic peak in appearance. The differential pulsed polarogram is initially recorded for the sample itself. The exact volume of the sample being analyzed must be known. This is most easily accomplished by pipetting the solution into the polarograph. At the conclusion of the scan, a small amount of a concentrated solution of the analyte is added to the sample and stirred to ensure complete dispersion of the standard through the sample. The differential pulsed polarogram is again recorded. This is the easiest situation, but the addition of the standard should be repeated at least three times for good statistical treatment of the data.

After the differential pulsed polarogram is obtained for each of the additions of standard, the concentration of the added analyte must be calculated. Note that it is the concentration of the analyte that has been added and not the concentration of the analyte in the sample that is calculated. The concentration of the added analyte is calculated from the following equation:

$$[A]_{\text{added}} = \frac{[A]_{\text{standard}} \cdot n_{\text{addition}} \cdot \text{Volume}_{\text{addition}}}{\text{Volume}_0 + n_{\text{addition}} \cdot \text{Volume}_{\text{addition}}} \qquad (29)$$

where n_{addition} is the addition number not the total number of additions, n_{addition} will range from 0 to the number of aliquots added. $\text{Volume}_{\text{addition}}$ is the volume of the standard added, Volume_{o} is the volume of the sample taken for analysis.

The height of the differential pulsed polarogram for each addition, including the *zero*th addition, and the concentration of the added analyte are recorded. These pairs are plotted— height *vs* added concentration. The concentration of analyte in the original sample may be thought of as being the amount of sample that must be removed from the sample to have a height of zero for the differential current. The mathematical nomenclature for this is the x-intercept. The x–intercept may be found by first calculating the regression equation for the data (height vs added concentration), then calculating that x–value ([Analyte] in the sample), which will give a y –value (height of differential current) of zero:

$$[A] = |x\text{-}intercept| = \left|\frac{-b_1}{b_0}\right| \tag{30}$$

Although this is a powerful technique, there are certain criteria that should be adhered to so as to minimize errors. The volume of each addition should be constant from addition to addition, the volume of addition should be as small as practical and should be considerably smaller than the volume of sample taken for analysis, the height of the differential pulsed polarogram after the first addition should be approximately 1.25 to 2 times the height of the original differential pulsed polarogram, and each successive added aliquot should yield a similar increase in height of the polarogram.

It should be noted that standard addition is not limited to the techniques of polarography, but is equally applicable to almost any analytical procedure.

Example

Suppose that a determination of the amount of copper in a solution was to be made using differential pulsed polarography. 10 mL of a solution of unknown copper concentration was used for the determination. 50 μL of a 200 μg copper/L solution was added incrementally 4 times. The solution was analyzed after each addition. The height of the resulting copper peak in the polarogram for the blank was 0 mm. The heights of the copper peak in the polarogram for the four additions were 12.00, 23.00, 33.00, 43.00, and 52.00 mm. Calculate the concentration of the copper in the unknown solution. The concentrations corresponding to the four additions are shown in Table 14-4. From the data, the

Table 14-4. Determination of Copper in an Aqueous Sample Using the Standard Additions Method with Differential Pulsed Polarography

Addition number	Added Concentration	Height
blank	0.000	12.000
1	0.995	23.000
2	1.980	33.000
3	2.956	43.000
4	3.922	52.000

y–Intercept = 12.50, the slope = 10.20 and the correlation coefficient (r^2) = 1.000. From the slope and the y–intercept, the x–intercept (and therefore the concentration of the unknown) is calculated to be 1.225 mg/L The data is graphically shown in Figure 14-5.

Effect of Volume for the Standard Addition Method

Although, the previous discussion suggested the use of addition of a constant amount to the sample, a situation may exist where the addition of the extra volume could destroy any advantage of the standard addition method because the matrix is no longer constant from one addition to the next.[10] The problem would be especially evident if the volume of the spike and the original volume of the sample are comparable, rather than the volume of the spike being considerably less than the volume of the sample. This would be particularly applicable to the situation of the analysis of soil or other solids, where it would be difficult if not impossible to add a very small amount of a highly concentrated addition.

Generalized Standard Addition

Rarely is a single analyte determined when analyzing a sample, but usually the interest is in obtaining concentrations of multiple analytes. Several analytical methods are capable of simultaneous determination of signals that are indicative of the concentration of several species in the sample. Inductively coupled plasma atomic emission is an example of such a technique. Unfortunately such a technique is often subject to the effect of interferences in such a manner that the presence of one analyte will subdue or enhance the signal of another analyte.[11] The single component standards addition technique described above will often not be able to overcome the interference effect. Any standards addition method necessitates the measurement of a signal before and after the addition of the additional, accurately known quantities of the analyte. It is assumed that the response of the instrument will be in the form of the following:

$$\text{Response}_i = \text{concentration}_i \cdot k_i \qquad (31)$$

where Response_i is the response of the instrument, concentration_i is the total

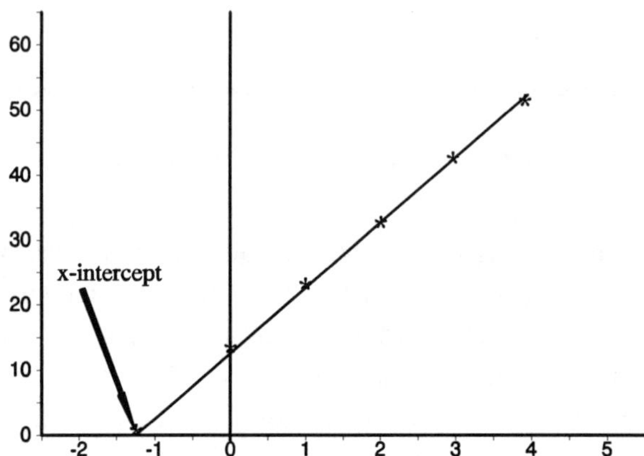

Figure 14-5. Graphic display of the determination of copper in an aqueous solution using standards addition and differential pulsed polarography.

concentration defined by

$$\text{concentration}_i = \Delta\text{concentration} + \text{concentration}_0 \tag{32}$$

and k_i is the linear response factor for the analyte. Δconcentration$_i$ is the effective concentration of the analyte that has been added. However, if, for example, the instrument is an absorbance spectrophotometer, the total analytical response is, more accurately, the sum of the analytical responses for each analyte:

$$R_i = \sum_{s=1}^{r} c_s \cdot k_{si} \tag{33}$$

Thus the total response at any given wavelength, for example, would be

$$R_i = \sum_{s=1}^{r} \Delta c_s \cdot k_{si} + \sum_{s=1}^{r} (c_0)_s \cdot k_{si} \tag{34}$$

This, of course, assumes that there are r analytes in the system. One can then measure several analytical responses in which the concentration of at least one analyte has changed through addition of a known amount of the analyte to the sample. The number of analytical responses that are measured must be greater than the number of analytes in the sample. This will require that at least the equivalent number of additions of standard be made. Using the notation of Jochum et al.[12] p analytical responses must be measured of n additions of multicomponent standard additions.

Equation (33) may be written as a matrix:

$$R = C \cdot K \tag{35}$$

where R is an n x p matrix of analytical responses and C is the n x r matrix of the total concentrations of the analytes as expressed by Equation (32) and K is the matrix of the response factors for each analyte. The K matrix will be a combination of the effect of adding the contribution of many analytes together. If there is little interference between analytes, the K matrix may contain many elements that are near, or possibly even equal to, zero.

The matrix that must be obtained is the K matrix and may be obtained from the expression developed by Saxberg and Kowalski[13] to be

$$K = (\Delta C^T \cdot \Delta C)^{-1} \cdot \Delta C^T \Delta R \tag{36}$$

Once the K matrix is found, the matrix of initial concentrations, *concentration*$_0$, may be found from

$$c_0^T = r_0^T \cdot K^{-1} \tag{37}$$

r_0 may be easily found as the transpose of the column vector of the initial responses of the analytes:

$$r_o = [(R_1)_0, (R_2)_0, \cdots, (R_i)_0]^T \tag{38}$$

The unfortunate situation here is that the matrix ΔC cannot be known unless the original concentrations are known. The problem is rather easily corrected since if the concentrations are multiplied by the volume, the absolute quantities of analyte is found. If Equation (34) is recast with the concentration expressed as a ratio of moles per volume, the following is obtained:

$$R_i = \sum_{s=1}^{r} \frac{N_s}{V} k_{si} \tag{39}$$

which leads to obtain a new matrix Q:

$$Q_i = V \cdot R_i = \sum_{s=1}^{r} N_s \cdot k_{si} \tag{40}$$

But N_s is merely the sum of N^0_s and ΔN_s, which allows the Q matrix to be separated and expressed as

$$Q_i = \sum_{s=1}^{r} \Delta N_s \cdot k_{si} + \sum_{s=1}^{r} N_s^0 \cdot k_{si} \tag{41}$$

The K matrix may now be expressed in terms of the change in volume corrected responses, ΔQ, and the total changes of quantities of analytes:

$$K = (\Delta N^{\mathrm{T}} \cdot \Delta N)^{-1} \cdot \Delta N^{\mathrm{T}} \cdot \Delta Q \tag{42}$$

Thus, the total response of the sensors can be separated into two components: a component from the original amount and a component from the added amount. By knowing these two components we can calculate the original amount. We were able to calculate the K matrix from the amount added and the additional response of each sensor from the added amount. The same K matrix may then be used to calculate the original amount from

$$K^{\mathrm{T}} \cdot N_0 = Q_0 \tag{43}$$

which must be solved for N_0 by left multiplying both sides of the equation by $(K^{\mathrm{T}})^{-1}$ to give

$$N_0 = (K^{\mathrm{T}})^{-1} \cdot Q_0 \tag{44}$$

where N_0 is the matrix of amounts present in the original sample. The matrix so obtained is thus determined from the addition of accurate and precise amounts each of the analytes to the original sample and determining the effect on the instrumental responses for each of the analytes. Let us now examine an example that uses this methodology.

Example

An aqueous sample, which may contain four different metals, (chromium, copper, cobalt, and nickel), was analyzed with ICP-AES. The response at four different wavelengths for the original sample before any additions were made is given by the following:

λ, nm	351.0	511.5	394.5	820.0
Q_0	0.117	0.500	0.546	0.861

Table 14-6 shows the amount added for each increment.The response at each of the wavelengths was measured after each of the additions and the incremental difference in the response of the instrument was recorded (Table 14-7).

From the first matrix, the added incremental concentrations, the following matrix was obtained

$$(\Delta N \cdot \Delta N^{T})^{-1}$$

$$\begin{bmatrix} 0.003906 & 0 & 0 & 0 \\ 0 & 0.015625 & 0 & 0 \\ 0 & 0 & 0.25 & 0 \\ 0 & 0 & 0 & 0.0625 \end{bmatrix}$$

Table 14-6. The Concentrations of 4 Metals Added Incrementally in a Generalized Standard Addition Method Analysis

Increment Number	Quantity Added (mg/L) (Increments not total) Cr	Co	Ni	Cu
1	0	0	1	0
2	0	0	1	0
3	0	0	1	0
4	0	0	1	0
5	0	0	0	2
6	0	0	0	2
7	0	0	0	2
8	0	0	0	2
9	0	4	0	0
10	0	4	0	0
11	0	4	0	0
12	0	4	0	0
13	8	0	0	0
14	8	0	0	0
15	8	0	0	0
16	8	0	0	0

Table 14-7. Incremental Change in Instrument Response on Addition of Increments of Metal Analytes

Increment Number	Δ(Instrument Response), ΔQ Cr λ=351.0 nm	Co λ=511.5 nm	Ni λ=394.5 nm	Cu λ=820.0 nm
1	0.042	-0.010	0.491	0.069
2	0.040	-0.001	0.488	0.035
3	0.069	0.016	0.497	0.105
4	0.058	0.013	0.516	0.068
5	0.056	-0.012	-0.007	1.540
6	0.060	-0.003	0.018	1.470
7	0.087	0.015	-0.008	1.600
8	0.094	0.027	0.048	1.560
9	0.058	1.950	0.115	0.079
10	0.063	1.860	0.015	-0.129
11	0.104	1.920	0.152	0.233
12	0.128	2.030	0.218	0.089
13	5.980	0.075	1.167	-0.125
14	5.930	0.026	1.630	-0.389
15	6.270	0.165	1.850	0.192
16	6.500	0.389	2.300	0.202

The change in instrument response matrix is then multiplied by the transpose of the quantity added matrix to yield:

$\Delta N^{\mathrm{T}} \cdot \Delta Q$

	Cr	Co	Ni	Cu
Cr	197.440	5.240	55.576	-0.960
Co	1.412	31.040	2.000	1.088
Ni	0.209	0.018	1.992	0.277
Cu	0.594	0.054	0.102	12.340

These two last matrices are multiplied together to obtain the K matrix

KMatrix

	Cr	Co	Ni	Cu
Cr	0.771	0.020	0.217	−0.004
Co	0.022	0.485	0.031	0.017
Ni	0.052	0.005	0.498	0.069
Cu	0.037	0.003	0.006	0.771

The original instrument response matrix is left multiplied by the inverse of the transpose of the K matrix to obtain

$N_0 = ((K^T)^{-1} \cdot \Delta Q$	
Cr	0.006
Co	1.014
Ni	1.018
Cu	1.002

The sample had been prepared so that the concentration was 0.000 for chromium, and 1.000 for the other three metals.

Discussion About Addition Methodologies

Addition without Dilution to a Constant Volume

Several different methods of adding the standards can be conceived. For example, the standards can be added without dilution of the sample to a constant volume. This type of methodology will work adequately if the volume of the addition is considerably smaller than the original volume of sample THAT is analyzed. A major reason to dilute the sample to a constant volume after addition is to minimize the potential matrix effects. The aim when using the generalized standard additions method is to minimize potential matrix effects. This type of addition methodology will work adequately in liquid samples when the concentration of standard is considerably higher than the concentration of the analyte in the sample.

Addition with Dilution to a Fixed Volume or Mass

When a matrix effect is evident, adding a standard without attempting to recreate the matrix of the original sample can be disastrous. To overcome that eventuality, the sample (to which a standard has been added) should be diluted with the original sample to a constant volume or mass before analysis. This type of addition methodology works particularly well with solid samples to which a solid standard is being added. Additions of a *pure* solid standard will need to be made to increase the total concentration of the analyte that will be comparable to the original concentration of the analyte. Such additions will often affect the concentration of the sample in a slight way compared to the addition of a concentrated liquid as in the last paragraph. By diluting the sample to a constant mass or volume, a minimization of matrix effects is obtained.[14]

Sample Partitioning

Successive additions of a standard to a sample, followed by dilution to either a constant volume or constant mass, can be difficult to adjust the mass after each addition. An alternative methodology is to split the original sample into several aliquots of equal size. To each of the aliquots is then added a standard that will give approximately 2, 3, 4, etc. times the original concentration of the analyte. Each of these additions is then diluted to constant mass with the original sample. A major disadvantage of this methodology is illustrated if the sample is liquid and will then be analyzed in a spectrophotometer. Successive additions of a standard to the original sample would allow one to use only one cuvette, whereas the use of partitioning may force the analyst to use as many cuvettes as additions have been made.[15]

References

1. Eisenhart, C. in John K. Taylor, Seminar materials. **Quality Assurance of Chemical Measurements**, National Bureau of Standards, March 1983, p. III-5.

2. Dean, J.A. and Rains, T.C. **Flame Emission and Atomic Absorption Spectrometry**, Vol. 1, Marcel Dekker, New York, 1969.

3. Skogerboe, R.K. and Freeland, S.J. *Appl. Spectrosc.*, **1985**, 39, 925.

4. Kalivas, J.H., and Kowalski, B.R. **Anal. Chem.**, 1981, **55**, 532.

5. Zagatto, E.A.G., Jacintho, A.O., Krug, F.J., Reis, B.F., Bruns, R.E., and Araújo, G. **Anal. Chim. Acta**, 1985, **145**, 169.

6. Bertin, E.P. **Principles and Practice of X-Ray Spectrometric Analysis**, 2nd ed., Plenum Press, New York, 1979, p. 506

7. See Bertin, E.P. *Principles and Practice of X-ray Spectrometric Analysis*, 2nd ed., Plenum Press, New York, Chapter 12 for a further discussion of matrix effects in X-ray fluorescence analysis.

8. Bader, M. *J. Chem. Ed.*, 1980, **57**, 703.

9. Cardone, M.M. *J. Assoc. Off. Anal. Chem.,* 1983, **66**, 1257.

10. Kalivas, J.H. **Talanta**, 1987, **34**, 899.

11. Kalivas, J.H., and Kowalski, B.R. *Anal. Chem.* 1981, **53**, 2207–2212.

12. Jochum, C., Jochum, P., and Kowalski, B.R. *Anal. Chem.*, 1981, **53**, 85–92.

13. Saxberg, B. E.H. and Kowalski, B.R. *Anal. Chem,*, 1979, **51**, 1031–1038.

14. Kalivas, J.H. and Kowalski, B.R. **Automated multicomponent analysis with corrections for interferences and matrix effects,** *Anal. Chem.*, 1983, **55**, 532–535.

15. Jochum, C., Jochum, P., and Kowalski, B.R. *Anal.Chem.*, 1981, **53**, 85–92.

15

MEASURES OF PERFORMANCE OF ANALYTICAL METHODS

General

Analytical chemistry has several goals—detection of analytes, identification of analytes, quantitation of analytes, and validation of results. Each of these is, however, inextricably related to the others. Each carries certain requirements and assumptions. The concepts of detection and identification may at first seem to be redundant, but as one examines many of the analytical techniques that are used today, the individual recognizes that many of the techniques are not analyte specific. For example, chromatographic techniques rely mainly on retention times for the identification of a chemical compound. The detectors that are used are, in general, compound nonspecific. The presence of a chromatographic peak at the retention time of a compound is at best circumstantial evidence of the existence of that compound. It is probable, especially in environmental analyses, that more than one compound may have the same retention time. Some analytical techniques are compound specific and are able to identify compounds when they are present. An example of a specific technique is mass spectrometry. When coupled with another technique such as gas or liquid chromatography, the coupled technique is able to identify specific compounds.

An additional requirement of analytical chemistry is to validate the results produced by the analytical results. This will require that additional performance measures be reported for the analytical method. Two of the most important measures that are to be validated and reported are the accuracy and precision of the method. These two will also be discussed in this chapter.

Detection Limit

Detection as opposed to identification is the process of deciding whether a bump in a chromatographic trace is a peak or merely detector noise. This process is important to both scientists and also lay individuals and, unfortunately, has entirely different meanings to both. The public has become sensitized to the issues of the presence of environmental contamination and pollution. Many individuals demand the absence of harmful compounds. To the lay person that means ZERO; to the chemist, it means that the compound is present at less than a given concentration. One of the goals that individual scientists should have is to be able to explain in familiar terms that zero concentration is not possible.

Many issues of societal nature are affected by advances in analytical technology. As analytical technology advances, the amount of material that we are able to detect in a sample will decrease and society will demand that the level of that compound in the water, air, and soil around us be below that level.Two papers have recently appeared that deal with issues of society and detection limits.[1,2]

Within the concept of detection limit, it is difficult from the literature to obtain a consistent picture as to what is meant by the term. There have been many terms used in the literature that seem to indicate that the authors are discussing "detection limit." Some of the terms that have been used are limit of detection, decision limit, detection limit, sensitivity, lower limit of concentration, limit of detectability, and minimum detectable level, among others. A detection limit should address the concepts of Type I and Type II errors. Any operational calculation of the detection limit should involve use of both α and β parameters that characterize the Type I and Type II errors. Calculation of the detection limit will be shown to be dependent not only on the response of a blank but also on the functional dependence of the response of an analytical instrument on the concentration (or mass) of analyte in samples.

Let us now examine some of the methods that have been used and are still being used to determine detection limits. Before we can examine these methods, we should try to define what is meant by the detection limit. For a first definition, let us state that the detection limit is the maximum concentration of an analyte that cannot be distinguished from zero. Several operational questions should arise from this definition. One is “How do we translate analytical data into such a hard, fixed number?" The second, in view of statistical measures, is "What degree of significance do we assign a detection limit?" The third is "What probability do we assign of risking not detecting the material when it is present?" Each of these questions may be answered using statistical techniques that we have examined.

Signal-to-Noise Ratio

The signal-to-noise ratio has been given a place in deciding whether an increase of a signal is the result of the presence of an analyte or whether it is a normal fluctuation of the background signal (noise). The signal-to-noise ratio has been computed by two different practices. The first method of computation is to average the background signal for a given period of time using a nonweighted technique (i.e., the common arithmetic mean). The signal is then monitored to determine if the signal varies from this average value by an arbitrary factor of usually 2 or 3. The method is amenable to automated detection of analytes by comparison of a digitized signal with the calculated noise level. When the ratio, signal/noise, exceeds the arbitrary factor, a decision is made that the increase in signal indicates the presence of analyte. Because of the random fluctuations inherent in digital signals, it is recommended that before a firm decision is made that analyte is present, the signal/noise ratio should continue to increase for several data points. An alternative method is to determine the derivative of the signal and use the derivative as a means to indicate the beginning of the analyte signal. This method has been successfully used to determine the endpoints of thermometric titrations.[3]

The second method of calculating the signal-to-noise ratio has been to subtract the minimum value of the background signal from the maximum background signal. This range is then compared to the *signal* from the suspected presence of a chemical phenomenon. The signal is then divided by the range of the background

$$\text{Signal-to-noise} = \frac{\text{Signal}}{\text{Max}_{\text{background}} - \text{Min}_{\text{background}}} \tag{1}$$

If the ratio is greater than a fixed number (usually at least 5), the phenomenon is deemed significant.

The signal-to-noise ratio may be enhanced using the data smoothing

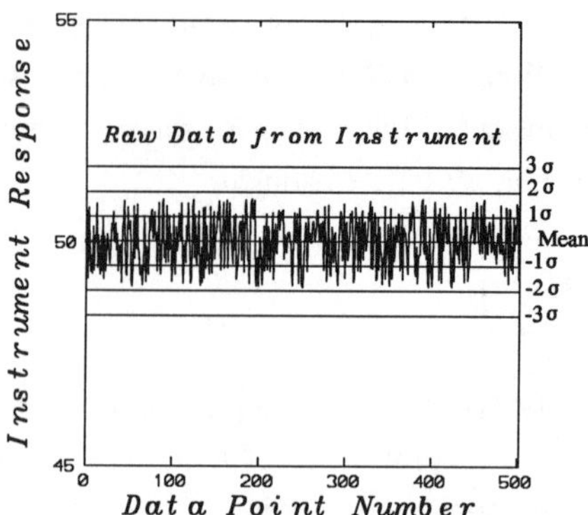

Figure 15-1. Graphic representation of data for a baseline. On this graph are also shown the ±1, 2, and 3σ confidence limits.

techniques discussed previously. These data smoothing techniques can also be used to detect the changes in slope that may indicate the beginning of the analyte signal.

The data collected for the determination of the noise level (the baseline) should be for a flat baseline or a sloping baseline. The simple concept of signal-to-noise ratio presupposes that a flat baseline is inherent. A test should be made on the data used to calculate the noise level to ensure that the noise may be described by the equation

$$\overline{\text{noise}} = \sum \frac{\text{background}}{\text{n}} \qquad (2)$$

where *background* is the digital data values obtained from an instrument and n is the number of data values used to compute the average noise level. This is in comparison to the baseline being described by the equation

$$\overline{\text{noise}} = b_0 + b_1 \cdot \text{background} \qquad (3)$$

where b_0 and b_1 are the slope and intercept of the background data values. The confidence limits on the slope should be calculated to decide whether the slope is not significantly different from zero at a desired level of confidence. If the slope is zero, the data are described by Equation (1) above. If not, care must be taken to ensure that the level of the noise that is used in the calculation of the signal-to-noise ratio is appropriate for the location in the data stream.

One of the major problems with the use of this approach to calculate a detection limit is that significance levels are very difficult to assign. Without significance levels, it is impossible to calculate confidence intervals, thus making it difficult to compare results from one or more individuals or laboratories.

Deviation from a Baseline

Some individuals have taken the concept of the signal-to-noise ratio one step further to calculate confidence limits about a baseline.[4,5,6] This is an extension of Equation (1). But the extension is an important one, because with the confidence limits, the need to take a ratio now becomes obsolete. The method of calculating a signal-to-noise ratio was limited because the variability of the baseline was not taken into account. Once the variability is known or at least estimated, limits may be calculated that allow us to obtain action limits. For example, let us look at the data stream of a baseline shown in Figure 15-1. In this figure individual data points, the average of the data points, and the ± 1, 2, 3, and 4 s confidence limits of the baseline are shown. The confidence limits can now give quantitative action levels. With this type of a formulation, the hypothesis is made that the mean of the baseline is zero:

$$H_0\colon \bar{x}_{\text{baseline}} = 0$$
$$H_a\colon \bar{x}_{\text{baseline}} \neq 0 \qquad (4)$$

If the signal crosses any of these limits, the null hypothesis may be rejected at the appropriate significance level. This gives a quantitative measure of the departure from the baseline. A limitation of this technique is there is no measure given of the probability of committing a Type II error (rejecting a sample as having no analyte when the analyte is present).

Comparison to a Blank

This same approach has also been used to detect the analyte when blanks are used in the detection process. Kaiser[4], Wilson[5] and Currie[6] indicated the need to determine the variability of a blank in regards to detection of analytes. Wilson[5] developed a procedure to determine the criteria of detection based almost entirely on the successive analysis of blanks. The technique is to analyze a series of blanks recording the difference between two successive blanks. A statistical distribution is made of the differences by assuming a normal distribution and calculating a mean and a standard deviation for the set of blank differences. For a significance level of 5% (α=0.05), the z-value is 1.645. If only two blanks are considered as a means to calculate the criterion of detection, then the action level is given as

$$1.645 \cdot \sqrt{2} \cdot \sigma_B = \text{criterion for detection} \tag{5}$$

where σ_B is the population standard deviation of the difference between blanks. The calculated mean should be zero for the difference between two blanks. The mean can be tested to determine if it is different than zero. A confidence interval for a desired significance level is calculated about this mean. The upper confidence limit is used as the lower confidence limit from which to calculate the mean to give a desired probability of committing a Type II error, i.e., accepting a decision as false when it should have been accepted as true.

Example

Calculate the limit of detection using the following values for required variables. The manner in which the variables are given or obtained will, of course, influence the way in which confidence intervals are calculated. The units on all required variables will be in concentration units. Confidence intervals could be given in response units from which concentration units would have to be calculated. Let the calculated difference between two blanks be 0.0001 mg/L (obtained from analyzing blank solutions and comparing the instrument responses to the calibration function), $\sigma_B = 0.00005$ mg/L and $n = 2$. A significance levels of $\alpha = 0.05$ and $\beta = 0.05$ are also assumed. The t-value for 1 degree of freedom and $\alpha = 0.05$ of 6.31 is used to calculate the confidence interval on the mean to be $-0.000216 \leq \mu \leq 0.000416$. Therefore, the null hypothesis that $\mu = 0$ cannot be rejected. From this a criterion of detection becomes 0.000416 mg/L which means that for a sample whose analysis gives a concentration less than 0.000416 mg/L, the concentration would be reported as not detected at the 5% level of significance. Only 1 out of 20 times (on the average) would a Type I error be committed for rejecting the hypothesis that the concentration is lower than the criterion of detection. However, concern should be voiced to determine at what concentration a Type II error might be committed. Since the criterion of $\beta = 0.05$ is desired, the concentration that will give 0.000416 mg/L as the lower limit of the confidence interval for the Type II error may be calculated from

$$0.000416 = x - \frac{t_{\beta/2,\nu} \cdot s}{\sqrt{n}} \qquad (6)$$

by substituting in the values of β (0.05) $t_{\beta/2,\nu}$ (6.31), σ_B (0.00005) and n (2) to obtain the lower confidence interval limit for indicating that we would be able to statistically reject a hypothesis that a sample has a concentration less than the detection limit is 0.000639. For a sample whose concentration is less than 0.000416 we can conclude that we would not commit a type II error by rejecting this sample as having a concentration less than the detection limit, and we would commit an error of 1 time out of 20 by rejecting the hypothesis that the sample's concentration is less than the criterion of detection.

Calculation of Detection Limits in Atomic Absorption Spectrophotometry

A method of calculating detection limits for atomic absorption is to alternatively measure the absorbance of a blank and a low concentration standard until at least 10 standards and 11 blanks have been measured. The series of analysis begins and ends with a blank. The blanks on either side of the standard are averaged and subtracted from the response of the standard. The mean and standard deviation of the set of 10 corrected standards are then calculated and used in the formula

$$\text{Detection limit} = \frac{2 \cdot s_{\text{corrected}}}{\bar{x}_{\text{corrected}}} \cdot [\text{Standard}] \qquad (7)$$

to calculate the detection limit. The standard deviation and the mean that are calculated are those of the corrected instrumental responses. This method is comforting in that a standard and a blank are both used in the calculation of the detection limit. But it leaves a bit to be desired since the method makes an assumption that the confidence interval of the standard can be directly transformed to be the confidence interval of a sample having "zero" concentration. As was seen in Chapter 10, the confidence limits on a regression curve were not constant over the range of the regression curve, but were curved with the degree of deviation from the regression line being dependent on the distance from the mean of the abscissa or the mean of the ordinate.

Calculation of Detection Limits from a Calibration Curve

These foregoing discussions of detection limits have largely ignored the dependence of instrumental response on the concentration of the sample. We have earlier shown how to calculate the confidence intervals for the slope, intercept, regression line, and individual values about the regression line. Let us use this concept to indicate how we will be able to calculate a detection limit based on the regression line. The main assumption that is made in using this method to calculate the detection limits is that the response of the instrument is linear to a zero concentration. As we have seen earlier, this assumption may not be very valid. A way to increase our confidence (not statistical but analytical confidence) in such a detection limit is to include samples with as low a concentration as feasibly possible. Hubaux and Vos[7] have formulated a similar approach to the following, but with confidence intervals built totally about the instrumental response and not by building confidence intervals about the concentrations.

Calibration data are obtained in the normal manner by analyzing standards of known concentration to obtain the functional dependence of the instrumental response on concentration. A regression equation is calculated from the data. The regression equation is usually a first-order polynomial, although, in principle, any type of a fitting model could be used. It is easier to calculate the confidence of the first order polynomial than for other fitting modcls, however. The model to be used then will be

$$\text{Response} = b_0 + b_1 \cdot [\text{Analyte}] \tag{8}$$

Figure 15-2 shows a graphic representation of such a calibration curve that would be obtained from this process. Also depicted in Figure 15-2 are the confidence limits on the regression line. The upper confidence limit (UCL) of the confidence interval of the intercept is shown as the intersection of the upper confidence limit of the regression line with the y axis. The confidence limits on the intercept, b_o, may be calculated using the following equation:

$$b_0 - t_{\alpha/2,\nu} \cdot s_{b_0} < \beta_0 < b_0 + t_{\alpha/2,\nu} \cdot s_{b_0} \tag{9}$$

where s_{bo} is given by

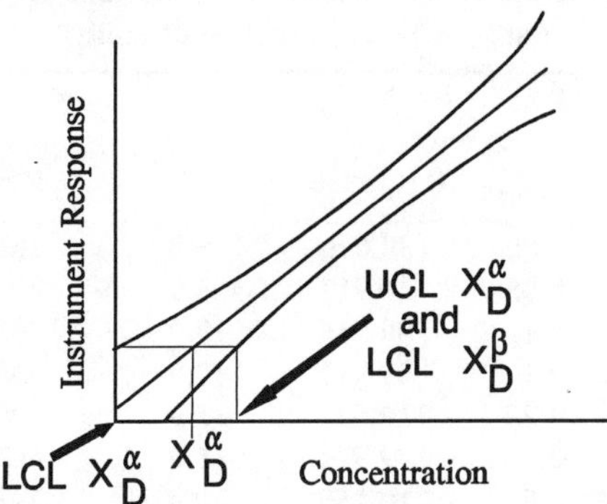

Figure 15-2. Graphic depiction of a calibration curve showing the 95% confidence limits on the regression and the UCL and LCL used to calculate the detection limit.

$$s_{b_0} = s_{y/x} \cdot \sqrt{\frac{\sum_i x_i^2}{n \cdot (n-1) \cdot s_x^2}} \tag{10}$$

and $s_{y/x}$ is given by

$$s_{y/x} = \sqrt{\frac{n-1}{n-2} \cdot \left(s_y^2 - b_1^2 \cdot s_x^2\right)} \tag{11}$$

But this is also the upper confidence limit of the regression line at zero concentration.

X_D may be used to denote the concentration of analyte that cannot statistically be distinguished from zero concentration. As can be seen from Figure 15-2, the lower limit of the confidence interval of X_D is zero. Solving the equation for confidence intervals about individual concentrations for the concentration that will have zero as the lower limit of the confidence interval will allow us also to calculate the upper limit of the confidence interval. The lower limit for the confidence interval (LCL) for the individual concentration is given by the equation

$$LCL \quad X_D = 0 = X_D - \frac{t_{\alpha/2,\nu}\, s_{y/x}}{b_1} \cdot \sqrt{\frac{1}{m} + \frac{1}{n} + \frac{(X_D - \bar{x})^2}{s_x^2}} \tag{12}$$

As indicated, this equation must be solved for X_D, which will be the concentration of analyte that will give zero as the lower limit of the confidence

Table 15-1. Data for the calculation of detection limit based on a calibration curve. [ ] concentration of analyte.

[ ]	Response	[ ]	Response
0.05	30.076	5.0	2016.392
0.05	29.913	5.0	2013.422
0.1	50.216	10.0	4024.712
0.1	49.922	10.0	4003.756
0.25	110.027	15.0	5984.713
0.25	109.972	15.0	6033.275
0.5	210.159	20.0	7972.900
0.5	209.196	20.0	7975.376
1.0	410.699	25.0	10038.470
1.0	410.269	25.0	10050.400

limit:

$$X_D^2 \cdot \left(b_1^2 - \frac{t^2 \cdot s_{y/x}^2}{s_x^2} \right) + \left(\frac{2 \cdot t^2 \cdot s_{y/x}^2 \cdot \bar{x}}{s_x^2} \right) \cdot X_D - t^2 \cdot s_{y/x}^2 \cdot \left(\frac{1}{m} + \frac{1}{n} + \frac{\bar{x}^2}{s_x^2} \right) = 0 \qquad (13)$$

After X_D is obtained, the upper confidence limit of the regression line is obtained from

$$UCL \quad X_D = X_D + \frac{t_{\alpha/2,\nu} \cdot s_{y/x}}{b_1} \cdot \sqrt{\frac{1}{m} + \frac{1}{n} + \frac{(X_D - -\bar{x})^2}{s_x^2}} \qquad (14)$$

This in turn is used to calculate a second concentration, which will give the desired confidence limit for Type II errors (β):

$$LCL \quad X_D^\beta = X_D^\beta - \frac{t_{\beta/2,\nu} \cdot s_{y/x}}{b_1} \cdot \sqrt{\frac{1}{m} + \frac{1}{n} + \frac{(X_D^B - \bar{x})^2}{s_x^2}} \qquad (15)$$

This second concentration should be considered as the concentration above which a null hypothesis that states that the concentration of an analyte in a sample is less than the detection limit may safely be rejected without thought to either a Type I or a Type II error. This concentration may safely be used to report a minimum detectable level for an analyte and the chemist can feel confident that the risks of making either a Type I (accepting a hypothesis that the analyte is present when it is in actuality absent) or a Type II error (accepting the alternative hypothesis that the analyte is absent when it is in actuality present) are minimized.

Example

Let us give an example of how this will work in practice. Table 15-1 gives data for the calibration of a nonspecific analyte. The concentrations range from 0.05 to 25.00 mg/L. The pertinent regression variables calculated from this data are

$$\begin{aligned}
\bar{x} &= 7.79 \\
m &= 2 \\
n &= 20 \\
b_0 &= 9.297 \\
r^2 &= 0.9999 \\
\nu &= n - 2 = 18 \\
b_1 &= 400.1816 \\
t &= 2.09 \\
s_x^2 &= 78.4964 \\
s_y^2 &= 12571156
\end{aligned}$$

From these variables, $X_D{}^{\alpha}$ is calculated from Equation (13) to be 0.024669 mg/L. This value for $X_D{}^{\alpha}$ is then substituted in Equation (14) and solved for the upper confidence limit of $X_D{}^{\alpha}$ to be 0.13566 mg/L. This value of the upper confidence limit of $X_D{}^{\alpha}$ is numerically equal to the lower confidence limit of $X_D{}^{\beta}$:

$$X_D^{\alpha}\,(\mathrm{UCL}) = X_D^{\beta}\,(\mathrm{LCL}) = 0.13566 \tag{16}$$

The value of $X_D{}^{\beta}$ is obtained by solving Equation (15) for $X_D{}^{\beta}$ to be 0.24488 mg/L. Some explanation is needed for these numbers that have just been calculated. LCL $X_D{}^{\alpha}$ is the lower limit of the confidence interval for $X_D{}^{\alpha}$. This confidence interval is a concentration interval and not in terms of response of the instrument. The two limits, LCL $X_D{}^{\beta}$ and UCL $X_D{}^{\alpha}$ numerically describe the same point, but differ in their interpretation. UCL $X_D{}^{\alpha}$ is the upper limit of the concentration confidence interval for the lowest concentration whose signal cannot be distinguished from zero. While LCL $X_D{}^{\beta}$ describes the lower limit of the confidence interval describing the distribution for the Type II error. $X_D{}^{\alpha}$ is the lowest concentration whose signal cannot statistically be distinguished from a zero signal. $X_D{}^{\beta}$ is the lowest concentration that will allow us to detect samples whose concentrations are greater than the detection limit within the β level of significance. The remaining variables have the usual meanings as previously described.

Although this method of calculation is longer than other methods, if programmed in a computer, the results are more statistically palatable. The one key to gaining good detection limits for this approach is to ensure that standards are analyzed with as low a concentration as feasible. The standards should also encompass the entire range of analysis. Unfortunately, the existence of standards with extremely high concentrations will tend to skew the standard deviations of x, y, and y/x. The value for X_D^{β} is not easy to calculate analytically and thus should be calculated using a computer program.[8]

Precision and Accuracy

Values that are obtained as the result of analyzing a chemical sample should be assessed for accuracy and precision. Even among chemists, the two terms are often used synonymously; however, the terms have greatly differing meanings. First, let us discuss accuracy. Accuracy refers to the ability of an analytical technique or analytical method to portray the real or population mean of an object. We have discussed sampling in terms of the need for a sample to faithfully represent the object being sampled. So the analysis portion of the method must faithfully represent the object that was sampled. This measure is called accuracy. The analogy given in Chapter 1 to explain accuracy was the concept of shooting a firearm at a bullseye target. How close the marksman is to the center of the target is termed accuracy. Another term for accuracy is relative error. Mathematically, we can think of accuracy as being the average relative deviation of the analysis of a set of samples from the mean of the population:

$$\text{Relative Error} = \frac{\sum \frac{\bar{x} - \text{True}}{\text{True}}}{N} \tag{17}$$

where N is the number of determinations. The percent relative error is

$$\%\text{Relative Error} = \frac{\sum \frac{\bar{x} - \text{True}}{\text{True}}}{N} \cdot 100 \tag{18}$$

It should be noted that the mean accuracy is a signed quantity, that is, it denotes a positive or negative deviation from the mean. As we shall learn later, it is also dependent on the concentration of the analyte being measured.

The other quantity that should be reported for a method is the precision. The term precision requires several components to adequately describe its behavior. Several synonyms have been given for precision, including *REPEATABILITY* and *REPRODUCIBILITY*. To adequately describe precision, one must accumulate data as to the values obtained from the analysis over time of the same "sample." As the data are accumulated, several components will begin to emerge, e.g., several different analysts analyze the sample, different instruments are used to analyze the sample, and different laboratories are used to analyze the sample. Each of these is a component of precision and must be considered in the final number reported as the precision. Repeatability, as defined by ASTM, is the ability of an analytical method when repeated several times in a single day by a single analyst to give the same answer. Reproducibility is defined as the ability of an analytical method to give the same answer on different days by different analysts and possibly even in different laboratories. Mathematically the precision may be described as the relative standard deviation from the mean of all analyses that are performed on a single sample:

$$\text{Relative standard deviation} = \frac{s}{\bar{x}} \quad \textbf{(19)}$$

The term coefficient of variation is often applied as a measure of the precision of a method and is another name for the relative standard deviation. In the literature, each of these is often expressed as a percentage:

$$\text{Coefficient of variation} = \frac{s}{\bar{x}} \cdot 100 \quad \textbf{(20)}$$

REFERENCES

1. McCormack, M. Realistic detection limits and the political world in ***Detection in Analytical Chemistry, Importance, Theory, and Practice***. Currie, L., ed., American Chemical Society, Washington, D.C., 1988, pp. 64–69.

2. Moss, T.H. Scientific measurements and data in public policy-making in ***Detection in Analytical Chemistry, Importance, Theory, and Practice,*** Currie, L.A., ed., American Chemical Society, Washington, D.C., 1988, pp. 70–77.

3. Graham, R.C. Design of software for thermal titrations *TALANTA*, 1987, **34(4)**, 381–384.

4. Kaiser, H. ***Two Papers on the Limit of Detection of a Complete Analytical Procedure***. Adam Hilger, London, 1968.

5. Wilson, A.L. ***Talanta***, 1973, **20**, 725.

6. Currie, L.A. ***Anal. Chem.***, 1968, **40**, 586.

7. Hubaux, A., and Vos, G. ***Anal. Chem.***, 1970, **42**, 849-855.

8. Examples of a program that can be used to numerically find solutions to equations are *Formula One*, *RS/1* (BBN Software), *MathCAD*, *Eureka*, and *TK Solver*. *Formula One* was used to obtain the values reported here.

16

Measurement of Data Quality

General

Before we can discuss what is meant by data quality, we must define what is meant by the word *quality*. Quality is generally meant to be the absence of defects that will affect the use of the product. It is accepted that when a product is purchased, the product will perform in the manner stated. Although quality is accepted to be the absence of defects, there are two subaspects to consider: deficiency and defectiveness. Deficiency is a general inadequacy for the product to perform, whereas, defectiveness is a wanting of a property. How is this applied to analysis of chemical data? Quite simply, the chemical data must be of sufficient quality and not lack essential elements of performance such that the analytical concentrations reported by the method are suspect.

The usual aim of routine analytical chemistry is to record concentrations of analytes in the samples. A task that is faced by users of such analytical data is to determine the quality of the data before the data can be used to make decisions. The quality of the numbers must often be proved, not just as a basis to make decisions, but as a basis of defense in legal actions. The defense of numbers in court goes beyond just indicating that the instrument gave a number for the concentration. Additional aspects such as the degree of calibration must be proved. An aspect that is really beyond the scope of this book to discuss, but is certainly part of the overall quality of the data, is the extent to which the sample history can be traced. The history of the sample, particularly in legal actions, must be documented from the time of sampling to the disposal of the remains of the sample after analysis. Such history is called a chain of custody.

What are the items that an investigator should determine about the quality of the data from a chemical laboratory and used in a decision-making process? Some of the items discussed in previous chapters include the accuracy and precision of the analysis, the calibration data, and the minimum detectable level. Implementation of procedures to obtain these numbers vary widely. The methods of interpretation of the results of such methods also vary widely. To obtain the calibration curve, the analyst must be aware of the concentration of the standards and use that accordingly to construct the calibration curve, from which the minimum detectable level can then be deduced. Unfortunately, the calibration curve is usually prepared with "ideal," well-behaved solutions.

When real samples are used, the applicability of the calibration curve to the analysis of the samples must be verified. A combination of "real" samples must be presented to the analyst for determination of concentrations. The combination of samples must include ones for which the concentrations are known to the analyst and ones for which the analyst has no knowledge of the actual concentrations (blind samples). The latter is to prevent a bias (although usually unintentional) from clouding the results of the analysis. Different methods of implementation of this scheme are used by different agencies and companies. A general aspect of the implementation of such a scheme is that analysts are given samples for which they have no knowledge of the concentration. These blind samples either are generated in the laboratory by the quality control manager or are from external sources.[1] These samples should be submitted to the analyst, who will not be aware which are the check samples. The samples are then analyzed in the same manner as all other samples. The interpretation of the results from analysis of such samples is the subject of this sample.

Quality Assurance

Two terms that are often used interchangeably but that have greatly different meanings are *Quality control* and *Quality assurance.* Both are extremely important in the overall production of quality results from the analytical laboratory. *Quality assurance* deals mainly with the attitude of the entire organization for whom the analyses are being performed. *Quality control* is just one small aspect of quality assurance. An organization that is dedicated to production of quality results will provide the means necessary to produce quality results. Elements that could be provided may be broken into three categories: quality control, facilities, and personnel. These aspects are shown in Figure 16-1.

The categories shown in Figure 16-1 indicate that the major aspects of

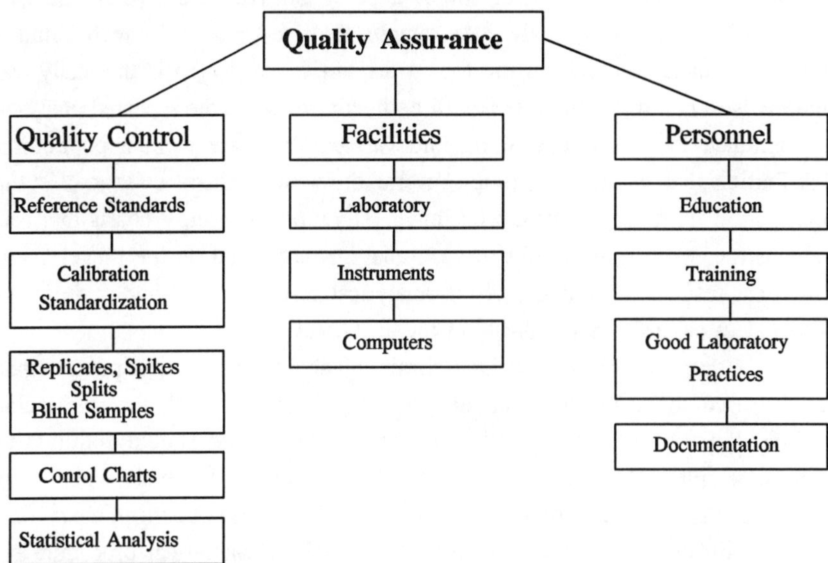

Figure 16-1. Aspects of quality assurance. Of main interest in this chapter are the aspects of quality control.

quality control are

1. Availability of reference materials and standards.
2. The proper use and preparation of calibration charts
3. The use of replicate analyses, spikes, splits and surrogate samples.
4. The results from the analysis of the samples indicated in (3) are used to prepare control charts.
5. The control charts are subjected to statistical analysis to determine a degree of confidence in the overall analytical results.

Before control charts can be prepared, analytical results must be obtained by the analyst. Samples are submitted to the analyst from the laboratory manager who either prepares the sample or receives the samples from the quality assurance officer of the company. The concentration and composition of the samples are unknown to the analyst. The sample composition should be similar to the composition of the samples that are being analyzed by the analyst. Samples whose concentration or appearance is out of the ordinary indicates to the analyst that a quality control sample is being analyzed and special care will then be taken to carefully analyze the sample. Repeated analysis of such samples will allow management from the top to the analyst to decide if the analytical process is in control. The process will be in control when the repeated analyses are assumed to give values of random variables that are from a population distribution that is stable in time. Several different methods exist to use the information from such control procedures. The types of control charts that can be constructed from the results of such analyses are based on the range of the analytical results, the average of the analytical results, the deviation from an accepted value, and the variability of the analytical results.

Control charts can provide information about the analytical process that can be used to decide whether analyses should continue for that particular analyte. If the process is in statistical control, the analyst should continue to analyze samples. However, often a decision will need to be made whether to discontinue analysis or to recalibrate a protocol. Proper use of control charts can also be valuable indicators of potential troubles in an analytical procedure by indicating potential trends or sudden shifts in means or variability.

A valuable, but often overlooked, tool of the control chart is to provide a better estimate of the variability of an analytical method than by the estimation that comes from repeated analysis of a single sample. This estimate of the variability can then be used in the tests that can be performed when the standard deviation is *known* as opposed to estimated. The variability in the control chart can provide that extra measure of confidence in the standard deviation or variance. The measure of variability, usually the sample standard deviation, can then be replaced in many statistical tests by the population standard deviation, thus giving a greater measure of confidence in the analytical results.

Control charts are characterized by a center line, which is the expected (or true) value of the check samples. Confidence limits are then calculated about this center line and used to indicate warning and action limits on the process. As long as the results remain within the bounds, the process is determined to be in statistical control. When the results fall outside of either of the limits (warning or action), the analyst is alerted that some type of remediation is needed. When the results fall outside the limits, it can be determined when the analytical system no longer provided results in statistical control. Such delineation of time when the process was not in control will provide the information needed to determine how many (and presumably which) samples will have to be reanalyzed. The control charts that are kept and maintained will result in long-range records used as historical and legal basis to assess the quality of the analytical data arising from the analytical system.

Each of the types of control charts will be presented along with its strong and weak points.

Variability of Analytical Results

Before a control chart can be constructed that can be used to assess the performance of an analytical method, the method must be brought under statistical control. If the variability of the analytical method is quite large, the results coming from the method cannot be assumed to be coming from a population distribution that is stable in time and that has a fixed mean. John Taylor, formerly of the National Bureau of Standards, has indicated that "until a measurement system has attained a state of statistical control, it cannot be assumed as measuring anything."[2]

Construction of control charts using the variability of the analytical results implies that data have been added to the control chart in groups and not added one data value at a time. Grouping may be acceptable if the length of time of acquisition of the data point is short and a large number of data points are collected in a relatively short period of time. Such a grouping might be applicable for analytical processes that do not have a long analysis time such as pH, specific conductance, or possibly even some of the ion-specific electrode measurements. Grouping data has several advantages including calculation of range (variability) and central tendency from the same data set.

Depending on the number of data points being added at a time, there are two choices that can be made to estimate the variability of the analytical method and consequently two measures of variability to use when constructing a control chart. Either the range or the standard deviation of the individual groups may be used. The estimation of variability using the range is presented because of the historical significance, however, much of the useful information in the data set would not be used by such a calculation. The information about variability is much richer and fuller when the standard deviation is calculated. The range was historically used because of the general lack of computers in the laboratory and, consequently, the calculations were performed manually. Today, the calculation can and should be done by a computer. Thus, it is much more important and acceptable to perform calculations for control charts using the standard deviation.

Range Control Chart

Limits for range control charts are expressed as 3σ limits about the range:

$$\bar{R} \pm 3 \cdot \hat{\sigma}_R \tag{1}$$

The upper and lower confidence limits are estimated from the values of two constants, D_3 and D_4 which are derived as follows.*

If the values of the random variable, Y, are normally distributed then the expected value of the range is

$$E(R) = \int_{-\infty}^{\infty} (y_{\max} - y_{\min}) f(R)\, dR = d_2 \sigma_y \tag{2}$$

d_2 depends on the number of elements in the group. Thus

$$\mu_R = d_2 \cdot \sigma_y \tag{3}$$

Then an estimate of σ_y would be given as

$$\hat{\sigma}_y = \frac{\hat{\mu}_R}{d_2} = \frac{\bar{R}}{d_2} \tag{4}$$

but it also follows that since the standard deviation of the y values can be obtained from the average of the ranges, then

$$\hat{\sigma}_R = d_3 \cdot \hat{\sigma}_y \tag{5}$$

Substituting Equation (4) into Equation (5), the following is obtained:

$$\hat{\sigma}_R = \frac{d_3 \cdot \bar{R}}{d_2} \tag{6}$$

But since the confidence limits on $\bar{R}$ may be calculated from Equation (1), it follows that the confidence limits can be expressed in terms of the two

---

* Tabulated values for the two constants are found in Appendix C.

constants, d_2 and d_3 as

$$\bar{R} \pm 3 \cdot \left(\frac{d_3}{d_2}\right) \cdot \bar{R} \tag{7}$$

which, after factoring out $\bar{R}$, results in being able to multiply $\bar{R}$ by D_3 and D_4, where D_3 and D_4 are equal to

$$D_3 = 1 + \left(\frac{d_2}{d_3}\right)$$
$$D_4 = 1 - \left(\frac{d_2}{d_3}\right) \tag{8}$$

D_3 and D_4 are tabulated in Appendix C for various values of the size of the group. The final useful equations for the control limits on the range are

$$\text{Upper control limit} = D_3 \cdot \bar{R}$$
$$\text{Lower control limit} = D_4 \cdot \bar{R} \tag{9}$$

s-Chart

Although the range chart has been used most extensively in the past, a large amount of data are ignored in the calculation of the ranges. Use of the entire amount of data in each group would be more desirable. One such way is to construct variability control charts based on the standard deviation of the groups. The use of the standard deviation to construct such charts, though, is calculation intensive if done by hand. However, if the data have been entered in a computer, the calculation should not be formidable. As with the range control charts, the control limits for the s-chart are based upon a three σ_s limit:

$$\bar{s} \pm 3 \cdot \sigma_s \tag{10}$$

If the values of the random variables in the subgroups are normally distributed then the ratio of the sample variance to the population variance is distributed as a χ^2 distribution:

$$\frac{(n-1) \cdot s_y^2}{\sigma_y^2} \sim \chi^2 (n-1) \tag{11}$$

Taking the square root leads to an expression for the expected value of the standard deviation having a value

$$E(s) = \frac{\sigma}{\sqrt{n-1}} \cdot E(\chi_{n-1}) \tag{12}$$

But, the expected value of χ_{n-1} equals

$$E(\chi_{n-1}) = \sqrt{2} \cdot \frac{\Gamma[n/2]}{\Gamma[(n-1)/2]} \tag{13}$$

where Γ is the gamma function. The expected value of s is then

$$E(s) = \bar{s} = \sqrt{\frac{2}{n-1}} \frac{\Gamma[n/2]}{\Gamma[n-1/2]} \cdot \sigma = k\,\sigma \tag{14}$$

But we also need an estimate of the standard deviation of s. This can be obtained by remembering that the variance of any random variable is calculated from

$$V(s) = E(s^2) - (E(s))^2 \tag{15}$$

and $E(s^2)$ is equal to σ^2 and $[E(s))^2$ is $(k\ \sigma]^2$. Since σ^2 can be estimated from Equation (14) to be

$$\sigma^2 = \left(\frac{\bar{s}}{k}\right)^2 \tag{16}$$

it follows that an estimate for the variance of s is

$$\hat{\sigma}_s = \frac{\bar{s}}{k} \cdot \sqrt{1-k^2} \tag{17}$$

Fortunately, the values of

$$1 \pm \frac{3 \cdot \sqrt{1-k^2}}{k} \tag{18}$$

have been tabulated and are available in Appendix C as B_3 and B_4. Equation (10) then becomes

$$\begin{aligned} \mathrm{UCL}_s &= \bar{s} \cdot B_4 \\ \mathrm{LCL}_s &= \bar{s} \cdot B_3 \end{aligned} \tag{19}$$

Example

Let us now examine an example of constructing control charts for the range and standard deviation of a chemical analysis. The data that will be used represent the analysis of a standard solution that has a conductance of 25.00 μS/cm. The analysis is rather easy to perform and many determinations can be accomplished rather quickly. The data are grouped into sets of seven data points. A total of 71 data sets (each consisting of seven data points) were taken. The grouped data are shown in Table 16-1. The range and standard deviation are calculated for each group. Before the data are plotted, the confidence limits to determine if the variability of the analysis is in a semblance of control are calculated using Equation (9). D_3 and D_4 for a group size of seven are 0.076 and 1.924, respectively. The average range is 0.690, therefore, the upper and lower confidence limits on the range are 1.328 and 0.052. The confidence limits on the standard deviation are calculated as the product of the average of the standard deviations (0.244) and the values for B_3 and B_4 are 0.118 and 1.882. Thus the confidence limits are 0.0288 and 0.460. The plots of the values for the standard deviations and the ranges for the data in Table 16-1 are shown in Figure 16-2 and Figure 16-3.

Control on Group Averages

Once the variability of an analytical process has been shown to be in control or at a minimum to be known, the control charts to determine if future values determined from the analytical process are valid may be constructed. One of the control charts which is in wide usage for the control of the mean of the analytical method is the *X* chart. This chart is constructed from the same data for which the range and *s* charts were constructed. The intent of the *X* chart is to provide a basis against which future values of the results of analysis for water samples using the conductance method can be compared. The further intent of the *X* chart is to be able to determine very quickly if the analytical method is still in control or whether modifications to the procedure need to be made. The modifications will often require such actions as recalibration of the method or further analyst training. The *X* control chart should be sensitive enough to be able to detect shifts, yet not so sensitive that many false "out-of-control" situations are detected. This brings us back to the former discussions about Type I and Type II errors. We do not want to have a large number of rejections of the analytical data that should not have been rejected. But on the other hand, neither do we want to have a large number of acceptance of invalid analytical results. One of the methods of accomplishing this seemingly paradoxical task is to construct the control chart in such a manner as to have warning limits and action limits. The level to which we wish to control an analytical technique can

Table 16-1. Data from Analysis of a Sample Having a Conductance of 25.00 μS/cm[a]

Group Number	Values of conductance						
1	24.81	24.71	24.69	24.92	24.58	25.04	25.12
2	24.60	25.00	24.66	24.70	24.90	25.40	24.83
3	24.79	25.13	24.98	24.89	24.93	25.16	25.39
4	24.99	25.16	25.06	24.96	25.07	25.18	25.11
5	25.09	25.09	24.83	25.28	25.09	25.14	24.80
6	24.97	24.95	25.06	24.99	24.95	25.61	24.84
7	25.02	24.63	25.01	24.84	24.97	25.11	24.92
8	25.03	25.22	25.07	25.03	24.98	24.81	25.16
9	24.73	25.58	24.90	25.11	25.06	24.99	25.01
10	24.56	25.09	25.09	24.51	25.15	24.95	25.06
11	24.62	24.84	24.66	25.08	25.31	24.93	24.94
12	25.08	24.71	25.01	24.74	24.60	24.85	24.96
13	24.90	24.70	24.91	25.16	24.82	25.16	25.13
14	25.43	25.45	24.95	24.94	24.99	25.01	24.49
15	24.61	24.86	25.06	24.98	24.80	25.15	25.06
16	25.08	25.13	25.00	24.74	24.64	24.77	24.93
17	24.81	25.36	25.28	25.04	25.32	24.78	25.36
18	24.67	25.06	25.47	24.95	25.16	25.29	25.21
19	25.35	24.57	24.78	24.80	24.92	24.81	24.88
20	25.10	24.93	24.78	25.13	24.78	24.96	24.80
21	25.16	25.07	25.17	24.70	25.42	24.86	24.79
22	25.18	24.79	25.05	24.92	25.59	25.50	24.92
23	24.63	25.17	25.24	25.51	24.89	24.92	25.59
24	25.17	24.79	25.33	24.68	24.74	24.97	24.67
25	25.24	24.95	25.18	25.10	25.07	25.51	25.16
26	24.79	25.04	25.49	24.80	25.09	24.92	24.86
27	25.42	24.69	24.61	25.05	24.97	24.98	24.84
28	24.95	24.57	25.26	25.19	24.84	25.05	24.97
29	24.41	25.19	25.17	25.15	25.08	25.11	25.25
30	24.92	24.86	25.26	25.12	25.08	25.11	24.80
31	24.66	24.83	24.94	24.89	25.29	25.24	25.30
32	25.02	25.36	25.54	25.28	25.36	24.69	24.98
33	25.24	24.82	25.02	24.81	25.28	24.80	25.76
34	24.95	24.75	25.08	24.68	25.20	25.08	25.58
35	25.17	25.27	24.90	25.62	25.03	25.04	25.26
36	25.12	24.86	25.24	25.33	25.23	24.46	25.00

Continued

be adjusted depending on the amount of risk that is acceptable to commit either a Type I or a Type II error. A common set of values for warning and action limits is 90 and 95%. The warning limit is the value above (or below for a

(Continued from Table 16-1)

37	25.34	25.15	24.95	24.97	25.52	24.70	24.95
38	25.08	24.95	24.89	25.02	25.07	24.72	25.17
39	24.45	24.76	24.71	24.65	24.72	25.66	25.09
40	24.59	24.83	25.04	25.47	24.94	24.60	24.73
41	25.04	24.87	24.58	24.85	25.20	25.01	24.84
42	25.05	25.22	24.81	25.13	24.75	24.84	25.31
43	25.02	24.78	25.02	25.02	25.21	24.95	24.95
44	24.73	24.54	25.49	25.17	24.96	24.61	25.10
45	24.66	25.26	24.88	25.31	25.06	24.70	25.23
46	25.17	24.62	25.00	25.25	25.18	25.46	24.68
47	24.53	24.63	25.18	25.09	24.92	24.95	24.88
48	24.99	25.28	24.86	25.34	25.33	24.88	25.03
49	24.67	25.20	25.22	24.99	25.11	25.17	24.99
50	25.15	25.06	25.00	25.30	25.05	25.30	24.93
51	24.73	24.67	24.84	24.75	25.26	25.03	24.87
52	25.17	24.63	25.18	25.40	25.12	24.77	24.48
53	24.80	25.22	24.97	25.20	24.59	25.27	25.55
54	25.10	25.12	24.76	24.68	25.13	24.90	25.72
55	25.04	25.34	25.15	24.58	24.84	25.08	24.80
56	25.02	25.15	24.49	24.76	25.21	24.98	25.16
57	24.72	24.83	25.41	25.43	24.96	24.62	25.40
58	25.01	24.32	25.21	25.32	25.04	24.95	25.00
59	25.12	24.78	25.16	24.59	24.74	24.74	24.85
60	25.08	25.22	25.02	25.10	24.95	24.94	24.98
61	24.67	24.85	25.50	25.62	24.72	24.85	25.82
62	25.02	25.13	25.34	24.81	24.64	24.92	24.99
63	25.04	25.19	25.09	25.32	24.73	24.90	25.32
64	25.11	24.82	25.32	24.92	24.79	24.72	25.09
65	25.45	24.67	24.88	25.18	25.02	24.69	24.89
66	24.96	24.94	24.91	25.42	25.06	24.93	24.95
67	25.28	25.12	25.05	25.04	25.09	25.32	24.97
68	24.53	24.95	25.46	25.38	25.34	25.18	24.78
69	25.04	25.21	24.88	24.76	24.98	25.09	25.44
70	25.44	25.04	25.08	24.54	24.83	25.03	25.00
71	24.81	24.85	24.81	24.97	25.21	25.39	24.73

[a] Data has been grouped in sets of seven for purposes of constructing control charts.

lower warning limit) which the result of analysis of a standard of known concentration must fall for the analyst to begin to be concerned about the overall quality of the data. Generally, the analyst should be aware of the results only after the fact. The analyst should not know that a standard for quality assurance purposes is being analyzed. The laboratory manager or a quality control manager should be assessing the results of the analysis. If the results fall in the band between acceptable and the limit where remedial action is required, the analyst should be apprised of the situation and be cautioned to examine

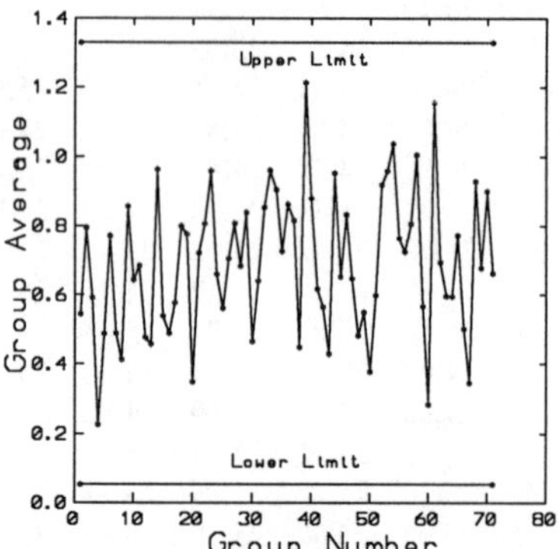

Figure 16-2. Ranges of groups (and upper and lower confidence limits) of seven determinations of conductance.

calibration. The results from the analysis are still acceptable, however, caution needs to be exercised. The use of the action and the warning limits gives a compromise between the commitment of Type I and Type II errors.

Let us now examine how a control chart is set up using the X chart. The same set of data from which the range and standard deviation charts are constructed should be used to construct the initial X chart. The arithmetic average of group is calculated and the average of the group averages is also calculated. The volume of data that should be available will allow the calculation of a standard deviation in which we should have a high degree of confidence. This standard deviation can then be used with either the Z-confidence interval

$$\mu \pm Z \cdot \sigma \tag{20}$$

or Student's t confidence interval

$$\mu \pm t \cdot \frac{s}{\sqrt{n}} \tag{21}$$

In Equation (20), σ is the overall standard deviation of the entire data set. In Equation (21), s is the standard deviation of the entire data set. The use of either σ in Equation (20) or s in Equation (21) will depend on the number of samples that have been analyzed. Recall that the central limit theorem allows us to estimate the expected value of a distribution by the arithmetic average if n is greater than 30. If, however, the number of samples is less than 30 and the

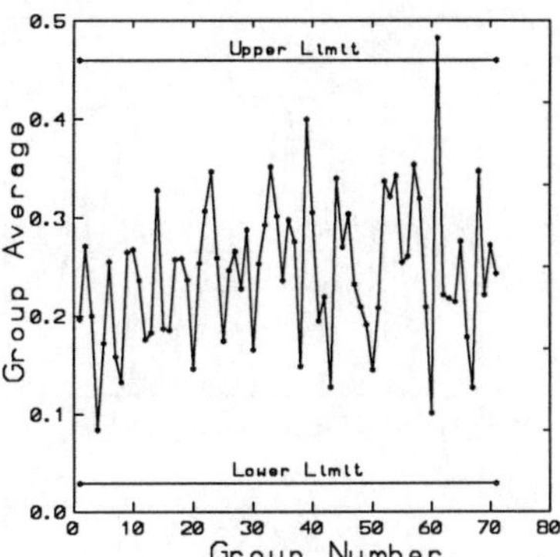

Figure 16-3. Standard deviation (and upper and lower confidence limits) for the conductance of groups of seven determinations.

results of the analysis are normally distributed (a very good assumption for this type of analysis), then the Student's t Equation (21) should be used. If it is desired to calculate the action limits based on the standard deviation of the group averages, the following can be used:

$$\mu \pm t \cdot s_{\bar{x}} \tag{22}$$

where $s_{\bar{x}}$ is the standard deviation of the group averages. These expressions are strongly influenced by the size of the groups. The larger the group, the closer to the mean will be the warning and action limits.

Example

The data in Table 16-1 may be used to calculate an initial control chart. The average of each group is calculated and plotted versus the analysis group number. The average of the group averages is 25.013. The desired warning limit is 95%, for which the Student's t value is 1.99. The action limit is chosen as 99%, for which the Student's t value is 2.65. The overall standard deviation of the data set is 0.252 and the overall mean is 25.013. The upper and lower warning limits are calculated to be 25.188 and 24.838. The upper and lower action limits are calculated to be 25.245 and 24.780. The data are plotted in Figure 16-4 which shows the average value for the analysis of the standard conductance solution (25.013) and the upper and lower action and warning limits. These

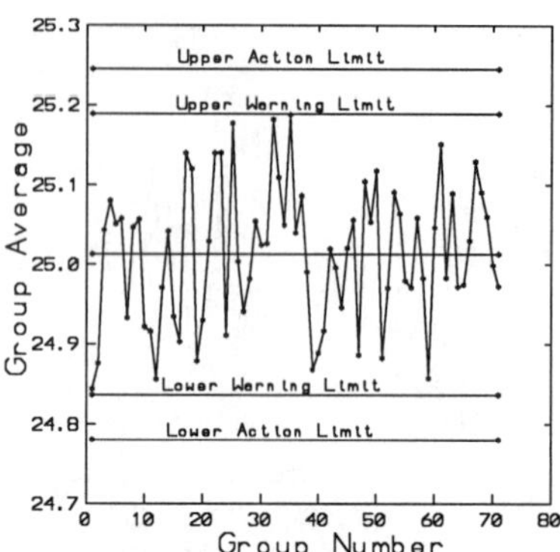

Figure 16-4. ***X*** **control chart for the analysis of conductance in water samples. The upper and lower action and warning limits are also shown.**

upper and lower action and warning limits can be used to check future determinations of the conductance of the standard solution.

As the control chart is used, the variability of the analytical method will likely decrease as the analyst becomes more familiar with the method and also realizes that the management of the laboratory is serious about quality control. As the variability decreases, the need to revise the control limits (both on the *X* chart and on the *s* chart) may become necessary. The question should be asked as to how long the control chart can be used before such modification is necessary. It can also be asked why is it necessary to revise the confidence limits. As the variability decreases, the detection of an out-of-control set of analyses will become more difficult to detect since the confidence limits are much wider than they should be. The distance between the confidence limits is very sensitive to the variability of the method. This is true whether the *X* chart or the *s* chart is being used.

A way to compensate for the decreased variability is to continually update the confidence limits. After a new group of analyses has passed the quality control check, that new group can be added to the set of data from which the confidence limits are calculated. The oldest member of the data set can then be dropped. This will ensure that the variability is always being updated. It should be noted that the minimum number of data points in the quality control set should always be at least 20. This number is the number of groups that are in the data set. Of course, the group size should also be kept constant as new analyses are performed.

An additional point to remember is that even though a set of control analyses appears to be in control, the average of the group averages should agree with the theoretical or expected value of the concentration of the control samples. For example, in the example above, the theoretical value of the conductance was 25.00 μS/cm. If the average of the group averages had been 27.53 μS/cm and the standard deviation the same as before, the analytical method would have appeared to be in control. However, it is clear that the true value of 25.00 μS/cm is clearly outside the 99% confidence interval for the data set. The value of the concentration of the control samples found by analysis must agree with the theoretical value of the concentration.

Another aspect that can be considered about the *X* chart is that not only can values that exceed the confidence limits signal an out-of-control condition, but runs of values that exceed the expected value of the control sample can also signal an out-of-control condition.

References

1. Sources for standards that are prepared and certified by external agencies include the Environmental Protection Agency, the National Institute for Standards Technology, the American Society for Testing and Materials, and the Occupational Safety Health Administration. Each of these prepares standards that span a large range of materials and concentrations. Many others prepare check samples for use within their programs, for example, standards for the National Atmospheric Deposition Program/National Trends Network are prepared by the U.S. Geological Survey.

2. Taylor, J. Seminar on ***Quality Assurance of Chemical Measurements***, National Bureau of Standards, Gaithersberg, Maryland, 1983.

17

NON-PARAMETRIC TESTS

Introduction

All of the tests which have been discussed to this point have generally assumed that the data is normally distributed. For example, the ANOVA test assumes the following:

1. The residuals calculated from the statistical model are normally distributed with a mean of zero.
2. The residuals have homogeneous variance throughout the range of the residuals.
3. The data points are also normally distributed.
4. The variance of individual columns are equal. (For one way and two way ANOVA).
5. The variance of individual rows are equal. (For two way ANOVA).

If these assumptions are not valid, the ANOVA should not be used.

Another situation in which many of the previous tests should not be used is if the data sets have very large standard deviations. As can be seen by examining the equations for the t-test, and so on, the confidence intervals that are obtained are directly affected by the size of the standard deviation. The standard deviation is sensitive to the presence of even one outlier in the data set, which can often lead to an inflated standard deviation. When such standard deviations are noted, it may be useful to first examine whether the outlier is the result of experimental or analytical difficulties and represents an out of control situation, or whether the data actually represents a real situation. Some analytical situations are not best treated as a continuous distribution. An extreme example can be seen in the mining industry when certain metals are mined, such as gold, which appears in elemental form in nature and which occurs as lumps. If the goal of the analysis program is to determine the average concentration of the gold in a rather limited sample, the presence of even one nugget can drastically affect the outcome of the analyses. In these situations, it is necessary that alternate methods of data analysis be available. A class of statistical techniques which do not assume that the underlying distribution of the data is normal nor in fact that knowledge about any underlying distribution is required are distribution free or nonparametric techniques. Several nonparametric techniques will be presented that can be used when assumptions such as normality are not valid.

Wilcoxon Rank Sum Test*

Let us examine a data set that is not normally distributed. The data in Table 17-1 depict sulfate concentrations in rainwater that were measured at two different locations. The mean and standard deviation for each location are given at the bottom of the table. The natural inclination is to use a *t* test to determine if the mean sulfate concentration at both locations is the same. The variance of x is 18.35 and the variance of y is 26.33. The F-ratio is 1.43 and the critical value for $\alpha=0.05$ is $F_{\text{critical}}=2.67$. Therefore, the hypothesis that the variances are equal would not be rejected. The *t* test that we should use is the *t* test for unknown means and equal variances. Recall from Chapter 9, that the test statistic is calculated from

$$t_\nu = \frac{(\bar{x}_1 - \bar{x}_2)}{s_p \cdot \sqrt{1/n_1 + 1/n_2}} \tag{1}$$

where the pooled variance s_p^2 is calculated from

$$s_p = \sqrt{\frac{s_1^2 \cdot \nu_1 + s_2^2 \cdot \nu_2}{\nu_1 + \nu_2}} \tag{2}$$

from which we obtain a pooled variance of 23.31 or a pooled standard deviation of 4.83. The calculated *t* test statistic is 0.146. This would be compared to $t_{\text{critical}}=1.70$ for 29 degrees of freedom ($n_1 + n_2 - 2 = 12 + 19 - 2$) and $\alpha = 0.05$. On first glance, we would be inclined to not reject the hypothesis that

* The Wilcoxon rank sum test is also known in some texts as the Mann-Whitney Test.

Table 17-1. Data Set for Concentration of Sulfate in Rainwater from Two Different Sampling Locations

Location					
X		Y			
14.8	10.6	12.7	16.9	7.6	2.4
7.3	12.5	14.2	7.9	11.3	
5.6	12.9	12.6	16.0	8.3	
6.3	16.1	2.1	10.6	6.7	
9.0	11.4	17.7	5.6	3.6	
4.2	2.7	11.8	5.6	1.0	
$mean_x = 9.45$		$mean_y = 9.189$			
$s_x = 4.284$		$s_y = 5.122$			

the means are equal. However, if we construct the confidence interval about the values for sulfate concentration at sampling location *y*, we obtain the following:

$$9.18 - 2.1 \cdot \frac{5.12}{\sqrt{19}} \leq \mu \leq 9.18 + 2.1 \cdot \frac{5.12}{\sqrt{19}} \tag{3}$$

or the confidence interval is calculated as

$$6.72 \leq \mu \leq 11.66 \tag{4}$$

As can be seen the standard deviation is quite large as is the range of the data (17.7-1.0 = 16.7). The number of extreme data points in this set indicate that the data is probably not normally distributed. The frequency histograms with a normal probability distribution function overlain for the set of sulfate data from sampling locations X and Y (Figure 17-1) verify that the data are probably not normally distributed. The tailing of the probability distribution indicates that the distribution is not symmetric, at a minimum. The implication is that the distribution is thus not normal. If the criterion is made that it is not important that the distribution be normal, but that whatever the distribution, it must be symmetric, a test is needed that will allow us to determine if the distribution is symmetric. A distribution free, or nonparametric, test would be desirable.

If we desire to formulate a null hypothesis that the mean of the sample is equal to the median of the sample, then the distribution must be symmetric.

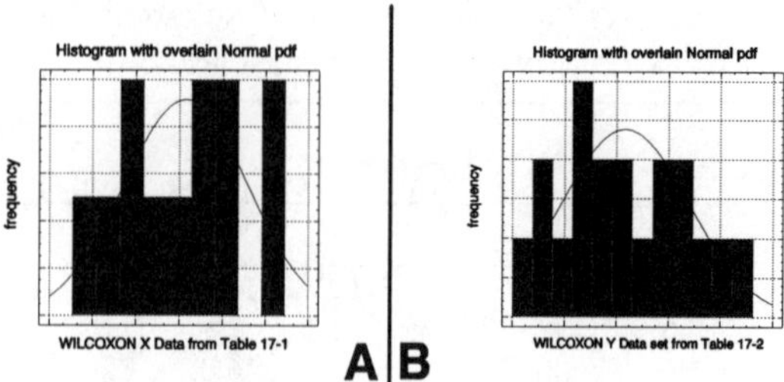

Figure 17-1. Histogram of sulfate concentration data for A sampling location X and B sampling location Y.

Recall that the definition of the median is that 50% of the area under the probability distribution function lie to the left of the median and 50% of the area under the probability distribution function lie to the right of the median. Recall also that the mean, or expected value, of a sample that is symmetrically distributed is the midway point of the probability distribution function. We would also like this test to depend on the relative distance from the median and the number of data points that are greater than (or less than) the median.

A test to use that meets the criteria of being distribution free and tests the symmetry of the data is the Wilcoxon rank sum test. To use this test, the set of data is sorted into rank order, regardless of whether the data belong to sampling location X or sampling location Y, as in Table 17-2. If there is a tie in the data, i.e., two or more data points have the same value, the rank that is assigned to each is the average of the ranks that otherwise would be assigned. The ranks of either the data points belonging to X or Y are summed [$\Sigma Rank(x_i)$=204.5; $\Sigma Rank(y_i)$=273.5] to give the test statistic, T

$$T_{test} = \sum \text{Rank}(y_i) \tag{5}$$

The null hypothesis of this test can be formulated in one of three ways: (1) the expected value of locations X and Y are the same; (2) the expected value of location X is less than or equal to the expected value of location Y; and (3) the expected value of location X is greater than or equal to the expected value of location Y. Alternative hypotheses would be formulated accordingly: (1) the expected value of locations X and Y are not equal, (2) the expected value of location X is greater than the expected value of location Y, or (3) the expected value of location X is less than the expected value of location Y. (1) is a two-

Table 17-2. Ranks of Sulfate Concentration Data for Sampling Locations X and Y[a]

X	Y	Rank	X	Y	Rank	X	Y	Rank
1.0		1	6.7		12	11.8		22
	2.1	2	7.3		13	12.5		23
	2.4	3		7.6	14		12.6	24
2.7		4		7.9	15		12.7	25
	3.6	5		8.3	16	12.9		26
4.2		6	9.0		17		14.2	27
5.6		8	10.6		18.5	14.8		28
	5.6	8		10.6	18.5	16.1		29
	5.6	8		11.3	20		16.9	30
	6.0	10	11.4		21		17.7	31
6.3		11						

$\sum \text{Rank}(X_i) = 204.5$ $\qquad$ $\sum \text{Rank}(Y_i) = 273.5$

tailed test; (2) and (3) are one-tailed tests. These are expressed as

$$H_0\text{: } E(X) = E(Y) \qquad H_a\text{: } E(X) \neq E(Y) \tag{6}$$

$$H_0 = E(X) \leq E(Y) \qquad H_a\text{: } = E(X) > E(Y) \tag{7}$$

$$H_0\text{: } E(X) \geq E(Y) \qquad H_a\text{: } E(X) < E(Y) \tag{8}$$

The critical value, T_{critical}, against which the test statistic is compared, is taken from the table of values of the cumulative distribution function of the F–Distribution in Appendix C for the number of elements in each data set and the appropriate significance level. For $n_1 = 12$, $n_2 = 19$ and $\alpha = 0.05$, $T_{\text{critical}} = 144$. This is compared with the sum of the ranks of the values from sampling location X (204.5). If $T_{\text{critical}} > \sum R(x_i)$, then H_0 is rejected. If the null hypothesis is that $E(X) = E(Y)$, then the upper critical value must also be calculated:

$$T^{\text{upper}}_{\text{critical}} = n \cdot (N+1) - T_{\text{critical}} \tag{9}$$

where the calculated sum of the ranks must be greater than the upper critical value of T for the null hypothesis to be rejected. The upper limit of the acceptance region is calculated to be 12•(31+1)=240. These two critical values (144 and 240) are compared with the calculated sum of the ranks for location X (204.5) and it is seen that the null hypothesis for a two tailed test [Equation (6)] cannot be rejected. The conclusion is that there is not sufficient reason to suspect that the two sampling locations are different.

To summarize, T_{critical} is the lower limit and $T^{\text{upper}}_{\text{critical}}$ is the upper limit of the acceptance region. To reject the null hypothesis for a two-tailed test, the sum of the ranks must be less than $T_{critical}$ or greater than $T^{\text{upper}}_{\text{critical}}$.

Kruskal–Wallis Test

When data have been collected that potentially represents more than two distributions and it is hypothesized that the independent samples represent identical populations, a test is needed that will allow unequal sample sizes without the requirement to estimate missing elements to be accomplished. Situations such as this occur frequently in chemistry, such as comparison of analytical methods, interlaboratory testing, evaluation of analysts in a laboratory, or estimation of differences in sampling methodologies. If each of the samples of a random variable is independent and the elements in each sample can be considered random, the Kruskal–Wallis test can be used. The Kruskal–Wallis test is similar to the Wilcoxon rank sum test in that the rank of each element of each sample is determined in relation to the entire set of all elements of all samples. Let the following represent the elements of the independent samples:

$$
\begin{array}{cccc}
\text{Sample 1} & \text{Sample 2} & \cdots & \text{Sample } l \\
X_{11} & X_{21} & \cdots & X_{l1} \\
X_{12} & X_{22} & \cdots & X_{l2} \\
\cdots & \cdots & \cdots & \cdots \\
X_{1,n_1} & X_{2,n_2} & \cdots & X_{l,n_l}
\end{array}
\qquad \textbf{(10)}
$$

where n_l, n_2, etc. are the number of elements in each sample and are not necessarily equal. Two quantities must be calculated before obtaining the test statistic, the grand number of all elements

$$N = \sum_i n_i \tag{11}$$

and the sum of the ranks for each of the independent samples

$$Rank_i = \sum_{j=1}^{n_i} Rank\,(X_{ij}) \tag{12}$$

where $Rank(X_{ij})$ is the rank of the individual element in relation to all N elements being considered in the test. $Rank(X_{ij})$ is calculated for each of the l samples. After calculation of the l sum of ranks, the test statistic, T, can be calculated:

$$T = \frac{1}{S^2} \cdot \left(\sum_{i=1}^{l} \frac{R_i^2}{n_i} - \frac{N \cdot (N+1)^2}{4} \right) \tag{13}$$

where S^2 is

$$S^2 = \frac{1}{N-1} \cdot \left(\sum_i \sum_j R(X_{ij})^2 - \frac{N \cdot (N+1)^2}{4} \right) \tag{14}$$

A null hypothesis that the distributions of all populations are the same is formulated with an alternate hypothesis that at least one of the l populations yields a larger mean. Extensive tables for many of the possible combinations of number of samples and number of elements in each sample are given by Imam et al.[1] Alternatively to using the exact tables, approximate significance levels can be obtained using the χ^2 distribution with l-1 degrees of freedom. The null hypothesis would be rejected if the test statistic, T, exceeds the value obtained from the χ^2 table.

If the null hypothesis is rejected, comparisons may be made to see which of the pairs of samples are the cause for the significant difference by using the following:

$$\left| \frac{R_i}{n_i} - \frac{R_j}{n_j} \right| > t_{\alpha/2} \cdot \sqrt{S^2 \cdot \frac{N-1-T}{N-k}} \cdot \sqrt{\frac{1}{n_i} + \frac{1}{n_j}} \tag{15}$$

Friedman Test

Several months of data for a set of rain samplers have been collected. The samples have been collected weekly and chemical analysis has been performed

Table 17-3. Sample Weight Data for 10 Collectors, August–November 1984

Collector									
1	2	3	4	5	6	7	8	9	10
281.7	279.7	325.5	315.0	305.2	294.0	283.9	20.4	123.3	281.8
2558.2	2550.7	2563.5	2572.4	2507.1	2530.4	2516.4	1629.2	397.7	2657.1
309.2	321.3	344.5	329.8	343.6	304.5	292.2	173.4	187.1	312.3
636.2	633.7	765.5	767.0	764.9	677.0	618.3	195.4	163.0	652.8
41.0	44.5	31.7	33.4	42.1	38.5	38.5	0.0	0.0	37.0
1145.0	1145.8	1190.8	1188.5	1192.2	1156.7	988.9	651.1	89.0	1197.5
618.8	730.2	823.1	818.1	785.4	738.2	724.5	335.7	370.2	744.4
1366.9	1366.2	1393.8	1405.5	1412.0	1376.9	1368.3	1367.2	355.5	1398.1
473.4	466.9	532.9	538.2	527.8	466.8	463.5	335.7	227.4	436.0
3507.6	3545.8	1256.9	3503.7	3528.2	3528.3	2563.3	2632.5	2504.7	3433.7

on the samples. Among the determinations performed on the samples was the measurement of sample weight for each sample for each week that precipitation was collected. A portion of the data collected in 1984 is shown in Table 17-3. Since this is a complete randomized block design, the analysis of variance would be the logical first method of analysis to determine if there is a difference in the amount of precipitation collected in each sampler. The analysis of variance table is shown in Table 17-4 for the entire data for the complete year of 19846. As can be seen, the within-groups mean square is significantly larger than the mean square for among groups as reflected in the calculated *F*-statistic. The assumptions of ANOVA are that the residuals of the data analysis will be normally distributed with a mean of zero and constant variance. Figure 17-2 and

Table 17-4. ANOVA Table for a Set of Sample Weights from Precipitation Collectors, January–December 1984

Source	SS	DF	MS	F
Among Groups	4.93E+06	5	985834	0.451
Within Group	4.850E+08	222	2185000	
Total	4.899E+08	227		

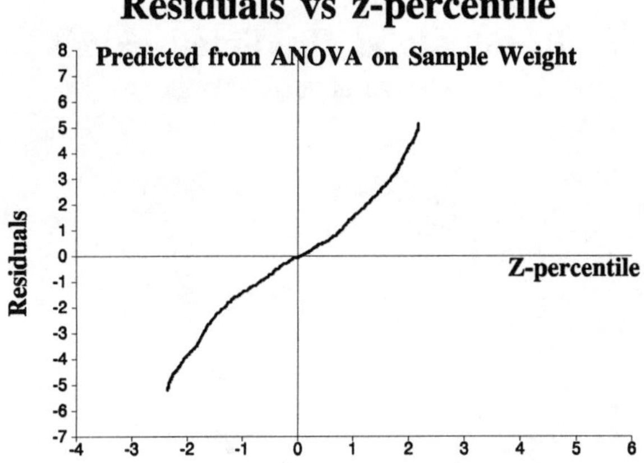

Figure 17-2. A plot of the residuals vs the normalized residuals. A linear plot would indicate normality.

Figure 17-3 show that the residuals are not normally distributed nor do they have constant variance. In addition, Bartlett's test for homoscedasticity (homogeneity of variance) between columns shows that the variances are not homogeneous. Therefore all of the major assumptions of ANOVA are violated. The data set must be analyzed by an alternative method.

The nonparametric analog to the ANOVA is the Friedman test.[2,3,4,5] The assumptions in the Friedman test are that the data in the blocks may be ranked by some criterion and that the data in a block do not influence the data in another block, i.e., independence of data between blocks. In this test, the data ARE ranked in each block according to sample weight (for this data set). The lowest sample weight in each block (i.e., week of data) will receive a rank of 1, the next lowest sample weight will receive a rank of 2, and so forth with the highest sample weight receiving a rank of 10. If there are ties in the data, the tying data will be assigned the average of the ranks. For example, if two samples had the same sample weight and they would normally have been assigned ranks 6 and 7, then each would be assigned 6.5 as the rank. The ranking procedure is repeated for each block of data. The matrix for the ranked data is shown in Table 17-5.

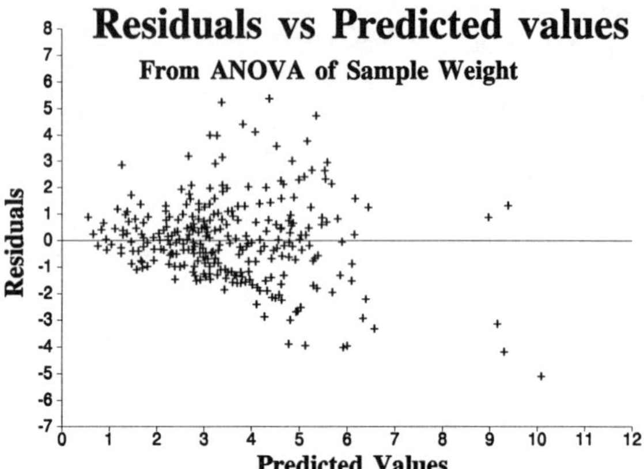

Figure 17-3. Plot of the residuals vs the predicted values from the model. Note the fanning, between 1 and 4 which indicates heteroscedasticity.

The hypotheses for the Friedman test are:

H_0: The mean sample weight of all samplers is the same
H_a: At least one sampler yields a different mean sample weight.

The parameters necessary to calculate the test statistic are as follows:

$$R_j = \sum_{i=1}^{b} \text{Rank}\,(x_{ij})$$
$$A_2 = \sum_{i=1}^{b}\sum_{j=1}^{k} \text{Rank}\,(x_{ij})^2 \tag{16}$$
$$B_2 = \frac{1}{k}\sum_{j=1}^{k} R_j^2$$

The test statistic, T_2 is calculated as

$$T_2 = \frac{(b-1)\cdot\left[B_2 - \dfrac{b\cdot k\cdot(k+1)^2}{4}\right]}{A_2 - B_2} \tag{17}$$

where b = number of weeks and k = number of samplers. T_2 is then compared to a critical value of F from the table of values of the cumulative distribution function for the F-distribution in Appendix C. If the value of T_2 exceeds the

Table 17-5. Ranked Data, Sum of Ranks, and Rank² Matrices for the Sample Weight Data Shown in Table 17-3

Rank array									
4.	3.	10.	9.	8.	7.	6.	1.	2.	5.
7.	6.	8.	9.	3.	5.	4.	2.	1.	10.
5.	7.	10.	8.	9.	4.	3.	1.	2.	6.
5.	4.	9.	10.	8.	7.	3.	2.	1.	6.
8.	10.	3.	4.	9.	6.5	6.5	1.5	1.5	5.
4.	5.	8.	7.	9.	6.	3.	2.	1.	10.
3.	5.	10.	9.	8.	6.	4.	1.	2.	7.
3.	2.	7.	9.	10.	6.	5.	4.	1.	8.
7.	6.	9.	10.	8.	5.	4.	2.	1.	3.
7.	10.	1.	6.	8.	9.	3.	4.	2.	5.

Rank Sum array									
53.	58.	75.	81.	80.	61.5	41.5	20.5	14.5	65.

Rank squared array									
16.	9.	100.	81.	64.	49.	36.	1.	4.	25.
49.	36.	64.	81.	9.	25.	16.	4.	1.	100.
25.	49.	100.	64.	81.	16.	9.	1.	4.	36.
25.	16.	81.	100.	64.	49.	9.	4.	1.	36.
64.	100.	9.	16.	81.	42.25	42.25	2.25	2.25	25.
16.	25.	64.	49.	81.	36.	9.	4.	1.	100.
9.	25.	100.	81.	64.	36.	16.	1.	4.	49.
9.	4.	49.	81.	100.	36.	25.	16.	1.	64.
49.	36.	81.	100.	64.	25.	16.	4.	1.	9.
49.	100.	1.	36.	64.	81.	9.	16.	4.	25.

critical F, pairwise comparisons may be calculated as the absolute value of the difference between the sum of the ranks for the collectors $|R_i - R_j|$, where R_i and R_j are calculated using Equation (16). A significant distinction in performance between collectors is noted if the absolute difference is greater than

$$|R_i - R_j| > t_{\alpha/2} \cdot \left[\frac{2 \cdot b \cdot (A_2 - B_2)}{(b-1) \cdot (k-1)} \right] \quad \textbf{(18)}$$

The parameters calculated for the set of data in Table 17-3 are $A_2 = 3849$, $B_2 = 3511.9$, $T_2 = 13.00$, F_{critical} (α=0.01) = 5.35, F_{critical} (α=0.05) = 4.03, and significant difference value = 18.61. The sum of ranks and the square of the ranks are shown in Table 17-5. The pairwise differences are shown in Table 17-6.

As can be seen, the nonparametric tests are much less sensitive to extreme data values. However, a major limitation is that it is difficult to tell by how much treatments differ in their effects. In the example given, it is difficult to say by how much the collectors differ in their collection efficiency. An indicator of ranking of collectors, i.e., the collectors that collect most and that collect least, is shown in Table 17-5 as the sum of the ranks array. The smallest value would indicate the collector that on the average collects the least amount of precipitation. The highest rank would indicate the collector that on the average collects the most precipitation.

Table 17-6. Rank Difference Matrix[a]

Pairwise Difference Array

	Collector								
Collector	**2**	**3**	**4**	**5**	**6**	**7**	**8**	**9**	**10**
1	5.0	22.0*	28.0*	27.0*	8.5	11.5	32.5*	38.5*	12.0
2		17.0	23.0*	22.0*	3.5	16.5	37.5*	43.5*	7.0
3			6.0	5.0	13.5	33.5*	54.5*	60.5*	10.0
4				1.0	19.5*	39.5*	60.5*	66.5*	16.0
5					18.5	38.5*	59.5*	65.5*	15.0
6						20.0*	41.0*	47.0*	3.5
7							21.0*	27.0*	23.5*
8								6.0	44.5*
9									50.5*

[a] Significant difference value = 18.61. * Significant difference between collectors.

References

1. Imam, R.L., Quade, D., and Alexander, D.A. *Exact Probability levels for the Kruskal-Wallis test* ***Selected Tables Math. Stat.***, 1975, **3**, 329-384.

2. Graham, R.C. Nonparametric statistical evaluation of data 1-- The Friedman test", ***Access— J. Microcomp. Appl.***, July-August 1987, 24-29.

3. Graham, R.C., and Obal, J. The Nonparametric statistical evaluation of precipitation chemistry intercomparison data. ***Atmos. Environ.***, 1989, **23(5)**, 1117–1130.

4. Graham, R.C., Robertson, J.K., Schroder, L.J., and LaFemina, J. Atmospheric deposition sampler intercomparison. ***Water, Air, Soil Poll.***, 1988, **37**, 139–147.

5. Graham, R.C., Robertson, J.K., and Obal, J. An assessment of the variability in performance of wet atmospheric deposition samplers, ***U.S. Geological Survey Water Resources Investigation Report*** *87-4125*, **1987**.

18

MULTIPLE REGRESSION

General

The situation in which the values of a dependent variable (such as instrument response) are related to the values of an independent variable (such as the concentration of an analyte) has been discussed. The general form of the relation was given as the calibration curve and was mathematically expressed as

$$Y = \beta_0 + \beta_1 \cdot X \qquad (1)$$

The parameters (β_0 and β_1) were estimated as the unbiased predictors, b_0 and b_1 using linear regression techniques through the minimization of the sum of the squares of the difference between the dependent variable and the predicted instrument response. This minimization led to a set of normal equations:

$$\begin{aligned} b_0 \cdot \sum 1 + b_1 \sum x_i &= \sum R_i \\ b_0 \cdot \sum x_i + b_1 \cdot \sum x_i^2 &= \sum x_i \cdot R_i \end{aligned} \qquad (2)$$

from which b_0 and b_1 can be directly calculated. Alternatively, a matrix approach was also presented that calculated a vector of coefficients as the product of several matrices:

$$B = (X^T \cdot X)^{-1} \cdot (X^T \cdot Y) \qquad (3)$$

where B is the coefficient vector, X is an n x 2 matrix of the values of the independent variable in the second column and all 1s in the first column, X^T is the transpose of the X matrix, and Y is the column vector of the values of the dependent variable.

In nature and chemistry, rarely is the magnitude of a measured quantity linearly and only related to a single independent variable. Consider, for example, an aqueous solution that contains strong and weak acids and strong and weak bases. The overall $[H^+]$ will be dependent on the relative amounts of the acids and bases that have been added and also on the ionization constants of the individual acids and bases. Situations such as this must also account for all of the potential pairwise combinations of acids and bases with the $[H^+]$. We must also account for any interaction between the pairs of acid/[H^+] or base/[H^+]. A technique that could be used to express multiple relations is multiple linear regression.

General Polynomial Regression

One of the simpler examples of a multiple regression technique is to obtain the regression coefficients for an equation of the form

$$Y = \beta_0 + \beta_1 \cdot X + \beta_2 \cdot X^2 + \dots \tag{4}$$

or more succinctly as

$$Y = \sum \beta_k \cdot X^k \tag{5}$$

where the β_k are estimated with the unbiased estimators, b_k.

The parameters may be obtained by using the matrix approach. The matrix equation will be

$$Y = B \cdot X + \epsilon \tag{6}$$

where ϵ is the matrix of residuals, B is the parameter vector, Y is the dependent variable matrix, and X is the dependent variable matrix. It has been shown previously that the B vector may be calculated from

$$B = (X^T \cdot X)^{-1} \cdot (X^T \cdot Y) \tag{7}$$

if the criterion for the best fit is the minimization of the sum of squares of the residuals with respect to the parameters. Of course, some care must be taken to correctly formulate the X, B, and Y matrices. To fit a quadratic function to a set of data, the following matrices would be used:

$$X = \begin{bmatrix} 1 & x_1 & x_1^2 \\ 1 & x_2 & x_2^2 \\ 1 & x_3 & x_3^2 \\ \dots & \dots & \dots \\ 1 & x_n & x_n^2 \end{bmatrix} \qquad X^T = \begin{bmatrix} 1 & 1 & 1 & \cdots & 1 \\ x_1 & x_2 & x_3 & \cdots & x_n \\ x_1^2 & x_2^2 & x_3^2 & \cdots & x_n^2 \end{bmatrix} \qquad Y = \begin{bmatrix} y_1 \\ y_2 \\ y_3 \\ \dots \\ y_n \end{bmatrix} \tag{8}$$

And the parameter matrix that would result is

$$B = \begin{bmatrix} b_0 \\ b_1 \\ b_2 \end{bmatrix} \quad (9)$$

An example of an analytical technique that requires a second-order regression equation is the analysis of sulfur compounds by gas chromatography using a flame photometric detector for which the response of the detector is proportional to the square of the concentration of the sulfur-containing compound through the regression equation

$$A = b_0 + b_1 \cdot [S] + b_1 \cdot [S]^2 \quad (10)$$

The peak areas for a series of standard solutions of dimethyl sulfide (ng/L) are given in Table 18-1. The matrices $(X^TX)^{-1}$ and X^TY are

$$(X^T \cdot X)^{-1} = \begin{bmatrix} 0.1779 & -0.0128 & 0.00154 \\ -0.0128 & 0.00148 & -2.0119 x 10^{-5} \\ 0.00154 & -2.0119 x 10^{-5} & 2.938 x 10^{-7} \end{bmatrix} \quad (11)$$

$$X^T \cdot Y = \begin{bmatrix} 6663.5 \\ 348586 \\ 2122214 \end{bmatrix} \quad (12)$$

Table 18-1. Table of Integrated Peak Areas for Several Concentrations of Dimethyl Sulfide (DMS)[a]

[DMS]	Peak Area	[DMS]	Peak Area
1	5.28	1	5.28
3	17.26	3	17.27
5	31.28	5	31.29
7	47.28	8	56.03
10	75.02	10	75.03
12	96.03	15	131.26
20	200.01	20	200.02
25	281.26	35	481.28
50	875.02	50	875.03
65	1381.27		

[a] Concentrations, [DMS], given in ng/L.

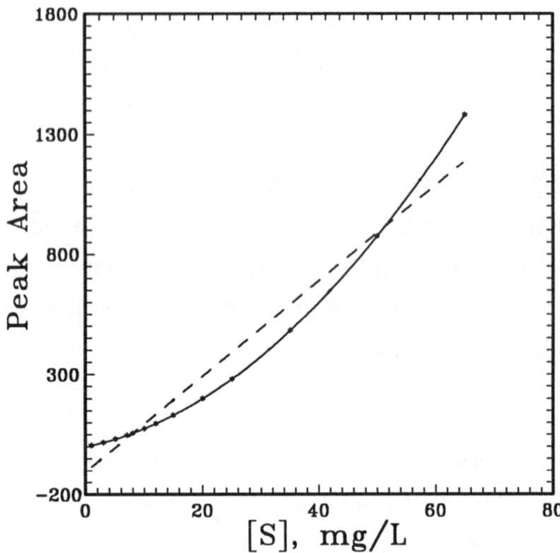

Figure 18-1. Relation between the concentration of a sulfur organic and peak area. - - - Linear first order, —— linear second order.

The parameter vector is found as the product of the matrices in Equations (11) and (12) to be

$$B = (X^{\mathrm{T}} \cdot X)^{-1} \cdot X^{\mathrm{T}} \cdot Y = \begin{bmatrix} 0.0238 \\ 5.0023 \\ 0.250 \end{bmatrix} \tag{13}$$

The data is plotted in Figure 18-1 with a linear first order and a linear second order polynomial fit both depicted.

General Multiple Regression with (l-1) Independent Variables

Although the general polynomial regression can fit many types of relationships, it is not an appropriate model to describe the relation between several independent variables and a single dependent variable. A model for such a relationship is mathematically described as

$$Y_i = \beta_0 + \beta_1 \cdot X_{i1} + \beta_2 \cdot X_{i2} + \cdots + \beta_{l-1} \cdot X_{i,l-1} + \epsilon_i \tag{14}$$

which is more succinctly written

$$Y_1 = \sum_{k=0}^{l-1} \beta_k \cdot X_{ik} + \epsilon_i \tag{15}$$

where it is explicitly stated that $X_{i\,0} = 1$.

An example of such a relationship that might be explored using multiple regression techniques is the optimization of chromatographic factors where the dependent variable is the height or area of a single peak and the maximum peak height is to be obtained for a single component. The independent variables might be the type of solvent used as the eluant, temperature, type of column (capillary versus packed), and type of detector. A similar approach could be used where the independent variable is Δt, the difference in retention time for the chromatographic analysis of two compounds. The independent variables could be the column length, the temperature, the column packing, and the flow rate.

The general multiple linear regression model assumes that the independent variables are not collinear nor is there interaction between the variables. Collinearity between two or more variables implies that there is a strong tendency for the two variables to behave quite similarly. The relation between collinear variables would lead to a linear relation between the two. If collinearity exists, then inclusion of only one of the variables or inclusion of the sum of the variables is usually appropriate.

A model that takes into account the potential interaction of two independent variables would be

$$Y_i = \beta_0 + \beta_1 \cdot X_{i,1} + \beta_2 \cdot X_{i,2} + \beta_3 \cdot X_{i,1} \cdot X_{i,2} + \epsilon_i \tag{16}$$

A multiple variable linear regression model for three independent variables with interaction would be

$$\begin{aligned} Y_i = {} & \beta_0 + \beta_1 \cdot X_{i,1} + \beta_2 \cdot X_{i,2} + \beta_3 \cdot X_{i,3} \\ & + \beta_4 \cdot X_{i,1} \cdot X_{i,2} + \beta_5 \cdot X_{i,1} \cdot X_{i,3} + \beta_6 \cdot X_{i,2} \cdot X_{i,3} \\ & + \beta_7 \cdot X_{i,1} \cdot X_{i,2} \cdot X_{i,3} + \epsilon_i \end{aligned} \tag{17}$$

The presence of the interaction term can be important when calculating the expected value of the dependent variable. Consider that the parameters of the model have been found to be expressed in the following model without any interaction between the two independent variables

$$Y_i = 16 + 3 \cdot X_1 + 4 \cdot X_2 + \epsilon_i \tag{18}$$

and if ϵ_i (the residual term) is 0, then the expected value of Y is

$$E(Y) = 16 + 3 \cdot X_1 + 4 \cdot X_2 \tag{19}$$

The residual term would be zero if each of the elements of the data array of the model exactly predicted the experimental value of the random variable. A graph of the response function (Figure 18-2A) with respect to X_1 and constant X_2, regardless of the value of X_2, will always have a slope of 3.

But, on the other hand, if an interaction term ($2 \bullet X_1 \bullet X_2$) is added, the expected value of Y becomes

$$E(Y) = 16 + 3 \cdot X_1 + 4 \cdot X_2 + 2 \cdot X_1 \cdot X_2 \tag{20}$$

then the response function will have different values of the slope for different values of X_2. For example, if $X_2 = 5$, then the expected value of Y will be

$$E(Y) = 36 + 13 \cdot X_1 \tag{21}$$

and the slope of the response function will be 13. But if $X_2 = 1$, then the expected value of Y will be

$$E(Y) = 20 + 11 \cdot X_1 \tag{22}$$

and the slope of the response function will be 11. The effects of these two situations are shown in Figure 18-2B. As can be seen from the figure, the slopes are quite different when the interaction term is included.

Let us now consider a set of rainfall data (Table 18-2) that represents the chemistry of rain at a single sampling site in southeastern New York* for several weeks. The chemistry data shown in Table 18-2 are only the data for those ionic species that would probably influence the concentration of H^+ in the sample of precipitation. The species chosen for this analysis are H^+ as the dependent variable and nitrate, sulfate, ammonium, calcium, magnesium, and chloride as the independent variables. We will attempt to fit the data to the model:

---

* The sampling site is NY99, West Point, New York. The data is furnished courtesy of the National Atmospheric Deposition Program/National Trends Network, Natural Resources Ecology Laboratory, Colorado State University, Fort Collins, Colorado.

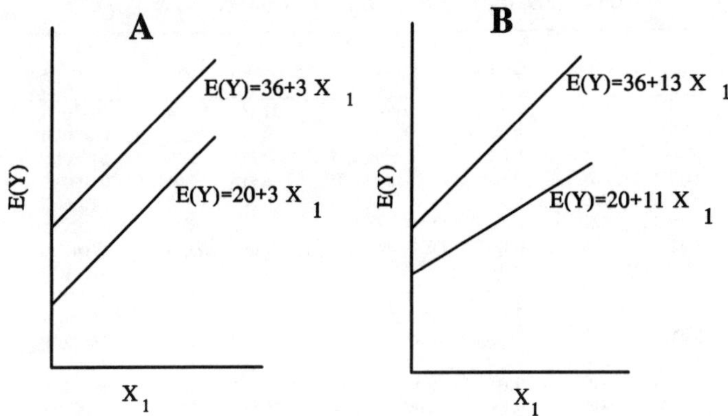

Figure 18-2. Plot of the response function for A. no interaction between independent variables and B. Interaction between independent variables.

$$[H^+] = b_0 + b_1 \cdot [NO_3^-] + b_2 \cdot [SO_4^{2-}] + b_3 \cdot [NH_4^+] \tag{23}$$
$$+b_4 \cdot [Ca^{2+}] + b_5 \cdot [Mg^{2+}] + b_6 \cdot [Cl^-] + \epsilon$$

which may be considered to be a form of the more general model

$$Y_i = \sum_{k=0}^{n-1} b_k \cdot X_{ki} + \epsilon_i \tag{24}$$

where k is the index of the variable and i is the index of the data block. Thus, we will predict the value of the dependent variable ($[H^+]$) based on a set of i observations of k variables. For the rainfall data set, there are 65 observations (weekly samples from 20 September 1983 to 22 January 1985) for six independent variables, $[NO_3^-]$, $[SO_4^{2-}]$, $[NH_4^+]$, $[Ca^{2+}]$, $[Mg^{2+}]$, and $[Cl^-]$ and one dependent variable, $[H^+]$. Since the equation is of the general matrix form

$$Y = B \cdot X + \epsilon \tag{25}$$

the equation may be solved to obtain the matrix of coefficients, B, as previously shown.

Table 18-2. Concentrations of Ionic Constituents of Precipitation from a Sampling Site in Southeastern New York

Dates	Ca^{2+}	Mg^{2+}	NH_4^+	NO_3^-	Cl^-	SO_4^{2-}	H^+
830920	0.53	0.984	0.10	3.47	11.48	7.64	0.0000115
830927	0.13	0.038	0.10	0.82	0.17	1.30	0.0000251
831004	0.05	0.035	0.03	0.57	0.38	0.79	0.0000182
831011	0.94	0.270	0.33	6.07	0.96	7.01	0.0000933
831018	0.03	0.054	0.02	0.34	0.77	0.66	0.0000145
831025	0.05	0.013	0.09	1.27	0.12	1.16	0.0000372
831101	0.48	0.171	0.36	7.03	1.03	5.08	0.0001230
831108	0.29	0.102	0.57	5.84	0.37	6.23	0.0001259
831115	0.04	0.017	0.23	1.79	0.19	1.65	0.0000447
831122	0.04	0.064	0.08	0.70	0.85	0.97	0.0000182
831129	0.04	0.025	0.09	1.11	0.29	1.65	0.0000389
831206	0.04	0.018	0.10	1.21	0.23	1.17	0.0000331
831213	0.05	0.118	0.08	0.79	1.71	1.33	0.0000288
831220	0.05	0.106	0.06	0.94	1.46	1.36	0.0000263
831227	0.02	0.020	0.00	0.49	0.24	0.89	0.0000195
840103	0.05	0.019	0.07	0.52	0.26	1.30	0.0000200
840110	0.65	0.142	0.52	15.11	1.32	5.00	0.0002512
840117	0.13	0.110	0.16	2.79	0.52	2.04	0.0000437
840124	0.13	0.087	0.09	3.13	1.02	2.12	0.0000676
840131	0.19	0.063	0.16	3.23	1.12	1.91	0.0000562
840207	0.06	0.040	0.10	1.27	0.41	1.51	0.0000372
840214	0.20	0.050	0.20	2.04	0.32	2.92	0.0000708
840221	0.12	0.239	0.10	0.99	3.54	2.36	0.0000427
840228	0.10	0.030	0.15	1.18	0.19	1.52	0.0000324
840306	0.12	0.041	0.12	1.33	0.17	1.93	0.0000363
840313	0.17	0.057	0.09	3.92	0.21	0.94	0.0000603
840320	0.06	0.017	0.00	0.78	0.05	1.18	0.0000302
840327	0.64	0.112	0.22	2.38	1.12	3.23	0.0000617
840403	0.25	0.069	0.03	0.17	0.29	1.05	0.0000155
840410	0.03	0.028	0.09	0.48	0.39	1.57	0.0000324
840417	0.09	0.022	0.17	1.18	0.12	1.45	0.0000380
840424	0.31	0.080	0.28	4.30	0.23	3.05	0.0000813
840501	1.52	0.266	0.66	8.24	0.62	10.16	0.0001820
840508	0.09	0.024	0.22	1.25	0.17	2.03	0.0000427
840515	0.40	0.093	0.57	3.68	0.29	3.87	0.0000794
840522	1.06	0.276	1.59	9.10	0.51	10.76	0.0001820
840529	0.06	0.030	0.21	1.21	0.28	1.87	0.0000398
840605	0.09	0.015	0.07	0.69	0.14	1.34	0.0000166
840619	0.23	0.082	0.99	5.53	0.28	8.86	0.0001738
840626	0.08	0.049	0.13	0.93	0.47	1.09	0.0000224
840703	0.04	0.015	0.18	2.32	0.10	2.30	0.0000794
840710	0.06	0.022	0.19	1.62	0.17	2.13	0.0000490
840717	0.07	0.016	0.22	1.66	0.11	3.59	0.0000741
840724	0.09	0.037	0.24	1.93	0.14	3.05	0.0000776
840731	0.13	0.033	0.13	4.13	0.19	3.39	0.0001202
840807	0.18	0.059	0.35	2.23	0.15	4.79	0.0001023
840814	0.13	0.064	0.23	1.26	0.20	4.03	0.0000741
840821	0.04	0.015	0.16	2.24	0.12	1.91	0.0000646
840904	0.09	0.024	0.29	2.22	0.11	2.78	0.0000724
840918	0.21	0.150	0.25	1.76	0.22	3.33	0.0000617
841002	0.13	0.087	0.40	1.11	0.12	1.88	0.0000214
841009	1.00	0.543	0.25	0.71	0.66	3.41	0.0000003
841023	0.18	0.067	0.50	2.78	0.29	2.99	0.0000661
841030	0.17	0.060	0.25	3.72	0.22	3.29	0.0001023

Continued

Table 18-2. (Continued) Concentration of Ionic Constituents of Precipitation from a Sampling Site in Southeastern New York

Dates	Ca^{2+}	Mg^{2+}	NH_4^+	NO_3^-	Cl^-	SO_4^{2-}	H^+
841106	0.03	0.041	0.00	0.29	0.65	0.94	0.0000195
841113	0.17	0.092	0.16	1.87	0.64	2.36	0.0000269
841204	0.05	0.079	0.08	0.58	1.24	1.42	0.0000347
841211	0.04	0.024	0.00	0.43	0.05	1.03	0.0000224
841218	0.44	0.340	0.31	5.20	2.17	5.89	0.0001122
841226	0.05	0.028	0.33	2.67	0.36	3.27	0.0000832
850102	0.17	0.052	0.22	3.63	0.61	2.85	0.0000871
850108	0.12	0.323	0.00	0.87	0.18	3.69	0.0000501
850115	0.97	0.554	0.00	5.77	1.23	2.12	0.0000023
850122	0.06	0.023	0.00	2.80	0.29	0.55	0.0000447

The formulation of a general multiple variable linear regression is very difficult to portray graphically. In fact, for anything beyond two independent variables or predictors, it is essentially impossible to depict on a sheet of paper or computer monitor. As the number of predictors increases, the complexity of representation increases. For one predictor, the graphic indicator is a line; for two predictors, the graphic indicator is a plane or three–dimensional surface; for three predictors, the graphic indicator is a four–dimensional surface, etc.

To graphically portray the dependence of $[H^+]$ on any one of the variables, two–dimensional scatter plots could be constructed for each variable with $[H^+]$. Several are shown in Figure 18-3. Two of these graphs (A and B) show that there appears to be some relationship between $[H^+]$ and the concentration of sulfate or nitrate. However, to fully explore the effect, we should also consider the cross relations, such as the nitrate/sulfate or nitrate/ammonium plots in Figure 18-3C and D, which also appear to have a high degree of correlation. The possible cross–relationships will be ignored so as to concentrate on the direct relations for the model given in Equation (23).

Let us now find the set of parameters for the rainfall data set that will minimize the residuals for the model of Equation (23). The matrices required to solve for the parameters are the X and the Y matrices. The X matrix has for the first column all ones and the middle six columns of Table 18-2. The response matrix, Y is column six of Table 18-2. The other matrices [$(X^TX)^{-1}$, (X^TY), and B) that are required are products or transpose of these two.

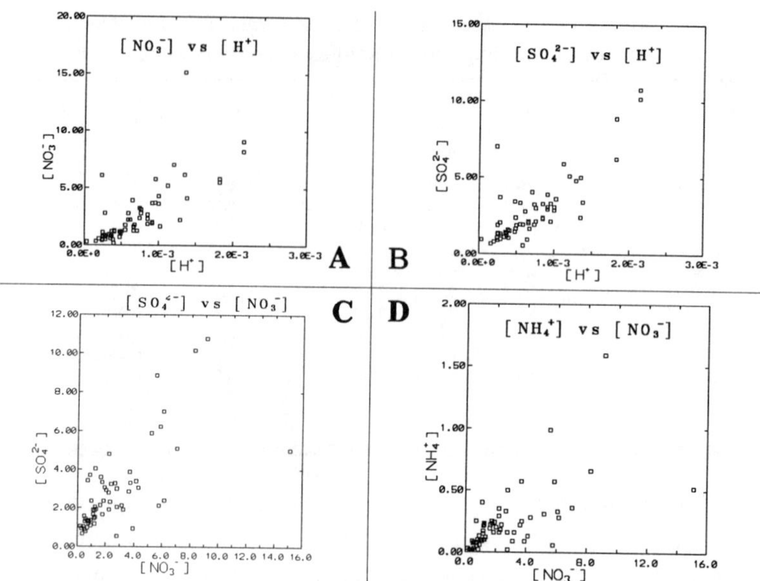

Figure 18-3. **A.** H^+/SO_4^{2-}, **B.** H^+/NO_3^-, **C.** SO_4^{2-}/NO_3^-, **D.** NH_4^+/NO_3^-

There are several assumptions that must be satisfied when using this approach to obtain relationships between variables:

1. For a given combination of independent variables, the dependent variable is assumed to be a variable that follows a normal distribution.
2. The dependent variable in each block is independent of the dependent variable in another block. In the rainfall example, this means that the $[H^+]$ for any week is independent of the $[H^+]$ in any other week. Another way of saying it is that Y_i does not depend on Y_j, where $i \neq j$. Another way of expressing this condition is that the measurements of the dependent variable are not correlated.
3. The mean of Y for a specific combination of X_ks is also a linear function of the X_ks, i.e., $\bar{Y}$ also fits the model in Equation (14).
4. The variances of the Y_is are fixed for any particular and specific combination of the X_ks. This is the condition of homoscedasticity, i.e., homogeneous variance over the entire region of the predicted values of the dependent variable.
5. The independent variables are not collinearly related. This can be thought of as the condition of independence of independent variables.

One of the stated assumptions was that the independent variables must not be linearly correlated. Thus we must find a means to measure the correlation between the independent variables. This is most easily accomplished by examination of the correlation and covariance matrices. The covariance between any two variables is a measure of the degree to which the two variables are related. The covariance is given as

$$C_{jk} = \frac{\sum_{i=1}^{n} (x_{ij} - \bar{x}_j) \cdot (x_{ik} - \bar{x}_k)}{n - 1} \tag{26}$$

Two properties of the covariance matrix should be evident. The covariance matrix (C_{jk}) is symmetric, i.e., $c_{jk} = c_{kj}$. The diagonal elements are the variances of the individual independent variables:

$$c_{jj} = \frac{\sum_{i=1}^{n} (x_{ij} - \bar{x}_j)^2}{n - 1} \tag{27}$$

A major limitation of the covariance matrix is that it is very dependent on the scale of the individual variables. If the magnitude of one of the variables is much greater than the magnitude of a second variable, the first variable will dominate the covariance and make it appear to have little correlation between the two. To account for the sometimes disparity in scale, the covariance matrix may be normalized by dividing each element of the covariance matrix by the product of the standard deviations of the two independent variables to obtain the correlation matrix:

$$r_{jk} = \frac{c_{jk}}{s_j \cdot s_k} \tag{28}$$

where s_j and s_k are the standard deviations of the jth and kth variables, respectively, e.g., if the jth variable represented $[NO_3^-]$ and the kth variable represented $[SO_4^{2-}]$, then r_{jk} would be the correlation between nitrate and sulfate.

Although the model for the multiple regression of the precipitation data set was obtained, the data were never tested for the correlation between the independent variables. The covariance matrix is shown in Table 18-3 and the correlation matrix is shown in Table 18-4. In these tables, it is seen that several variables have rather high correlations, e.g., sulfate and nitrate have a

Table 18-3. Covariance Matrix for Rainfall Data

	Ca^{2+}	Mg^{2+}	NH_4^+	NO_3^-	Cl^-	SO_4^{2-}	H^+
Ca^{2+}	0.090	0.030	0.040	0.504	0.162	0.479	1.0E-5
Mg^2	0.030	0.025	0.005	0.120	0.190	0.176	0.000
NH_4^+	0.040	0.005	0.063	0.409	0.023	0.449	1.0E-5
NO_3^-	0.504	0.120	0.409	6.507	2.467	3.973	1.0E-4
Cl^-	0.162	0.190	0.023	2.467	3.980	1.315	2.0E-5
SO_4^{2-}	0.479	0.176	0.449	3.973	1.315	4.930	8.0E-5
H^+	1.0E-5	0.000	1.0E-5	1.0E-4	2.0E-5	8.0E-5	0.000

correlation of 0.70 and calcium and magnesium have a correlation of 0.62. Each of these two pair of variables should be combined into two new variables:

$$Z_1 = \left[NO_3^-\right] + \left[SO_4^{2-}\right] \qquad (29)$$

$$Z_2 = \left[Ca^{+2}\right] + \left[Mg^{2+}\right]$$

The model that describes this reduced set of variables is given:

$$Y_i = \beta_0 + \beta_1 \cdot \left[NH_4^+\right] + \beta_2 \cdot Z_1 + \beta_3 \cdot Z_2 + \beta_4 \cdot \left[Cl^-\right] + \epsilon_i \qquad (30)$$

where $Z_1 = [NO_3^-] + [SO_4^{2-}]$ and $Z_2 = [Ca^{2+}] + [Mg^{2+}]$. The process of combining variables should continue until the minimal set of variables is obtained which adequately explains the dependent variable. The estimated regression equation is:

$$\left[H^+\right] = 8.63 x 10^{-6} \cdot \left\{\left[NO_3^-\right] + \left[SO_4^{2-}\right]\right\} + 2.1 x 10^{-5} \cdot \left[NH_4^+\right] \qquad (31)$$

$$-4.52 x 10^{-6} \cdot \left[Cl^-\right] + 5.15 x 10^{-6} \cdot \left\{\left[Mg^{2+}\right] + \left[Ca^{2+}\right]\right\} + 2.2 x 10^{-5}$$

This regression equation predicts that the $[H^+]$ will increase by $8.63x10^{-6}$ mg/L for an increase of 1 mg/L of $[NO^{3-}]$ + $[SO_4^{2-}]$ while the other variables are held constant. Similarly, $[H^+]$ will increase by $5.15x10^{-6}$ mg/L for each mg/L increase in $[Ca^{2+}]$ + $[Mg^{2+}]$ as the other independent variables are held constant. Although the model predicts quite a dependence on the sum of calcium and magnesium, the significance is very low. The change in $[H^+]$ is seen to be relatively not affected by the magnitude of the significance level of the $[Cl^-]$ ion as shown in Table 18-5. The significance level of 0.34 indicates that the

Table 18-4. Correlation Matrix for the Full Set of Variables

	Ca^{2+}	Mg^{2+}	NH_4^+	NO_3^-	Cl^-	SO_4^{2-}	H^+
Ca^{2+}	1.0000	.6232	.5280	.6601	.2714	.7203	.4591[a]
	.0000	.0000	.0000	.0000	.0301	.0000	.0001[b]
Mg^{2+}	.6232	1.0000	.1241	.2956	.5991	.4989	.0070
	.0000	.0000	.3285	.0177	.0000	.0000	.9560
NH_4^+	.5280	.1241	1.0000	.6376	.0461	.8051	.7402
	.0000	.3285	.0000	.0000	.7176	.0000	.0000
NO_3^-	.6601	.2956	.6376	1.0000	.4849	.7015	.8645
	.0000	.0177	.0000	.0000	.0000	.0000	.0000
Cl^-	.2714	.5991	.0461	.4849	1.0000	.2968	.2549
	.0301	.0000	.7176	.0000	.0000	.0172	.0421
SO_4^{2-}	.7203	.4989	.8051	.7015	.2968	1.0000	.7506
	.0000	.0000	.0000	.0000	.0172	.0000	.0000
H^+	.4591	.0070	.7402	.8645	.2549	.7506	1.0000
	.0001	.9560	.0000	.0000	.0421	.0000	.0000

[a] Entries in this line correspond to the correlation between variables.
[b] Entries in this line correspond to the significance of the correlation (P-value)

dependence could result by chance 34% of the time. Similarly, the magnitude of the significance of combined variable, $[Ca^{2+}]+[Mg^{2+}]$, indicates that 55% of the time the results would arise by chance. Thus each of these two variables would be deemed as not significant in the interpretation of the results. The significance of the ammonium ion is marginal as shown in the same table. The low significance indicates these variables should not be included in the final

Table 18-5. Table of results of fitting a model to the rainfall data when variables are combined together.

Independent Variable	Coefficient	*t*-Value	Significance Level
Constant	0.000022	4.8377	0.0000
$[NO_3^-]+[SO_4^{2-}]$	0.00000863	7.1147	0.0000
$[Cl^-]$	0.00000452	-0.9590	0.3416
$[NH_4^+]$	0.000021	1.1667	0.2482
$[Mg^{2+}]+[Ca^{2+}]$	0.00000515	0.6016	0.5499

regression equation because of the magnitude of the significance level to the contribution of the predictive power of the model. For this particular model, the combined variable, $[NO_3^-]+[SO_4^{2-}]$, and ammonium ion should be included. This is congruous with the expectation that sulfuric and nitric acids and ammonium ion are the probable major sources of hydrogen ion in rainfall.

The result for just the variables that are significant in the predictive model are shown in Table 18-6. As can be seen the only two variables which are significant in the prediction of H^+ are the constant and the sum of the nitrate and sulfate.

Determination of the Aptness of the Model

Methods were presented in earlier chapters to ascertain the aptness of a model. Among the methods were the fit to an Empirical Distribution Function and the analysis of residuals. For such a model as this, *i.e* a multivariate regression, the analysis of residuals is most appropriate.

We require the calculation of the residuals from the model used in the analysis. After the residuals are calculated, they are subjected to the tests given in chapter 10. The condition that the variance of the residuals be constant can be determined by the tests that have been given in earlier chapters such as Bartlett's or Hartley's test. The condition of homoscedasticity can often be observed graphically by plotting the residuals of the model against the predicted values of the random variable. The residuals for the data set are plotted in Figure 18-4. This figure shows that the variance is probably not constant over the range of the predicted values of hydrogen ion.

For further discussion about the use of multiple linear regression, see Draper and Smith.[1]

Table 18-6. Multiple linear regression model results for the prediction of $[H^+]$ by the sum of nitrate and sulfate.

Independent variable	coefficient	t-value	significance level
Constant	0.000019	4.7561	0.0000
$[NO_3^-]+[SO_4^{2-}]$	0.00000986	16.3801	0.0000

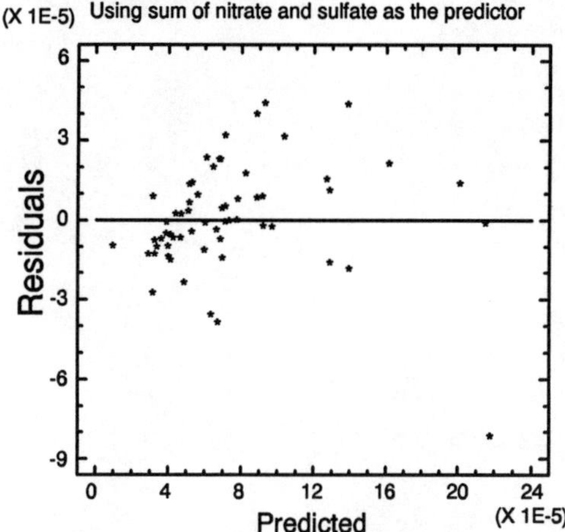

Figure 18-4. **Residuals resulting from the prediction of [H^+] using the sum of nitrate and sulfate.**

References

1. The best reference for further details on regression analysis is Draper, N., and Smith, H. ***Applied Regression Analysis***, 2nd ed., Wiley, New York, 1981.

19
MULTIVARIATE ANALYSIS

Introduction

Many data sets have hidden or not easily recognized similarities, patterns, or structures. The similarity may be exhibited as the interdependence of ions in an acidic solution, in the similarity of portions of a chemical spectrum such as infrared or nuclear magnetic resonance, or the mutual coexistence of minerals in geological specimens. One aim of data analysis is to uncover the hidden structure and similarities in a data set. The techniques given in previous chapters do not address the idea that patterns may exist in data sets. One of the techniques does address the notion that there may be relationships between a dependent and an independent variable— the linear regression model technique. However, uncovering the relations between more than one variable that may be correlated is not easy to do. Consider, for example, that a data set consists of 50 rows of data with values of 15 random variables. The plots shown in Chapter 18 concerning the multilinear regression indicated that relations existed between several of the variables. How many such bivariate plots would have to be examined to ferret out all of the relations between variables? The answer is of course given in the number of combinations of variables that can be taken two at a time:

$$C_{k,n} = \binom{n}{k} = \frac{n!}{(n-k)! \cdot k!} = \frac{15!}{13! \cdot 2!} = 104 \qquad (1)$$

which in itself is not a large number of plots to be made, especially with the modern easy to use desktop computers that are available. However, this is just the bivariate plots. The number rapidly increases if sums of variables are considered: certainly not the type of task that should be relished by anyone. How then can we ascertain the structure in the data sets? The answer lies in the use of the multivariate techniques discussed in this chapter. These techniques give the data analyst the mathematical tools to rapidly and relatively easily ascertain the relationships and patterns in the data. The first technique that will be discussed is cluster analysis.

Cluster Analysis

A powerful method of data analysis that can accomplish the goal of pattern recognition is cluster analysis. Cluster analysis is defined as the process by which objects can be classified into groups based on subjective or objective properties of the objects.

Scientists have long been interested in assigning order to a seemingly disordered world. Carl von Lagneia, a Swedish biologist, gave us the taxonomic classification system used in biology, wherein each plant or animal is assigned to a hierarchical order (Table 19-1). Decisions are made at each classification step to assign an organism to one of the classifying groups. As each classifying step is taken the groups get smaller as one progresses from the general to the specific.

Chemists have similar classification schemes to identify compounds, although not as rigidly used as the taxonomic schemes of the biologist. The potential criteria, however, for chemical classification are much more varied and can be very complex. Decisions to place a chemical compound in one group or another can be based on molecular weight, solubility in solvents, resistance to solvents, whether it contains a given element or not, or whether a particular functional group is present. The classification can also be accomplished on the basis of composition of samples. Kowalski has classified several obsidian samples by the content of 10 different metals present in the obsidian[1]. An example of a classification is the selection of a column to use in HPLC analysis. The realm of chemical compounds is divided into smaller sections by such classifying factors as molecular weight, elemental composition, and solubility in solvents (such as water or alcohol). The general scheme is to divide the totality of chemical compounds into smaller and smaller sections until a column is selected that is satisfactory for the separation task at hand.

Before the technique of cluster analysis is further pursued, means must be discussed that can be used to ascertain similarities in the data set. The idea of the distance between two or more groups can be used as a measure of the similarity.

Table 19-1. Hierarchical classification system for biological specimens[a]

Kingdom	Phylum	Class	Order	Family	Genus	Specie
Monera	Bacteria	...	...	...	*Bacillus*	...
Protista	Protozoa	...	...	...	*Euglena*	...
	Ameba					
Fungi	Molds					
	Mushrooms					
Plantae	Bryophyta	Moss				
	Tracheophyta	Filicineae				
		Gymnosperm				
		Angiosperm				
Animalia	Coelenterata	...	...	...	*Hydra*	...
	Annelida			Lumbricidae	*Lumbricus*	*terrestris*
	Arthropoda	Chilopoda				
		Diplopoda				
		Crustacea				
		Insecta	Lepidoptera			
			Homoptera			
	Chordata	Pisces				
		Aves				
		Amphibia				
		Mammalia	Rodentia			
			Primates	Hominidae	*Homo*	*sapiens*

[a] Only a few representative groupings are shown.

Distances

Euclidean Distance

We now wish to consider the case in which it is desirable to obtain a variable that will give the "distance" between two data sets. If we consider the situation in which two populations may each be described by a set of two variables that are binormally distributed, then the distance between the two populations may be described by the Euclidean distance shown in Figure 19-1. The Euclidean distance, d_{ij}, is given as the geometric straight line distance between the two data points:

$$d_{12} = \sqrt{(X_{21} - X_{11})^2 + (X_{22} - X_{12})^2} \tag{2}$$

If the two populations are described by three variables then the Euclidean Distance is given by the expression

$$d_{12} = \sqrt{\sum_{k=1}^{3} (X_{2k} - X_{1k})^2} \tag{3}$$

Calculation of the Euclidean distance may be extended to any number of potential data populations with any number of data variables (p) describing each population where the expression

$$d_{ij} = \sqrt{\sum_{k=1}^{p} (X_{jk} - X_{ik})^2} \tag{4}$$

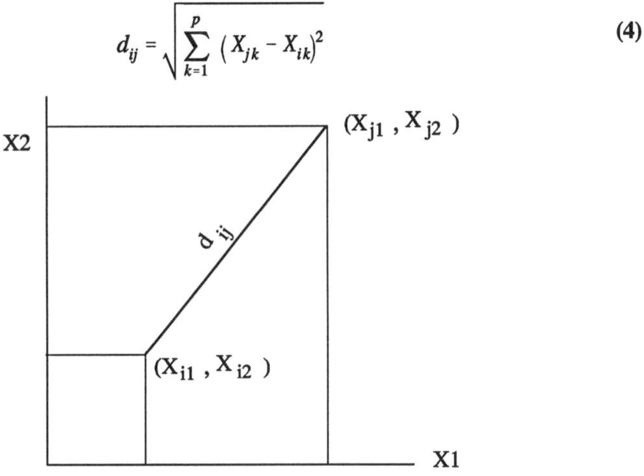

Figure 19-1. Geometric representation of the Euclidean distance. In this schematic there are two populations represented.

gives the distance between population i and population j.

There are problems with this particular type of description of a data system, however; if two or more of the variables are collinearly related, then the distance will be unduly large. Similarly if one of the variables is more variable than the other variables, then it will clearly dominate the distance.

Penrose Distance

Consider a situation in which we have r populations each of which has a multivariate distribution for p variables such as shown in Figure 19-2. If the difference between the means of two populations is squared and divided by the variance, one of the problems mentioned with the Euclidean distance (dominance of the distance by a single variable) can be minimized. Such a measure of distance is the Penrose distance. However, this distance still does not take into account the colinearity between variables.

The Penrose distance is calculated from the following

$$P_{ij} = \sum_{k=1}^{p} \frac{(\mu_{ik} - \mu_{jk})^2}{p \cdot V_k} \tag{5}$$

where V_k is the variance of the kth variable and is considered equal from population to population. μ_{ik} and μ_{jk} are the mean of the variable in the ith and jth populations for the kth variable. The full intent of this symbology will be clear in the example. For this measure of the distance, any number of

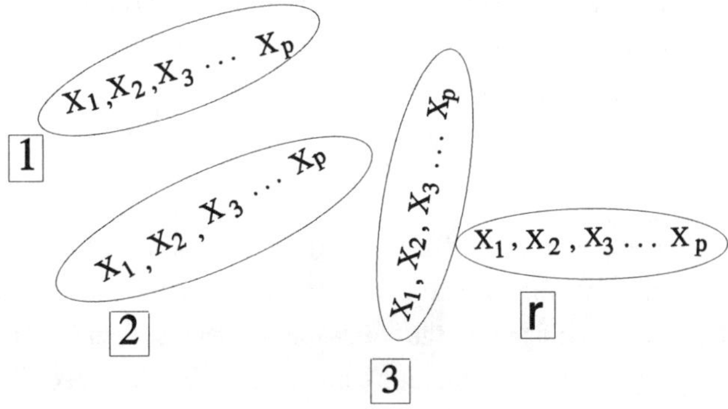

Figure 19-2. A schematic representation of *r* populations each having *p* variables associated with it.

populations can be compared pairwise. A requirement is that the variables of interest have been measured for each population. Of course, this assumption of equality of variance from population to population is a prerequisite for the use of this distance measure and should be checked using one of the equality of multivariate variances such as Bartlett's, Cochran's, or Hartley's test.

Mahalonobis Distance

A more satisfying measure of the distance between two populations is the Mahalonobis distance. This particular measure takes into consideration the possible correlation between two or more of the p variables in the r populations. This particular measure is a scalar and is expressed as the product of

$$D_{ij}^2 = \sum_{r=1}^{p} \sum_{s=1}^{p} (\mu_{ri} - \mu_{rj}) \cdot c_{rs} (\mu_{si} - \mu_{sj}) \tag{6}$$

where μ_{ri} is the mean of the rth variable in the ith population. The rest of the means are similarly described, except being described for the jth variable of the sth population. c_{rs} is the element of the rth row and the sth column of the inverse of the covariance matrix of the p variables. We must here assume that the covariance matrix for each of the populations is statistically the same and has been combined to form a single pooled covariance matrix.

This particular distance measure is very useful for determining whether a given data record is out of agreement with the rest of the data records for a given population. Consider that the questionable data record is described by the set of variables, x_k, where $k=1$ to p. The mean of this variable in the remainder of the data population is described by μ_k as the mean for that variable in the data population. Then D^2 is described by the expression

$$D_{ij}^2 = \sum_{r=1}^{p} \sum_{s=1}^{p} (m_r - \mu_r) \cdot V_{rs} \cdot (m_s - \mu_s) \tag{7}$$

where V_{rs} is the element in the rth row and the sth column of the inverse of the covariance matrix of the p variables. If the population is multinormally distributed, then D^2 will follow the ^{2} distribution with p degrees of freedom. If $D^2_{\mathrm{calc}} > D^2_{\mathrm{critical}}$ the sample population (the questionable data record) is either (1) genuine but highly unlikely or (2) there is an error in that particular record and the record should be critically examined for correctness.

Similarly, this concept may be used to determine how far a single population is from the center of a set of populations that has been determined to be close together.

Hierarchical Analysis

Two major different clustering techniques are discussed in the literature—hierarchical and nonhierarchical. Hierarchical is the more important of the techniques and will be discussed here. To use hierarchical cluster analysis, a measure of similarity between each of the p populations is calculated. The measure of similarity is usually one or more of the previous distances, e.g., as the Euclidean, the Penrose, or the Mahalonobis distance. The two closest populations are joined together to form a single new population. The distance to each of the remaining p-1 populations is then calculated as the average from the new population to each of the remaining populations. The two closest populations are again joined to form a new joint population. The distance to each of the p-2 remaining populations is again calculated as the average from the new population to each of the other populations. This process is continued until there are only two populations.

Example

The process is best explained by way of example. Consider a set of gas chromatographic data for five different columns and retention time of four different organics:

	Column				
	A1	**A2**	**A3**	**A4**	**A5**
a1	4.542	2.042	3.708	2.458	3.708
a2	3.708	1.208	2.875	0.792	3.292
a3	3.292	1.208	2.458	1.208	2.042
a4	2.875	0.792	2.042	0.792	2.458

Each of the columns, A1, A2, A3, A4, and A5, represents a population. The values of the retention times for the different compounds, a1, a2, a3 and a4, represent the values of the random variables which characterize each population. For the sake of demonstration, the Euclidean distance will be chosen as the best measure of similarity between populations. The Euclidean distance between each of the populations is calculated using

$$d_{ij} = \sqrt{\sum_{k=1}^{p} (X_{jk} - X_{ik})^2} \tag{8}$$

where $p = 4$ for the four variable retention times, a1, a2, a3, and a4. The distance, d_{ij}, is the distance between two of the populations, *e.g.* A1 and A2. The following table represents the Euclidean distance matrix:

Euclidean Distance					
A1	A2	A3	A4	A5	
0.000	4.602	1.667	4.640	1.614	A1
	0.000	1.909	1.998	2.125	A2
		0.000	4.983	2.125	A3
			0.000	1.559	A4
				0.000	A5

As can be seen the closest distance is between columns A4 and A5 with a Euclidean distance of 1.559. These two populations are combined into a new population, A4‡. The distance between A4‡ and A1 is the average of the distances between A1–A4 and A1–A5

$$D_{A5^{\ddagger}-A1} = \frac{D_{A1-A4} + D_{A1-A5}}{2} = \frac{4.640 + 1.614}{2} = 2.627 \tag{9}$$

The distances, A2–A4‡ and A3–A4‡ are found similarly. The resulting Euclidean distance matrix, for the populations A1, A2, A3, and A4‡ is

Euclidean Distance				
A1	A2	A3	A4‡	
0.000	4.602	1.667	3.127	A1
	0.000	1.909	2.061	A2
		0.000	3.5536	A3
			0.000	A4‡

The next closest distance is between columns A1 and A3 (1.667). These two are combined into one population, A1‡. The distance between A1‡ and A2 is calculated as

$$D_{A1^{\ddagger}-A2} = \frac{D_{A1-A2} + D_{A3-A2}}{2} = \frac{4.602 + 1.909}{2} = 3.256 \quad (10)$$

The matrix for the reduced set of columns, A1‡, A2, and A4‡ is

Euclidean Distance			
A1‡	**A2**	**A4‡**	
0.000	3.256	3.341	**A1‡**
	0.000	2.061	**A2**
		0.000	**A4‡**

The next closest set of populations is A2 and A4‡ with a distance of 2.061. As before, these are combined into a single population, A2‡. The distance between A1‡ and A2‡ is calculated as

$$D_{A1^{\ddagger}-A2^{\ddagger}} = \frac{D_{A1^{\ddagger}-A2} + D_{A1^{\ddagger}-A4^{\ddagger}}}{2} = \frac{3.256 + 3.341}{2} = 3.299 \quad (11)$$

which is the final distance between the reduced set of columns. The linkages can be shown graphically as in Figure 19-3.

Principal Component Analysis

General

Most of the data analysis techniques described so far measure whether two or more data sets are statistically indistinguishable. In each of these the techniques, the dependence between sets using the variability of random variables within each data set was used as the tool to identify similarities between the data sets. In this chapter, the technique of cluster analysis was described to determine potential groupings within a data set using the principle of a distance between

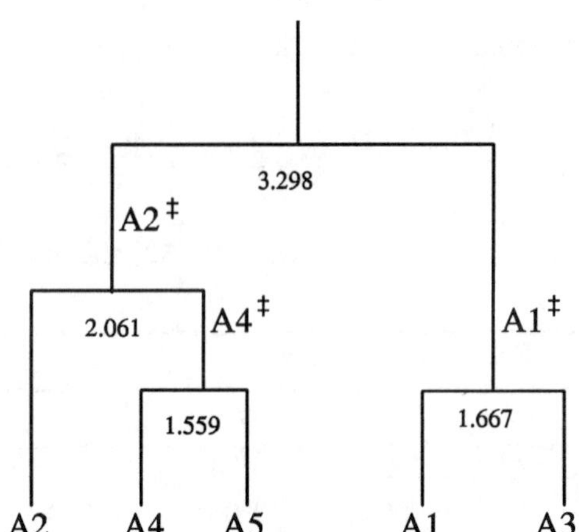

Figure 19-3. Linkages which can be formed from the gas chromatographic data of four compounds on five columns.

the two data sets. In Chapter 18, the technique of multiple linear regression described the potential dependence of a single dependent variable on several independent variables. A major limitation of multiple linear regression is that it may not be clear from the outset which (if any) of the measured random variables is the dependent variable. Nor may it necessarily be desirable to describe the data set using multilinear regression by limiting ourselves to only one dependent variable.

Consider a data set that is composed of p columns of random variables and m rows of values of the random variables, such as shown in Table 19-2. It is desired to determine which of the measured random variables in a data set are most important and which of the random variables may be grouped together in some fashion. One method of grouping variables together would be as a linear combination of the variables, which predict the value of a random variable by the values of all other random variables:

$$dv_{ij} = sc_{i1} \cdot ld_{1j} + sc_{i2} \cdot ld_{2j} + \cdots + sc_{ik} \cdot ld_{kj} = \sum_{k=1}^{n} sc_{ik} \cdot ld_{kj} \qquad (12)$$

where dv_{ij} is the data value in the ith column and the jth row of the data matrix. sc_{ik} is the true value of the ith random variable in the kth row as affected by the magnitude of the loading ld_{kj} for that particular variable. In this formulation of predicting values of random variables, the concept of independent and dependent

Table 19-2. Representative Data Matrix for *p* Variables and *m* Rows

	Variable				
Row	1	2	3	...	p
1	a_{11}	a_{12}	a_{13}	...	a_{1p}
2	a_{21}	a_{22}	a_{23}	...	a_{2p}
3	a_{31}	a_{32}	a_{33}	...	a_{3p}
...	...	...	...	...	...
m	a_{m1}	a_{m2}	a_{m3}	...	a_{mp}

random variables loses its meaning. All random variables are, at the same time, dependent and independent variables. This method looks at the interdependence of the random variables.

If there are p variables, then there would be p such linear combinations of variables. The goal of forming such linear combinations is to be able to reduce the number of rows of data in the data set and to be able to predict future values of random variables. A criterion to form the linear combinations is to make the variance of the expected value of the linear combination as large as possible and be consistent with the data set. Of course, one way to accomplish this restriction is to make the score coefficients, ld_{ij}, arbitrarily large, thus making the variance arbitrarily large. Remember this is true since the variance of a set of numbers each of which is multiplied by a constant, b, is b^2 times the original variance. A criterion by which the magnitude of the variance can then be limited is to require that the sum of the square of the loadings for each linear combination will sum to 1:

$$\sum_j ld_{ij}^2 = 1 \tag{13}$$

An additional criterion is that each of the linear combinations must be uncorrelated. This last criterion can be ensured if the linear combinations are orthogonal to each other.

The linear combinations formed with the foregoing criteria result in principal components of the data set. The method of finding the principal components of the data set is called *factor analysis*. The number of principal components can be either the number of rows or the number of columns in the data set and will be whichever is smaller, the number of rows or the number of columns of data. Generally, the number of columns of data is smaller than the number of rows of data, thus the number of principal components will usually be the number of columns of data.

The data set can be considered as a matrix equation in which the principal components (the linear combinations) represented by Equation (12) can be resolved into a product of two matrices:

$$[\text{Data Set}] = [\text{scores}] \cdot [\text{loadings}] \tag{14}$$

The matrix for the scores is

$$\begin{bmatrix} sc_{11} & sc_{12} & \cdots & sc_{1q} & \cdots & sc_{1p} \\ sc_{21} & sc_{22} & \cdots & sc_{2q} & \cdots & sc_{2p} \\ \cdots & \cdots & \cdots & \cdots & \cdots & \cdots \\ sc_{q1} & sc_{q2} & \cdots & sc_{qq} & \cdots & sc_{qp} \\ \cdots & \cdots & \cdots & \cdots & \cdots & \cdots \\ sc_{m1} & sc_{m2} & \cdots & sc_{mq} & \cdots & sc_{mp} \end{bmatrix} \tag{15}$$

and the matrix for the loading is

$$\begin{bmatrix} ld_{11} & ld_{12} & \cdots & ld_{1q} & \cdots & ld_{1p} \\ ld_{21} & ld_{22} & \cdots & ld_{2q} & \cdots & ld_{2p} \\ \cdots & \cdots & \cdots & \cdots & \cdots & \cdots \\ ld_{q1} & ld_{q2} & \cdots & ld_{qq} & \cdots & ld_{qp} \\ \cdots & \cdots & \cdots & \cdots & \cdots & \cdots \\ ld_{m1} & ld_{m2} & \cdots & ld_{mq} & \cdots & ld_{mp} \end{bmatrix} \tag{16}$$

The product of the two matrices should yield the original data matrix as shown in Equation (14).

The rows in the loading matrix are termed the principal components of the data set. Likewise, the columns in the score matrix could be used as the principal components. Either the score matrix or the loading matrix can contain the principal components. Which one is used will depend on whether the number of rows or the number of columns is less.

Effect of Error in the Data

If there is no experimental error in the data set, the number of principal components needed to completely span (i.e,. reproduce) the data space is the same as either the number of rows or the number of columns of the data set, whichever is smaller. Rarely will chemical data be free of experimental error. The presence of error will complicate the analysis of the data set. The original data set can be considered to be the sum of two different matrices, a matrix giving the "true" values of the random variables and a matrix that contains errors associated with the data.

$$[\text{data}] = [\text{true data}] + [\text{error}] \tag{17}$$

The error matrix will contain the obvious error that can be removed, but there will still be the inherent error in the data matrix, which will introduce a measure of uncertainty that cannot be removed.

In principle, it is desired to eliminate the error matrix and find only the matrix that contains the real data. The number of principal components that can span the matrix of the true data will be less than the number of principal components that can span the complete data matrix. The principal components required to span the data space are called the *primary principal components.* Those principal components required to span the remaining error space are the *secondary principal components.* We desire to determine only the primary principal components and discard the secondary principal components.

We should be concerned as to the type of data that this method can analyze. One of the first applications reported in the literature was to use factor analysis to reduce the number of mass peaks in a mass spectrum for the 22 different isomers of a common empirical formula, $C_{10}H_{14}$, so as to be able to use the principal components in an interpretation of the mass spectra.[2] Another area to which principal component was applied was the interpretation of the number of absorbing species in the infrared region of the electromagnetic spectra, particularly in the region where the carbonyl moiety[3] absorbs electromagnetic radiation. Several solutions of differing concentrations of acetic acid and chlorinated acetic acid were prepared in carbon tetrachloride and the absorbance measured at several wavelengths. The conventional theory was that the acetic acid would form cyclic dimers.[4,5] However, when the spectral data were analyzed by principal component analysis, it was indicated that four different species, probably in a chain, were responsible for the deviations from the idealized Beer's law plots of absorbance versus concentration.

Determination of the Principal Components

A theorem of principal component analysis that will aid us to determine the principal component (stated without proof) is that the principal components that span the space of the correlation or covariance matrix are the same principal components that span the original data matrix. This aids us because the principal components are much easier to determine for the covariance or correlation matrix than for the original data matrix. Two considerations are used to determine whether the correlation or covariance matrix should be used in the principal component analysis. Once the principal components have been determined for the correlation or covariance matrix, they can be applied to calculations pertinent to the original data matrix.

The first consideration is whether the data have been taken relative to an absolute standard. Also implied in this is whether the uncertainty in each of the data points is constant. If the decision is made that these criteria are fulfilled, then a covariance matrix should be used to determine the principal components. It should be recognized that by so doing, data points that are inordinately large or small can drastically affect the covariance matrix. Each of the data columns will not have the same statistical weight.

If it is desired that each of the data columns have the same statistical weight, the correlation matrix will be used. The correlation matrix is formed using normalized data. There are three reasons to use the correlation matrix when determining the principal components of a data set.[6]

1. When the experimental error of the measurement is the same order of magnitude as the measurement. When this prevails, the raw data are composed of two nearly equal terms. The decomposition of the matrix into the "real" data and the uncertainty in the measurement will be unduly influenced by the variability of the measurements. This situation is corrected by normalization of the data.
2. When the magnitudes of the properties represented by the data columns are very dissimilar. If the magnitudes of the measurements in different columns are very dissimilar, the larger of the data magnitudes will be unduly affected by the larger of the measurements. As a result any real contribution to the principal components by the smaller magnitude measurement may be obscured by the larger magnitude measurements. This is a consequence much the same way that the calculation of the standard deviation or variance is affected by the presence of very large values of random variables.
3. When each column of the data has been measured without respect to an absolute standard or scale. In this situation, the experimenter is not certain of the zero of the measurements and is therefore not certain if some of the variability between columns is not the result of the differing points of reference. This particular problem may be corrected by subtracting the mean for a particular column from each data value in the column.

There are means available to account for each of these limitations in the data set using either the covariance or the correlation matrix as a precursor to the principal component analysis procedures. Each of the matrices may be formed with respect to the origin or the mean of the data. If the data must be adjusted to account for the mean of the data, the mean of a particular column is subtracted from each data value in the column.

$$dv'_{ij} = dv_{ij} - \bar{dv}_i \tag{18}$$

In matrix notation, this matrix is denoted by DV' .

Normalization of Data

Normalization of data may need to be accomplished if the conditions given in the previous section exist. When the data are normalized, the correlation matrix results. There are two primary methods of normalizing the data. The first is normalization with respect to the origin of the data measurements. Each data value is divided by the square root of the sum of the squares of the data:

$$dv^*_{ij} = \frac{dv_{ij}}{\sqrt{\sum_j^n dv^2_{ij}}} \tag{19}$$

In matrix notation, this matrix is denoted by DV^*. The normalization is accomplished for each column of data represented by the first subscript.

The second is normalization with respect to the mean of the data in the column. The mean of the data in a column is first subtracted from each data value. Each of these data values is then divided by the square root of the square of the difference between the data value and the mean of the data in the column:

$$dv^{**}_{ij} = \frac{\left(dv_{ij} - \bar{dv}_i\right)}{\sqrt{\sum_j^n \left(dv_{ij} - \bar{dv}_i\right)^2}} \tag{20}$$

In matrix notation, this matrix is denoted as DV^{**}.

Calculation of the Covariance and Correlation Matrices

The covariance and correlation matrix may be formed with respect to either the origin or the mean of the data. Four matrices are obtained from the normalization procedure, DV (the original, nonnormalized data matrix), DV' (the data matrix from which the mean of the data in a particular column has been subtracted from the data in that column), DV^* (each element in a column of the data matrix has been divided by the square root of the sum of the squares of the data values in that column), and DV^{**} (each element in a column of DV' divided by the square root of the sum of the squares of the difference between the data values in a column and the mean of the data in the column). Each of these can be used to form either a correlation or a covariance matrix by forming the inner dot product of the matrix with itself:

$$\begin{aligned} \text{Cov}^{\text{o}} &= DV \cdot DV \\ \text{Cov}^{\text{m}} &= DV' \cdot DV' \\ \text{Corr}^{\text{o}} &= DV^* \cdot DV^* \\ \text{Corr}^{\text{m}} &= DV^{**} \cdot DV^{**} \end{aligned} \quad \textbf{(21)}$$

This matrix, regardless of the number of rows and columns in the data matrix, will be square. A square matrix may have eigenvalues associated with it. The eigenvalues of the square correlation or covariance matrix may be found by finding the roots to the determinant equation:

$$|C - \lambda I| = 0 \quad \textbf{(22)}$$

where λ are the eigenvalues of the matrix and I is the identity matrix in which all diagonal elements are 1 and all off-diagonal elements are zero. The roots of the determinant equation are the eigenvalues. From each of the roots, an eigenvector may be calculated. Many computer routines exist to find the eigenvalues and eigenvectors of a matrix. An alternative, iterative technique is presented by Malinowski et al.[7] Consult this book for tips on how to iteratively find the eigenvalues and eigenvectors. Alternatively, consult a text on matrix algebra[8] to learn other techniques on finding eigenvalues and eigenvectors.

The eigenvectors that are found are the principal components of the correlation, the covariance or the original data matrix. We must now determine which of the principal components are those that span the "true" or "real" data and which of the principal components are those that represent the error portion of the original data. One of the interesting properties of the eigenvalues is that each eigenvalue represents the variance of the eigenvector. If the eigenvalues and associated eigenvectors are sorted in decreasing sequence of the eigenvalues, the principal components are sorted in order of importance. Therefore, the first principal component represents the component that has the greatest variance.

Determining the Significant Principal Components

After the eigenvectors have been obtained, they must be interpreted as to their significance. The primary aim of interpretation is to determine how many principal components are significant to the "true" data and which of the principal components are those that represent the error portion of the data. Several methods have been proposed to determine which of the principal components are the most significant. Two of these methods will be presented—variance of the eigenvectors and residual standard deviation.

Variance of the Eigenvectors

It was stated above that the eigenvalue of the eigenvector is the variance of the eigenvector or the principal component. The total variance of all of the principal components is then the sum of the values of the eigenvalues:

$$\text{variance}_{\text{total}} = \sum_{i=1}^{p} \lambda_i \tag{23}$$

The fraction of the variance represented by any single eigenvector is then

$$f_j = \frac{\lambda_j}{\sum_{i=1}^{p} \lambda_i} \tag{24}$$

The sum of all f_js is, of course, 1. To determine which of the eigenvectors or principal components is significant, determine how much of the variance in the principal components represents variance in data, such as 90, 95, or 98%. Then form the cumulative sum of the fraction of the eigenvalues. When the cumulative sum exceeds the predetermined limit, the number of principal components has been found. Unfortunately, this particular method (although simple and easy to apply) does not lend itself to a quantitative determination of the number of significant principal components since the method is very arbitrary.

Residual Standard Deviation

The residual standard deviation is compared to the standard deviation of the measurements. The integer value of the eigenvector or principal component when the residual standard deviation is smaller than the standard deviation is taken as the number of significant principal components. The residual standard deviation is calculated from

$$S_k = \frac{1}{NS} \sqrt{\frac{\sum_{i=1}^{NW} \lambda_i - \sum_{j=1}^{K} \lambda_j}{NW - K}} \tag{25}$$

where *NW* is the number of rows in the data matrix, *NS* is the number of columns in the data matrix, *K* is the number of significant principal components, and λ_j is the eigenvalue of the *j*th principal component. S_k is compared to the standard deviation of the measurements.

Which standard deviation of the measurements to use when making this decision can be difficult to determine. Let us suppose that the data with which we are concerned are a set consisting of measurements of concentrations of several different analytes at a single sampling location for many different weeks. The standard deviation against which to compare the residual standard deviation certainly should not be the standard deviation of the measurements of the concentrations of the analytes of the natural samples. This standard deviation can be unduly influenced by the presence of very large data values in the data matrix. The methods that yielded the concentrations should have had the average precision of the method determined over the expected range of concentrations for the samples. It is thus more appropriate to use the precision of the analytical method as the comparison value for the residual standard deviation calculated using Equation (25).

Example

Let us examine a set of rainfall data to determine the principal components of the data. The data are shown in Table 19-3. To find the principal components, the covariance or the correlation matrix must first be obtained. The covariance matrix is shown in Table 19-4. This covariance matrix was calculated with regard to the mean of each data set. The eigenvalues of the covariance matrix are obtained by diagonalizing the matrix using computer techniques and are shown in Table 19-5. This table also shows a

Table 19-3. Ionic Composition of Rainfall from a Site in Southeastern New York

Ca^{2+}	Mg^{2+}	NH_4	$^+NO_3$	$^-Cl^-$	SO_4^{2-}	H^+
0.53	0.984	0.10	3.47	11.48	7.64	0.0000115
0.13	0.038	0.10	0.82	0.17	1.30	0.0000251
0.05	0.035	0.03	0.57	0.38	0.79	0.0000182
0.94	0.270	0.33	6.07	0.96	7.01	0.0000933
0.03	0.054	0.02	0.34	0.77	0.66	0.0000145
0.05	0.013	0.09	1.27	0.12	1.16	0.0000372
0.48	0.171	0.36	7.03	1.03	5.08	0.0001230
0.29	0.102	0.57	5.84	0.37	6.23	0.0001259
0.04	0.017	0.23	1.79	0.19	1.65	0.0000447
0.04	0.064	0.08	0.70	0.85	0.97	0.0000182
0.04	0.025	0.09	1.11	0.29	1.65	0.0000389
0.04	0.018	0.10	1.21	0.23	1.17	0.0000331
0.05	0.118	0.08	0.79	1.71	1.33	0.0000288
0.05	0.106	0.06	0.94	1.46	1.36	0.0000263
0.02	0.020	0.00	0.49	0.24	0.89	0.0000195
0.05	0.019	0.07	0.52	0.26	1.30	0.0000200
0.65	0.142	0.52	15.1	11.32	5.00	0.0002512
0.13	0.110	0.16	2.79	0.52	2.04	0.0000437
0.13	0.087	0.09	3.13	1.02	2.12	0.0000676
0.19	0.063	0.16	3.23	1.12	1.91	0.0000562
0.06	0.040	0.10	1.27	0.41	1.51	0.0000372
0.20	0.050	0.20	2.04	0.32	2.92	0.0000708
0.12	0.239	0.10	0.99	3.54	2.36	0.0000427
0.10	0.030	0.15	1.18	0.19	1.52	0.0000324
0.12	0.041	0.12	1.33	0.17	1.93	0.0000363
0.17	0.057	0.09	3.92	0.21	0.94	0.0000603
0.06	0.017	0.00	0.78	0.05	1.18	0.0000302
0.64	0.112	0.22	2.38	1.12	3.23	0.0000617
0.25	0.069	0.03	0.17	0.29	1.05	0.0000155
0.03	0.028	0.09	0.48	0.39	1.57	0.0000324
0.09	0.022	0.17	1.18	0.12	1.45	0.0000380
0.31	0.080	0.28	4.30	0.23	3.05	0.0000813
1.52	0.266	0.66	8.24	0.62	10.16	0.0001820
0.09	0.024	0.22	1.25	0.17	2.03	0.0000427
0.40	0.093	0.57	3.68	0.29	3.87	0.0000794
1.06	0.276	1.59	9.10	0.51	10.76	0.0001820
0.06	0.030	0.21	1.21	0.28	1.87	0.0000398
0.09	0.015	0.07	0.69	0.14	1.34	0.0000166
0.23	0.082	0.99	5.53	0.28	8.86	0.0001738
0.08	0.049	0.13	0.93	0.47	1.09	0.0000224
0.04	0.015	0.18	2.32	0.10	2.30	0.0000794
0.06	0.022	0.19	1.62	0.17	2.13	0.0000490
0.07	0.016	0.22	1.66	0.11	3.59	0.0000741

(Continued)

Table 19-3. (Continued) Ionic Composition of Rainfall from a Sampling Site in Southeastern New York

Ca^{2+}	Mg^{2+}	NH_4	$^+NO_3$	$^-Cl^-$	SO_4^{2-}	H^+
0.09	0.037	0.24	1.93	0.14	3.05	0.0000776
0.13	0.033	0.13	4.13	0.19	3.39	0.0001202
0.18	0.059	0.35	2.23	0.15	4.79	0.0001023
0.13	0.064	0.23	1.26	0.20	4.03	0.0000741
0.04	0.015	0.16	2.24	0.12	1.91	0.0000646
0.09	0.024	0.29	2.22	0.11	2.78	0.0000724
0.21	0.150	0.25	1.76	0.22	3.33	0.0000617
0.13	0.087	0.40	1.11	0.12	1.88	0.0000214
1.00	0.543	0.25	0.71	0.66	3.41	0.0000003
0.18	0.067	0.50	2.78	0.29	2.99	0.0000661
0.17	0.060	0.25	3.72	0.22	3.29	0.0001023
0.03	0.041	0.00	0.29	0.65	0.94	0.0000195
0.17	0.092	0.16	1.87	0.64	2.36	0.0000269
0.05	0.079	0.08	0.58	1.24	1.42	0.0000347
0.04	0.024	0.00	0.43	0.05	1.03	0.0000224
0.44	0.340	0.31	5.20	2.17	5.89	0.0001122
0.05	0.028	0.33	2.67	0.36	3.27	0.0000832
0.17	0.052	0.22	3.63	0.61	2.85	0.0000871
0.12	0.323	0.00	0.87	0.18	3.69	0.0000501
0.97	0.554	0.00	5.77	1.23	2.12	0.0000023
0.06	0.023	0.00	2.80	0.29	0.55	0.0000447

distinctive characteristic of eigenvalues obtained from covariance matrices. The eigenvalue and the variance of the principal component are equal. Also shown in this table is the cumulative sum of the variances giving the fraction of the total variance explained by the principal component. Associated with each of the eigenvalues is an eigenvector. The eigenvectors are shown in Table 19-6. This table shows the loading or eigenvector matrix and gives the relative weighting of each of variables to the true value of the variable. The last required matrix is the scores matrix. This matrix is shown in Table 19-7. The matrices may also be represented graphically. The usual presentation is to plot the first two principal components against each other. Some plot only the scores or the loadings. Gabriel[9,10] pioneered the use of the biplot in which both the scores and loadings are placed on the same plot. The biplot for the rainfall data is shown in Figure 19-4. A couple of interesting features are exhibited in this biplot. The first is that the two ions, magnesium and chloride, are by themselves in the biplot indicating that they are associated with each other. This is to be expected since magnesium and chloride are two of the major ions associated with sea salt. The location from which these samples

Table 19-4. Covariance Matrix Used to Compute the Eigenvalues and Eigenvectors of the Rainfall Data

	Ca^{2+}	Mg^{2+}	NH_4^+	NO_3^-	Cl^-	SO_4^{2-}	H^+
Ca^{2+}	0.0895	0.0295	0.0397	0.5039	0.1620	0.4786	0.00001
Mg^{2+}	0.0295	0.0251	0.0049	0.1196	0.1895	0.1757	0.00000
NH_4^+	0.0397	0.0049	0.0631	0.4086	0.0231	0.4491	0.00001
NO_3^-	0.5039	0.1196	0.4086	6.5065	2.4671	3.9726	0.00010
Cl^-	0.1620	0.1895	0.0231	2.4671	3.9795	1.3146	0.00002
SO_4^{2-}	0.4786	0.1757	0.4491	3.9726	1.3146	4.9292	0.00008
H^+	0.0000	0.0000	0.0000	0.0001	0.0000	0.0000	0.00000

were taken in southeastern New York state should have a prominent component of sea salt. The two ions, nitrate and sulfate, are also strongly associated. These two are the major contributors to the acidity of the rain. The degree of correlation between ions is related to the angle between the vectors drawn from the origin to the point in principal component space.

Another example of the application of principal component analysis has been to the analysis of data from three of the major acid rain monitoring networks on the North American continent, the National Atmospheric Deposition Program/National Trends Network in the United States, the MAP3S in the northeastern United States, and the CANSAP in Canada.[11,12] The results of several years of sampling efforts being subjected to the principal component process culminating in the biplot are shown in Figure 19-5. This figure shows three predominant groupings: a coastal grouping based

Table 19-5. Eigenvalues of the Covariance Matrix Computed for the Rainfall Data

Component Number	Percentage of Variance	Cumulative Percentage
1	58.57591	58.57591
2	21.30388	79.87979
3	11.33030	91.21008
4	5.11974	96.32983
5	2.22298	98.55281
6	1.14108	99.69389
7	.30611	100.00000

Table 19-6. The Loading or Eigenvector Matrix for the Rainfall Data

Ca^{2+}	Mg^{2+}	NH_4^+	NO_3^-	Cl^-	SO_4^{2-}	H^+
0.3988	-0.1645	-0.4444	0.5969	-0.2003	0.4578	0.1010
0.2579	-0.6416	-0.3047	-0.1898	0.1566	-0.5081	0.3319
0.3879	0.3705	-0.1960	-0.4756	-0.6495	-0.0886	0.1338
0.4424	0.0816	0.3363	0.4328	-0.1188	-0.5532	-0.4206
0.2367	-0.5304	0.6260	-0.2213	-0.2494	0.3980	-0.0341
0.4558	0.0650	-0.2255	-0.3749	0.5106	0.2364	-0.5291
0.4055	0.3629	0.3395	0.0669	0.4200	0.0634	0.6354

on the proximity of the sites (M6, M7, and C15) to the vectors corresponding to sodium and chloride, a CANSAP grouping based on the similarity of chemistries of nitrate, sulfate, and calcium, and a grouping of NADP and MAP3S sites. The C7 site in the NADP/MAP3S group is logical since it is located in Ontario very near to the U.S.

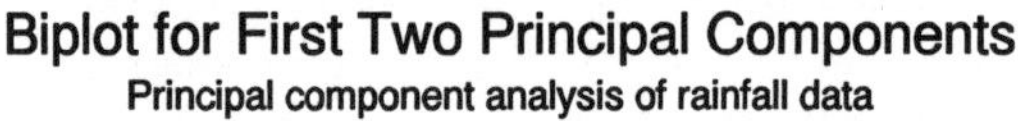

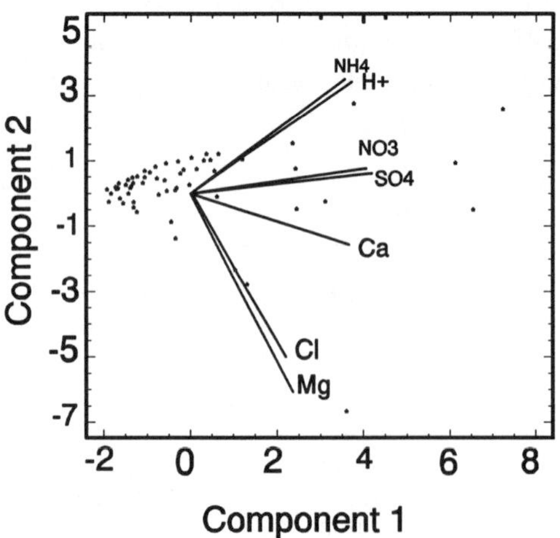

Figure 19-4. Biplot of loadings and scores for principal component analysis of rainfall data. This figure plots the first two components.

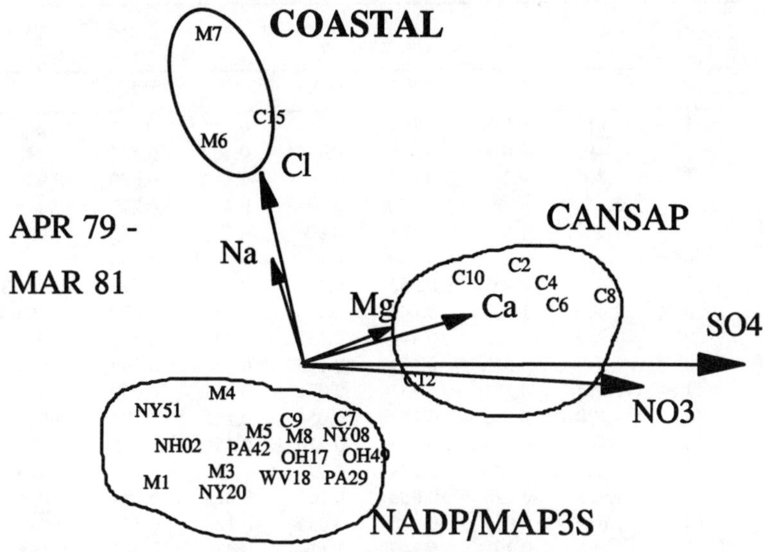

Figure 19-5. Biplot of the first two components from the analysis of rainfall data from three monitoring networks, NADP/NTN, MAP3S, and CANSAP.

sampling sites.

Conclusion

The multivariate techniques provide tools to rapidly (with the aid of a computer) analyze a set of data that has many random variables. It is especially applicable to data sets in which no one of the random variable can be determined to be a dependent variable and the remainder to be independent variables. If all of the variables are independent, these methods are even more applicable. They can be used quickly with the aid of computer software packages such as MINITAB, SYSTAT, or STATGRAPHICS. Graphical presentation of the data is easily achieved for the data sets.

Table 19-7. Score Matrix for the Principal Component Analysis of Rainfall Data

	Analytes						
row	Ca^{2+}	Mg^{2+}	NH_4^+	NO_3^-	Cl^-	SO_4^{2-}	H^+
1	3.655	-6.914	0.577	-2.111	0.275	0.065	-0.251
2	-1.402	-0.028	-0.169	0.117	-0.192	0.145	-0.043
3	-1.804	-0.206	0.045	0.104	-0.153	0.090	-0.046
4	3.166	-0.427	-1.150	0.954	0.471	0.277	-0.491
5	-1.867	-0.430	0.125	-0.004	-0.196	0.111	-0.005
6	-1.418	0.218	0.149	0.128	-0.077	0.000	-0.006
7	2.470	0.451	0.255	0.621	0.368	-0.479	-0.141
8	2.405	1.338	0.046	-0.390	0.313	-0.370	-0.265
9	-0.946	0.483	0.141	-0.152	-0.283	-0.113	-0.030
10	-1.577	-0.360	0.111	-0.105	-0.270	0.049	-0.034
11	-1.305	0.151	0.135	-0.031	0.054	0.069	-0.055
12	-1.438	0.155	0.143	0.054	-0.136	-0.003	-0.042
13	-1.194	-0.718	0.313	-0.276	-0.159	0.095	0.107
14	-1.263	-0.646	0.272	-0.179	-0.110	0.058	0.010
15	-1.913	-0.126	0.086	0.106	-0.015	0.104	-0.094
16	-1.671	-0.023	-0.038	-0.030	-0.122	0.170	-0.145
17	6.593	-0.791	5.149	1.114	-0.736	0.281	0.164
18	-0.499	-0.097	0.072	0.114	-0.077	-0.360	-0.119
19	-0.306	-0.045	0.535	0.296	0.231	-0.195	0.030
20	-0.268	0.005	0.422	0.337	-0.179	-0.090	-0.082
21	-1.239	0.051	0.130	0.006	-0.039	0.020	-0.029
22	-0.184	0.427	-0.005	0.034	0.214	0.177	0.079
23	-0.292	-1.561	0.557	-0.642	-0.017	0.258	0.270
24	-1.206	0.162	-0.065	0.004	-0.213	0.066	-0.057
25	-1.066	0.114	-0.092	0.051	-0.014	0.084	-0.114
26	-0.565	0.204	0.451	0.825	-0.096	-0.506	0.048
27	-1.686	0.013	0.058	0.225	0.126	0.119	-0.058
28	0.673	-0.298	-0.593	0.703	-0.195	0.751	0.102
29	-1.531	-0.455	-0.443	0.356	-0.188	0.380	0.061
30	-1.482	0.046	0.054	-0.168	0.005	0.184	-0.017
31	-1.176	0.289	-0.025	-0.016	-0.224	0.056	0.025
32	0.632	0.543	0.042	0.452	-0.007	-0.258	-0.040
33	6.180	0.684	-1.769	1.490	0.674	0.976	-0.047
34	-0.918	0.396	-0.068	-0.198	-0.186	0.094	-0.031
35	1.272	0.842	-0.501	-0.185	-0.611	-0.033	0.051
36	7.279	2.342	-1.813	-1.142	-1.299	-0.231	0.030
37	-1.016	0.316	-0.003	-0.242	-0.208	0.042	-0.030
38	-1.629	-0.017	-0.133	0.084	-0.166	0.183	-0.221
39	3.825	2.527	-0.148	-1.701	0.301	-0.152	-0.038
40	-1.416	-0.103	-0.029	-0.035	-0.340	0.033	-0.030
41	-0.516	0.742	0.407	-0.016	0.286	-0.107	0.161

Continued

Table 19-7 (continued)

row	Analytes						
	Ca^{2+}	Mg^{2+}	NH_4^+	NO_3^-	Cl^-	SO_4^{2-}	H^+
42	-0.306	0.754	0.077	-0.371	0.443	0.196	-0.074
43	-0.306	0.754	0.077	-0.371	0.443	0.196	-0.074
44	-0.245	0.699	0.116	-0.256	0.289	0.047	0.117
45	0.451	0.923	0.725	0.406	0.888	-0.213	0.187
46	0.700	0.969	-0.103	-0.520	0.569	0.231	0.117
47	-0.101	0.517	-0.181	-0.480	0.533	0.282	0.010
48	-0.765	0.580	0.352	0.050	0.119	-0.139	0.058
49	-0.242	0.795	0.121	-0.244	0.029	-0.032	0.064
50	0.015	0.051	-0.426	-0.278	0.216	-0.072	0.146
51	-0.727	0.227	-0.560	-0.557	-0.825	-0.133	-0.015
52	1.070	-2.588	-2.800	0.499	-0.453	0.119	0.562
53	0.379	0.809	-0.194	-0.483	-0.586	-0.178	0.066
54	0.493	0.808	0.358	0.145	0.412	-0.220	0.173
55	-1.841	-0.330	0.128	0.013	-0.031	0.183	-0.033
56	-0.697	-0.226	-0.188	-0.031	-0.168	-0.006	-0.294
57	-1.281	-0.394	0.246	-0.219	-0.056	0.189	0.125
58	-1.859	-0.078	-0.011	0.132	0.059	0.116	-0.061
59	2.506	-0.703	-0.013	-0.154	0.725	-0.368	0.117
60	0.072	0.904	0.306	-0.423	0.112	-0.102	0.029
61	0.244	0.561	0.443	0.206	0.202	-0.154	0.050
62	-0.393	-1.063	-0.678	-0.415	1.120	-0.468	0.252
63	1.359	-2.997	-1.590	1.916	-0.362	-0.992	-0.067
64	-1.303	0.082	0.555	0.661	-0.008	-0.337	-0.039

This table is used with the loadings matrix to model the original data set.

Bibliography

Anderberg, M.R. *Cluster Analysis for Applications*. Academic Press, New York, 1973.

Batchelor, B.G. *Practical Approach to Pattern Classification. Plenum, Lon*don, 1974.

Bock, H.H. *Automatische Klassifikation*. Vandenhoeck and Rupprecht, Gottingen, Germany, 1974.

Duda, R.O., and Hart, P.E. *Pattern Classification and Scene Analysis.* Wiley, New York, 1973.

Flury, B. *Common Principal Components and Related Multivariate Models*. Wiley, New York, 1988.

Hartigan, J.A. *Clustering Algorithms*. Wiley, New York, 1975

Malinowsky, E.R., and Howery, D.G. *Factor Analysis in Chemistry*. Wiley, New York, 1980.

Massart, D.L., and Kaufman, L. ***The Interpretation of Analytical Chemical Data by the Use of Cluster Analysis***. Wiley, New York, 1983.
Sneath, P.H.A., and Sokal, R.R. ***Numerical Taxonomy***. Freeman, San Francisco, California, 1973.
Spath, H. ***Cluster Analysis Algorithms. Ellis Ho***rwood, Chichester, 1980.
Steinhausen, D., and Langer, K. ***Cluster Analyse. Wal***ter de Gruyter, Berlin, 1977.
Tryon, R.C., and Bailey, D.E. ***Cluster Analysis***. McGraw-Hill, New York, 1970.

References

1. Kowalski, B.R., Schutzki, T.F., and Stross, F.H. **Anal. Chem.**, 1972, **44**, 2176.

2. Rozett, Richard W. and Petersen, E. M., ***Anal. Chem.***, 1975, **47(8)**, 1301–1308.

3. Bulmer, J.T. and Shurvett, R. ***J. Phys. Chem.***, 1973, **77(2)**, 256-262.

4. Harris, J.T. and Hobbs, M.E. ***J. Am. Chem. Soc.***, 1954, **76**, 1419.

5. Melnick, G. et al. ***Spectrochim. Acta***, 1964, **20**, 285.

6. Malinowski, E.R., and Howery, D.G. ***Factor Analysis in Chemistry***. Wiley Interscience, New York, 1980, p. 30.

7. Malinowski, E.R. and Howery, D.G. ***Factor Analysis in Chemistry***. Wiley Interscience, New York, 1980.

8. For example, Searle, S.R. ***Matrix Algebra Useful for Statistics***. Wiley, New York, 1982.

9. Bradu, D. and Gabriel, K.R. *The Biplot as a diagnostic tool for models of two-way tables*, ***Technometrics***, 1978, **20(1)**, 47.

10. Gabriel, K.R. *Analysis of meteorological data by means of canonical decompositions and Biplots*. ***J. Appl. Meteorol.***, 1972, **11(7)**, 1071.

11. Graham, R.C., Wilson, J. W, Robertson, J.K., and Mohnen, V. *A chemical climatology of the Northeast*, ***Proceedings of the Second New York State Symposium on Acid Deposition***, Albany, NY, November 1983.

12. Graham, R.C., Wilson, J.W., and Robertson, J.K. *Correlation of Intrastorm Sequential Precipitation Chemistry with Storm Meteorology*, ***Proceedings of the 4th International Conference on Precipitation Scavenging, Dry Deposition, and Resuspension***, Santa Monica, CA, November 1982

Appendices

Appendix A -- Listing of Software for Statistical Analysis, graphing and equation solving.

Appendix B -- Problems and Questions

Appendix C -- Statistical Tables

- Normal deviates
- *F*
- Student's *t*
- Poisson
- χ^2
- Quality Control
- Savitzky-Golay Convolution Integers
- Duncan's Separation of Means
- Kolmogorov-Smirnov

Appendix A
Sources of Statistical Analysis Software for Personal Computers

Program Name	Company	Type of Software
Axum	TriMetrix Inc. 444 NE Ravenna Blvd. Suite 210 Seattle, WA 98115 800-548-5633 ext 550	Graphics
SYSTAT	Systat Inc. 1800 Sherman Ave. Evanston, IL 60201 708-864-5670	Statistics
SYGRAPH	Systat Inc. 1800 Sherman Ave. Evanston, IL 60201 708-864-5670	Graphics
W.A.V.E	Electronic Decisions Inc. 1776 East Washington St. Urbana, IL 61801 217-367-2600	Data analysis
Table Curve	Jandel Scientific 2591 Kerner Blvd San Rafael, CA 94901 415-453-6700	Equation fitter

Sigma-Plot	Jandel Scientific	Graphing (also for the Macintosh)
Slide Write Plus	Advanced Graphics Software 333 West Maude Ave. Suite 105 Sunnyvale, CA 94086 408-749-8620	Graphics
Graph-in-the-Box	New England Software Greenwich Office Park #3 Greenwich, CT 06831 203-625-0062	Graphics
GB-STAT	New England Software Greenwich Office Park #3 Greenwich, CT 06831 203-625-0062	Statistics
EasyPlot	Spiral Software 6 Perry St. Suite 2 Brookline, MA 02146	Plotting
Statgraphics	Manugistics Inc. 2115 East Jefferson St Rockville, Maryland 20852 301-984-5094	Statistics
Plot-It	Scientific Programming Enterprises PO Box 669 Haslett, MI 48840 517-339-9859	Plotting and statistics
DaDiSp	DSP Development Corp. One Kendall Square Cambridge, MA 02139 617-577-1133	Plotting and data analysis

MathCAD	MathSoft Inc. 201 Broadway Cambridge, MA 02139 800-MATHCAD	Plotting and data analysis
Temple Graph	Mihalisin Associates 600 Honey Run Road Ambler, PA 19002 215-646-3814	Graphing
MINSQ	Micromath Scientific Software PO Box 21550 Salt Lake City, UT 84121 800-942-MATH	Nonlinear curve fitting
GRAPH	Micromath Scientific Software PO Box 21550 Salt Lake City, UT 84121 800-942-MATH	Plotting; data transformation
CSS	Statsoft Available from Micromath Scientific Software PO Box 21550 Salt Lake City, UT 84121 800-942-MATH	Statistics
SEGS	Advanced MicroSystems 3817 Windover Drive Edmond, OK 73013 800-284-3381	Graphing and plotting
STATA	Computing Resource Center 1640 Fifth Street Santa Monica, CA 90401 800-STATPAC	Statistics, plotting, and graphing
MathGraf	Caren Co. 12137 Midway Drive Tracy, CA 95376 209-835-0295	Plotting, graphing

Macsyma	Symbolics, Inc. 8 New England Exec. Park East Burlington, MA 01803 617-221-1250	Symbolic mathematics
TK Solver	Universal Technical Systems 1220 Rock Street Rockford, IL 61101 815-963-2220	Data analysis
MATLAB	The MathWorks Inc. 21 Eliot Street South Natick, MA 01760 508-653-1415	Data analysis
F-Curve II	LEDS Publishing PO Box 12487 Research Triangle Park, NC 27709 919-477-3690	Curve fitting
FLURP	LEDS Publishing PO Box 12487 Research Triangle Park, NC 27709 919-477-3690	Linear regression
ALLCOR	LEDS Publishing PO Box 12487 Research Triangle Park, NC 27709 919-477-3690	Equation fitting
DATA Doctor	LEDS Publishing PO Box 12487 Research Triangle Park, NC 27709 919-477-3690	Standalone statistical routines

GRAPHER	Golden Sofware Inc. PO Box 281 Golden, CO 80402 303-279-1021	Graphing and plotting
SURFER	Golden Sofware Inc. PO Box 281 Golden, CO 80402 303-279-1021	Surface and topographic plotting
Unistat	Adhoc 29 Brunswick Woods East Brunswick, NJ 08816 800-783-3210 908-254-7300	Statistical
Quick Statistics for Macintosh	Statsoft Tulsa, OK 918-583-4149	Statistics for Macintosh
Presentations	WordPerfect Inc. 1995 North Technology Way Orem, UT 84057 800-451-5151	Drawing and graphing
Formula One	MATHTEC Soft-Sense 12 Rockaway Lane Arlington, MA 02174 617-641-3808	Graphing and equation solving
MINITAB	Minitab, Inc. 3081 Enterprise Drive State College, PA 16801 814-238-3280	Statistics and graphing
SPSS	SPSS, Inc.	Statistics and graphing
SAS	SAS Institute SAS Campus Drive Cary, NC 27330 919-677-8000	Statistics and graphing

Statistica	Statsoft 2325 East 13th Street Tulsa, Oklahoma 74104 918-583-4149	Statistics and graphing
Origin	Microcal Software, Inc. 22 Industrial Drive East Northhampton, MA 01060 800-969-7720	Technical graphics and data analysis
StatView	Abacus Concepts 1984 Bonita Avenue Berkeley, CA 94704-1038 510-540-1949	Data analysis and graphing for the Macintosh

APPENDIX B

Problems and Exercises

1. One of the methods of constructing joints in metals is to use a technique called ultrasonic welding. A set of data [1] on the shear strength, expressed in pounds, of the welds produced by ultrasonic welding on aluminum clad sheets follows:

5434	5299	4848	4308	5095	4848	4925	4965	4698	5764
4948	5288	5378	5498	5273	5189	4918	4786	5227	5582
4521	4599	5260	4723	5069	4417	4173	4803	4493	5388
4570	4886	5055	5173	5207	5076	5078	5518	5069	5245
4990	5043	5828	5049	4974	5419	5205	5640	5333	4740
5702	4820	5218	5256	4568	5653	4774	4931	4951	5461
5241	4381	4859	5342	5275	4592	5364	4452	4500	5679
5112	5670	4780	5133	4681	5621	4986	4900	5170	5164
5015	4637	5027	4772	4823	4755	5089	5296	5248	5188
4659	4806	5008	4609	5042	4618	5001	5138	5555	5309

Prepare:

a. A stem-leaf diagram of the data.
b. A frequency distribution plot (histogram) of the data.
c. A cumulative frequency distribution plot of the data.

Compute:

d. The mean of the data.
e. The standard deviation of the data.
f. 95% confidence limits on the mean of the data.

Determine if the data are normally distributed using:

a. normal probability plot
b. χ^2 test
c. Kolmogorov-Smirnov test

2. A common experience in a freshman laboratory is to issue to a student a sample of a compound that will dissolve in water. If the pH of the resulting solution is measured by a randomly selected group of students ($n=25$), compute the probability:

$$P\left(-0.100 \le \frac{\overline{pH} - \mu_{pH}}{s} \le 0.100\right)$$

The data are assumed to be normally distributed with a mean of 5.00 and a variance of 0.04.

3. Sea scallops bioabsorb heavy metals such as cadmium when exposed in the natural environment. The amount they take up is quite variable. A number of scallops from different locations in North Atlantic waters were analyzed for cadmium content and found to have the following concentrations (mg of cadmium per kg of scallop):

13.1	8.4	16.9	2.7	9.6	4.5	12.4	5.5	12.7	17.1	10.8	18.9	27.0
18.0	6.4	13.1	8.5	7.5	12.1	8.0	11.4	5.1	5.5	9.6	5.0	10.1
4.5	7.9	7.9	8.9	3.7	9.5	14.1	7.7	5.7	6.5	10.8	14.7	14.4
5.1												

Compute:

a. The mean of the data.

b. The standard deviation of the data.

Construct:

c. A histogram of the data

d. A stem-leaf diagram of the data.

Compute:

e. 10% trimmed mean of the data.

f. The standard deviation of the trimmed data.

g. Is there a significant difference between the standard deviation of the trimmed data and the full data set?

4. Density of sediments in rivers and lakes varies greatly from one region to another across the world. In one particular region, the sediment was sampled (n=25) and the density determined. If the average density of the sample (n=25) was 2.8 g/mL, and the variance was 0.062 g^2/mL^2, what is the probability that the density is at most 3.2? What is the probability that the density is between 2.1 and 2.9 g/mL?

5. Polarographic techniques work best in a relatively high ionic strength aqueous solution. An attempt to evaluate seawater as the supporting electrolyte for direct current polarography led to the following data:

	Standard Number					
	1	2	3	4	5	6
Molarity x 10^5	1.0	2.0	3.0	4.0	5.0	10.0
mg/L	1.124	2.248	4.556	5.620	9.112	11.24
$E_{½}$ (V)	-0.560	-0.545	-0.573	-0.560	-0.578	-0.583
i_d (mA)	0.045	0.103	0.248	0.313	0.504	0.653

a. Prepare a calibration curve for the data using the concentration as the independent variable and the diffusion current (i_d) as the dependent variable.

b. Does the intercept indicate a possibility of a bias in the method? Justify your answer.

c. If analysis of a seawater sample gave a diffusion current reading of 0.320 mA, what would be the concentration?

d. What are the confidence limits on the concentration calculated in part (c)?

6. Chemical Engineers are often concerned about distillation rates in distillation towers. One of the methods of measuring the efficiency of a distillation tower is to measure the rate at which steam is produced. The following steam production rates were reported in the literature[2]:

1170	1620	1495	1530	1170	1710	1710	1530	1440	1260	1440
1800	1170	1260	1440	1170	1260	1530	1800	1540	1260	1440
1710	1440	1170	1170	1640	1800	1800	1530	1170	1350	1800
1530	1170	1440	1350	1530	1260	1350	1350	1350	1350	1440
1170	1710	1620	1440	1350	1730					

Prepare:

a.. A stem-leaf diagram of the data.

b. A frequency distribution plot (histogram) of the data.

c. A cumulative frequency distribution plot of the data.

Compute:

d. The mean of the data.

e. The standard deviation of the data.

f. 95% confidence limits on the mean of the data.

7. Verify that frequency data for the distribution of means of sample size, $n=25$ may be adequately described by a normal distribution.

8. Table C-1 has 3 columns of rainfall data (date sample was collected, mass of rain collected, and mm precipitation recorded on a rain gauge) from the Bradford Forest in Florida.[3] Propose a model that relates the mass of precipitation to the mm precipitation. Test the model and determine goodness of fit parameters. What is the theoretical value of the b_0 coefficient of the model? Test to determine if the theoretical value of the coefficient is statistically different than the actual value.

9. Use the mm precipitation and mass of precipitation data from Table C-1 to

a. Construct a normal probability plot. Are you justified to indicate that the data are normally distributed? If there is reason to believe that data are not normally distributed, transformations such as

$$x' = \sqrt{x}$$
$$x' = \log_{10}(x)$$

may yield data that is normally distributed. If you do not feel the data are normally distributed, suggest and test possible transformations of the data that might lead to a normal distribution of the transformed data.

10. A major portion of a quality control program in a laboratory is to ensure that all of the pieces of equipment in the laboratory are in calibration. One of the methods to calibrate a piece of equipment such as a balance is to weigh a specimen that has a known mass. Suppose that 25 weighings of a 100 g reference standard were made. A standard deviation of 0.01 g was obtained.

a. What would you propose as a hypothesis for this metrological exercise?

Table C-1. Rainfall data from the Bradford Forest in Florida

Date	mass	mm	date	mass	mm	date	mass	mm
860513	332.9	5.08	870602	16.5	7.00	880607	1224.0	17.78
860520	2180.3	30.48	870609	2784.4	40.64	880614	995.7	14.73
860603	110.3	8.64	870616	71.6	1.02	880621	0.0	0.00
860610	2209.4	33.53	870623	1808.8	25.40	880628	1156.2	17.78
860617	3322.6	48.77	870630	2733.6	38.61	880705	1943.6	29.72
860624	5012.2	71.12	870707	1356.0	20.07	880712	315.7	4.57
860701	1.3	0.02	870714	6814.0	116.33	880719	1526.5	21.59
860708	3915.1	58.42	870721	599.2	8.38	880726	2855.0	41.91
860715	436.4	6.35	870728	2148.7	32.51	880802	1835.3	27.43
860722	1081.6	16.51	870804	920.2	14.99	880809	10366.9	154.69
860729	2256.7	31.24	870811	1410.3	21.08	880818	6734.6	135.89
860805	1928.5	27.69	870818	8378.2	139.45	880823	75.3	1.27
860812	0.0	0.06	870825	1512.6	21.59	880830	5615.9	83.06
860819	5106.7	76.20	870901	42.5	0.51	880906	11008.9	160.27
860826	4027.9	57.40	870908	2893.3	43.18	880913	4460.4	66.04
860902	2267.0	34.29	870915	4762.3	70.87	880920	572.4	9.40
860909	1607.2	23.62	870922	752.4	10.16	880927	3957.2	59.18
860916	2100.9	30.23	870929	0.0	0.00	881004	2474.2	36.07
860923	915.6	12.70	871006	346.2	5.08	881011	0.0	0.00
860930	118.9	1.78	871013	41.5	0.12	881018	0.0	0.00
861007	0.8	0.00	871020	0.0	0.00	881025	361.9	7.62
861014	6113.1	86.61	871027	0.0	0.00	881101	45.0	0.76
861021	449.6	6.10	871103	98.2	1.27	881108	1266.3	19.56
861028	482.0	7.11	871110	7064.3	105.41	881115	0.0	0.00
861104	1542.3	22.61	871117	18.3	0.00	881122	142.5	1.52
861111	0.0	0.01	871124	5114.7	77.47	881129	3413.5	48.77
861118	139.2	2.29	871201	159.9	2.29	881206	5.2	0.00
861125	513.9	7.62	871208	0.0	0.00	881213	1416.6	19.56
861202	3456.6	52.07	871215	141.0	1.78	881220	71.9	1.02
861209	1115.5	16.51	871222	1022.5	15.24	881227	0.0	0.00
861216	2738.7	39.37	871229	170.2	2.54	890103	27.8	0.51
861223	177.5	2.54	880105	265.4	4.32	890110	0.0	0.00
861230	2065.5	31.24	880112	1125.6	29.72	890117	0.0	0.00
870106	2828.9	54.61	880119	81.8	1.52	890124	1560.4	25.65
870113	971.6	13.46	880126	8229.9	124.97	890131	0.0	0.00
870120	2036.2	29.21	880202	27.0	1.02	890207	1.5	0.00
870127	4721.6	66.29	880209	927.1	13.97	890214	0.0	0.00
870203	987.5	13.21	880216	1225.1	17.78	890221	1.1	0.00
870210	3277.9	50.55	880223	5565.6	91.95	890228	1241.0	17.78
870217	592.7	8.89	880301	0.0	0.00	890307	2425.1	36.83
870224	7637.7	113.54	880308	5134.5	76.96	890314	1.7	0.00
870303	1508.8	22.35	880315	1763.5	30.48	890321	0.0	0.00
870310	2254.6	32.77	880322	2242.8	34.29	890328	806.3	11.43
870317	50.5	1.02	880329	535.1	6.35	890404	402.0	5.59
870324	481.1	7.11	880405	8.7	0.25	890411	2892.5	43.18
870331	7192.6	107.19	880412	265.9	4.32	890418	1887.5	27.94
870407	375.7	6.10	880419	626.1	8.89	890425	112.7	1.27
870414	0.0	0.00	880426	26.2	1.52	890502	1855.3	27.18
870421	481.3	6.60	880503	1699.1	30.48	890509	4.5	0.00
870428	732.5	10.16	880510	0.0	0.00	890516	689.1	9.40
870505	2.1	0.00	880517	2072.2	29.97	890523	426.5	5.59
870512	1052.6	14.73	880524	0.0	0.00	890530	514.6	6.35
870519	1551.1	22.10	880531	0.0	13.46	890606	3925.2	57.15
870526	421.9	6.60						

b. What is the probability that recalibration will be accepted if recalibration is required if the average of the weighings is more than 0.1% from the true mass of the reference standard (100.000 g)? (Hint: calculate the probability of a Type II error.)

c. What is the probability that recalibration will be performed unnecessarily if the true mass of the reference standard is 100.11 g? What is the probability that recalibration will be performed unnecessarily if the true mass of the reference standard is 98.49 gram?

11. Flow in a stream is often measured with V-Notch and rectangular weirs. The following data were obtained from a stream for several summers (each value represents 1000s of acre feet for the period):

127.96	210.07	203.24	108.91	178.21	285.37	100.85
89.59	185.36	126.94	200.19	125.86	117.64	204.91
94.33	311.13	302.74	114.79	66.24	247.11	299.87
109.64	85.54	330.33	109.11	280.55	145.11	95.36
477.08	262.09	150.58				

a. Is the normal distribution a reasonable distribution for this data? Justify your answer.

b. A possible alternative distribution is the log-normal distribution. Is the log-normal a reasonable distribution for the data? If it is, estimate the parameters of the distribution. Remember that a log-normal distribution is that distribution of the logarithm of the random variable [$\log_{10}(X)$] is normally distributed with mean μ and variance σ^2.

12. An article in ***Journal of Chromatography*** reports a method to determine cibenzoline in blood using gas chromatography and mass spectrometry. Three repeated analyses of a blood sample that had 5 ng of the cibenzoline added resulted in a sample mean of 4.59 ng with a standard deviation of 0.08 ng. Is there sufficient evidence to indicate that too little cibenzoline was being detected by the method at the α=0.05 significance level?

13. Marijuana smoking leads to increased levels of two major cannabinols,

Δ^9 tetrahydrocannabinol (Δ^9 THC)

11-hydroxy-Δ^9 tetrahydrocannabinol(11-OH-Δ^9 THC)

Dose required to effect μg drug/kg of body weight	
Δ^9-THC	11-OH-Δ^9-THC
19.54	25.89
26.39	20.53
11.49	15.52
16.00	15.95
14.47	14.18
24.83	16.00

Calculate a 95% confidence interval for the difference between the two drugs, Δ^9-THC and 11-OH-Δ^9-THC.

14. High concentrations of metals in culinary and drinking water may pose health hazards and may also impart unusual taste characteristics to the water. In 1982, a study[4] was conducted to see if a difference would be seen if the water in a river was sampled at the surface and at the sediment-water interface. The samples were analyzed for zinc at each location.

[Zn], mg/L	
Surface	Sediment–water interface
0.430	0.415
0.266	0.238
0.567	0.390
0.531	0.410
0.707	0.605
0.716	0.609

Do the data suggest that the concentration is higher at the surface or at the sediment–water interface?

15. If a significance test is being performed at the $\alpha=0.10$ level, which of the following *P*-values would result in rejection of the null hypothesis?

0.001 0.50 0.35 0.09 0.05 0.78

16. Two aliquots of sewage were each treated with a different bacterial culture to determine the effect on the chemical oxygen demand (COD).[5] The average for one culture (n=14) was 18.1 mg/L with a standard deviation of 5.0 mg/L. The mean for the other culture (n=16) was 15.9 mg/L with a standard deviation of 6.0 mg/L. Is there justification to choose one of the cultures over the other? Hint: the null hypothesis is μ_1-μ_2=0 with an alternative hypothesis that μ_1-$\mu_2 \neq 0$.

17. The pH of soil may be measured by mixing an aliquot of the soil with a fixed volume of water and determining the pH with a glass electrode. Ten samples from two different locations on an experimental farm[6] were measured in this manner. The data yielded for the two locations:

pH			
Location 1	Location 2	Location 1	Location 2
8.53	7.23	7.89	7.27
8.52	7.35	7.85	7.40
8.01	7.58	7.82	7.27
7.99	7.73	7.93	7.30
7.80	7.85	7.87	7.82

Does the true mean pH of the soil differ in the two locations? Repeat the calculations based on ion concentration rather than pH. Do your conclusions differ? Why?

18. The vanadium concentration in a National Bureau of Standards orchard leaves (SRM 1571) was determined by five different methods.[7] The results of the analyses were:

Analytical Method	Mean Value µg/kg	Standard Error of the Mean	Number of Samples
Dry ashing	410	15	7
Wet ashing — preirradiation separation	408	16	0
Schöninger combustion	435	20	12
Wet ashing-radiochemical	377	10	2
No separation	598	32	6

Determine if there is reason to believe that the methods do not give comparable results at the 95% significance level.

19. The following data were obtained for the analysis of N-nitrosoamines by liquid chromatography with polarographic detection[8]:

pH	Peak Height, μA	pH	Peak Height, μA
2.57	9.90	5.06	10.00
3.06	10.20	6.06	10.00
3.56	9.90	7.02	9.90
4.06	10.00	8.07	9.90
4.56	9.90		

Is there sufficient reason to not reject a null hypothesis that the slope=0 at the 95% confidence level?

20. National Bureau of Standards SRM-1645 was analyzed to determine the amount of transition metals in the sediment standard.[9] Use Duncan's separation of means test to determine if the proposed method of extraction gives acceptable levels compared to the values reported by the National Bureau of Standards.

	Concentration of Transition metal μg/g of sediment unless stated otherwise								
	Cr	Mn	Fe	Co	Ni	Cu	Zn	Cd	Pb
Mean (n=16)	2.8%	750	9.8%	6	46	109	1740	8.2	690
s_x/mean (%)	7	2	3	8	8	6	2	6	4
NBS Values	2.96%	785	11.3%	8	45.6	109	1720	10.2	714

21. Many measurement techniques are based on physical or chemical relationships of the form:

$$c = k \cdot m$$

where c is a property of the sample, k is a proportionality constant and m is the measurement. For example, if k is Hook's constant, and m is mass, then c is the elongation of a spring. These type of relationships are often calibrated using a single point. Comment on the validity of this practice. Suggest a way to measure the validity of the calibration.

22. Should time be considered as a major factor in the analysis of trace metals using the following data?[10]

	Concentration of Metal Ion (μg/L) Mean $\pm$ Standard Deviation (n=3)		
Metal	Freshly Sampled and Extracted	10 Day Old Sample before Extraction	Freshly extracted Sample, 10 Days before Analysis
Cd	5.0±0.1	5.0±0.1	4.9±0.1
Co	25.0±0.2	24.7±0.3	24.4±0.5
Cu	25.0±0.4	25.8±0.2	25.2±0.5
Ni	25.0±0.6	24.5±0.4	24.2±0.6
Pb	25.0±1.1	25.3±0.4	24.7±0.4
Zn	25.0±0.2	25.2±0.2	24.6±0.6

23. An acid extraction method of analysis was compared to a fusion method to determine several metals in samples of Chesapeake Bay Mud.[11] Use a paired t-test to determine if there is reason to reject a null hypothesis that the methods give identical results at the 95% confidence level.

	Concentration of Metal μg/g unless otherwise stated					
	Cr	Mn	Fe	Ni	Cu	Zn
acid extraction (n=8)	75	1300	4.6%	53	59	450
%Coefficient of variation	10	5	3	3	7	3
Fusion (n=6)	100	1000	4.8%	91	67	420
%Coefficient of variation	6	5	2	2	9	3

24. A laboratory uses HPLC to assay the concentration of dioxamine in a liquid. The laboratory manager is concerned about the overall precision of the analytical method and has designed a factorial experiment with replication in an attempt to determine the source(s) of lack of precision. The analysts were chosen at random from all of the analysts in the laboratory, as were the days used in the analysis of the results. The results of the precision experiment are given below:

Analyst	Replicate	Dates of Analysis 3/22	4/2	4/4	4/9
A	1	28.1	26.9	26.9	26.9
	2	27.7	27.0	27.9	27.5
	3	28.8	26.8	27.4	26.9
B	1	26.6	26.1	25.9	26.3
	2	27.2	25.8	25.9	26.5
	3	26.9	26.7	26.5	26.4
C	1	28.1	27.0	27.2	27.5
	2	28.4	27.5	26.6	27.2
	3	28.4	27.4	26.9	27.8

a. Obtain the ANOVA table for such an experimental design and assign a percentage of the total variability of the method to the possible source(s) of variation. Be sure to calculate the standard deviation for each source of variability.

b. If the true value of the dioxamine content of the liquid that was analyzed was 28.5, comment on the overall accuracy of the method.

c. If the true value of the dioxamine content of the liquid that was analyzed was 28.5, calculate the probability that the control procedure would have accepted the hypothesis that the true mean was 27.0.

25. The following calibration data were obtained for the analysis of an Arochlor PCB.

PA, peak area C, concentration, μg/L											
PA	C	PA	C	PA	C	PA	C	PA	C	PA	C
5.0	3.5	10.0	6.4	20.0	12.0	30.0	18.2	50.0	31.0	70.0	43.1
5.0	3.7	10.0	6.4	20.0	12.1	30.0	18.2	50.0	30.8	70.0	43.6
5.0	3.6	10.0	6.3	20.0	12.0	30.0	18.3	50.0	30.7	70.0	43.3
5.0	3.8	10.0	6.3	20.0	12.0	30.0	18.1	50.0	31.0	70.0	43.9

b. Use the calibration data to obtain the minimum detectable limit (α=0.05) and the detection limit (β=0.05) for the analysis.

c. determine if the variance is constant over the range of the regression. What is the effect if the variance is not constant?

d. The precision may be expressed as a confidence interval about individual points. Calculate the *t*-interval about each of the responses. Compare this to the confidence interval (that includes the individual *y*-values) calculated about the regression line.

e. A subsequent analysis of a sample containing the Arochlor was performed. The average peak area was found to be 37.4. Calculate the amount of Arochlor in the sample and determine the 95% confidence limit on the concentration.

26. Calculate the first five smoothed data values (Y_1^*, Y_2^*, Y_3^*, Y_4^*, and Y_5^*) for the data set below. Use a five-point smooth for the first-order polynomial smooth. Show pertinent equations and calculations.

	Data point number						
	1	2	3	4	5	6	7
Signal	28.3	29.4	28.1	28.7	28.4	29.3	27.6
Time (seconds)	1	2	3	4	5	6	7
Smoothed data point	?	?	?	?	?	?	?

27. An analyst has obtained the following data for an experiment to determine the amount of arsenic in hair using arsine (AsH_3) generation in an atomic absorption spectrophotometer. The hair was dissolved in an acid solution. Hair obtained from an individual who was never exposed to arsenic was used as a blank. The method of standard additions was used to quantitate the amount of arsenic in the samples. The original sample of 100 mL volume was split into 5 equal aliquots. The following data was obtained from the experiments.

Concentration of the Standard, 200 mg/L Volume of Each Addition = 20 μL		
Sample	Absorbance	Corrected Absorbance
Blank	0.130	
Hair Sample	0.250	
Hair + 1 addition	0.370	
Hair + 2 addition	0.485	
Hair + 3 addition	0.604	
Hair + 4 addition	0.728	

Complete the Corrected Absorbance column in the table and determine the amount of arsenic in the hair sample.

28. A set of data for the content of gold in three placer deposits exhibits several extreme values. Determine if the content of gold in each of the deposits is the same. Suggest some possible reasons for the existence of extreme values.

	Placer deposit		
Sample number	**A**	**B**	**C**
1	8.3	6.4	10.3
2	6.7	3.4	2.1
3	14.5	3.5	11.5
4	5.6	2.3	9.4
5	4.5	2.1	8.7
6	8.3	6.8	13.5
7	18.4	22.0	3.4
8	10.3	3.4	2.3
9	2.5	18.2	8.4
10	3.4	3.6	9.3

29. You have been given a budget of $2000 to complete a sampling and analysis program to determine the average concentration of lead in the interstitial waters of sediments at the confluence of the Susquehanna River with the Chesapeake Bay. Presumably the major source for the lead has been through the incorporation of the lead into the sediments from lead aerosols that have been removed from the atmosphere. Each analysis will cost $75 to perform and it will

cost $100 to acquire each sample. What is the minimum variance that can be expected for such a sampling program? Assume $(\sigma_m)_A = 0.05$ and that $(\sigma_m)_S = 0.70$. You must take at least 12 samples.

30. The following values were obtained for the normality of a solution of $KMnO_4$ when pure KI and pure As_2O_3 were used as primary standards.

Normality using KI	*Normality* using As_2O_3
0.13110	0.13118
0.13126	0.13113
0.13120	0.13119
0.13113	0.13125
0.13129	0.13128
0.13108	

a. Calculate the variance, standard deviation, mean, coefficient of variation and standard error of the mean for each set of data.

Data Set	Variance	Standard Deviation	Mean	Coefficient of Variation	Standard Error of the Mean
KI					
As_2O_3					

b. Calculate the 95% confidence limits for each data set.

c. Do the two means differ significantly? Show all work to support your answer. Explicitly state any assumptions and hypotheses that are used.

d. If the true normality of the $KMnO_4$ is 0.13000 *N*, calculate the relative accuracy for the two determinations of the normality by KI and by As_2O_3.

31. Discuss briefly six requirements for a sample to be a "good" sample.

32. You have reason to suspect that a set of peaks in a mass spectrum is from a tetrachloro-substituted hydrocarbon. Calculate the integer coefficients for such a substitution and determine if it fits the experimental data ratio of

$$(M:M+2:M+4:M+6:M+8 \mid \mid 80:110:52:11:1).$$

33. The following data (Table 1) resulted from several years of analysis of methoxychlor using a gas chromatographic technique. The values in Table 2 were obtained from the data in Table 1.

a. Fill in the blank values in Table 2 and determine whether the data are normally distributed. (Show pertinent calculations.)

Table 1. Data for Several Years of Methoxychlor Analysis by a Gas Chromatographic Technique (units μg/L)

34.75	36.00	36.51	37.25	36.89	37.56	37.92	38.17	38.60	39.25	34.95
36.00	36.52	36.91	37.26	37.60	37.93	38.25	38.61	39.25	34.98	36.03
36.53	36.92	37.29	37.60	38.25	38.61	39.26	37.94	35.40	36.04	36.55
36.92	37.30	37.62	37.94	38.25	38.61	39.26	35.45	36.06	36.62	36.94
37.33	37.64	37.95	38.29	38.69	39.30	35.46	36.06	36.65	36.96	37.34
37.65	37.97	38.30	38.71	39.33	35.46	36.15	36.65	37.00	37.34	37.66
37.97	38.31	38.76	39.44	35.50	36.18	36.68	37.10	37.36	37.66	37.98
38.34	38.77	39.58	35.66	36.20	36.68	37.11	37.39	37.69	37.99	38.35
38.77	39.61	35.67	36.27	36.68	37.12	37.41	37.70	38.00	38.37	38.81
39.62	35.72	36.29	36.70	37.12	37.43	37.72	38.02	38.38	38.84	39.64
35.79	36.31	36.72	37.13	37.45	37.74	38.44	38.86	39.68	35.85	36.32
36.75	37.15	37.45	37.75	38.07	38.05	38.94	39.88	35.85	35.87	36.33
36.75	37.18	37.45	37.77	38.08	38.47	38.95	39.88	35.90	36.33	36.79
37.20	37.48	37.84	38.11	38.49	38.99	40.05	35.90	36.40	36.79	37.22
37.48	37.86	38.12	38.49	39.07	40.18	35.96	36.42	36.81	37.24	37.49
37.86	38.15	38.50	39.07	40.21	35.98	36.44	36.83	37.24	37.55	37.86
38.16	38.53	39.15	40.25	35.99	36.50	36.86	37.25	37.56	37.87	38.17
38.60	39.23	40.72	35.44	36.05	36.58	36.94	37.33	37.63	37.95	38.27
38.66	39.29									

b. For the data in Table 1 and Table 2, show schematically a histogram and a box whisker plot. Be sure to give numbers showing the pertinent calculations.

Table 2. Pertinent values to determine if values for the random variable, concentration of Methoxychlor, are normally distributed. $n=200$ $mean=37.50$ $s=1.25$

End of Interval	Observed Frequency	Z-transform	P(X≤x)	p	Expected Frequency
<35.0	3	-2.0	0.0228	0.0228	4.56
35.5	6	-1.6	0.0548	0.0320	6.40
36.0		-1.2	0.1151	0.0603	12.06
36.5	19		0.2119	0.0968	19.36
37.0	28	-0.4	0.3446		26.54
37.5	30	0.0		0.1554	
38.0		0.4	0.6554	0.1554	31.08
38.5	28		0.7881	0.1327	26.54
39.0	18	1.2	0.8849		19.36
39.5	12	1.6	0.9452	0.0603	
40.0	7	2.0		0.0320	6.40
>40.0	5		1.00	0.0228	4.56

34. An analyst obtained the data for the atomic weight of cadmium:

112.23 112.34 112.30 112.22 112.32 112.34

Calculate mean, variance, standard deviation, and coefficient of variation.

35. For the data in problem 34,

    a. Calculate the 95% confidence limits on the true mean.

    b. Does the experimental mean agree with the accepted value of 112.41?

36. For the data in problem 34, calculate the accuracy of the results based on an accepted value of 112.41.

37. A set of standards was prepared to calibrate a spectrophotometer for the analysis of iron (II) in water. Three absorbance readings were obtained for each standard.

$[Fe^{2+}]$, mg/L	Absorbance Readings		
2.0	0.195	0.204	0.205
3.0	0.295	0.300	0.305
4.0	0.388	0.398	0.398
5.0	0.485	0.490	0.495
6.0	0.585	0.590	0.590
7.0	0.680	0.685	0.590
8.0	0.785	0.790	0.785
9.0	0.875	0.880	0.880
10.0	0.970	0.970	0.970

a. Determine the linear least–squares best estimate fit of the data to a first–order polynomial model.

b. Is the intercept equal to 0? Justify your answer.

c. If the intercept is equal to zero, how should the regression equation (calibration curve) be expressed?

d. Determine the correlation coefficient. Is the correlation significant at the 95% confidence level?

e. Three determinations of a solution (250.00 mL) containing an unknown amount of iron were accomplished. The average of the three determinations was 0.587. Estimate the iron content of the sample. Estimate the 95% confidence limits for the average value and for the true mean value.

f. The iron for the liquid sample was obtained from a sample of ore that weighed 15.40 g. If the concentration of the iron in the ore must be 100 mg Fe/kg ore to be economically feasible to mine the ore, is the ore worth mining? (Hint, statistically test the calculated amount against the cutoff.)

38. A laboratory uses several different analysts and spectrophotometers to determine chloride in aqueous solutions. Three analysts and three spectrophotometers were chosen at random to determine which has the larger effect on the variability of the results. It was also desired to determine if there is a significant interaction effect between the analyst and the spectrophotometer; therefore replicate determinations were made for each combination of analyst and spectrophotometer.

Chloride Concentrations Determined from Spectrophotometer Analysis

Analyst	Spectrophotometer 1	2	3
1	2.3	3.7	3.1
	3.4	2.8	3.2
	3.5	3.7	3.5
2	3.5	3.9	3.3
	2.6	3.9	3.4
	3.6	3.4	3.5
3	2.4	3.5	2.6
	2.7	3.2	2.6
	2.8	3.5	2.5

a. Calculate the percent contribution to the total variability of the method for each of the possible effects. Be sure to give the entire applicable ANOVA table. Pool variances where possible. Report the variance (or standard deviation) for each of the effects. Identify any significant effects. Discuss the ramifications of the results.

39. A new technician in the laboratory has just designed a new detector to determine the water content in antifreeze (ethylene glycol). It is claimed that the detector is good over the range of water from 0.1 to 99.0%. If this is true, this

detector would be very useful in maintenance of the fleet of vehicles owned by the laboratory since it would decrease the amount of time needed to check the freezing point of the coolant in the vehicles. If it is true, the new technician should get a raise. From the data, decide if the technician should be given a raise or not. Base your decision on whether the precision is constant over the range claimed. To test the claim, a series of samples with known contents of water were analyzed with the detector. The following are the results of the those determinations.

Water Level, %	Determinations, %
0.1	0.1500 0.0692 0.0823 0.1033 0.0751
1.0	0.9669 1.0433 0.9999 0.9744 0.9681
10.0	10.0500 9.9950 9.991110.020210.0701
25.0	25.070224.991225.049924.995125.0203
50.0	50.050149.995050.020050.070049.9913
100.0	99.85 99.88 99.90 99.86 99.91

40. Determine, by calculating the Mahalonobis distance, whether the data record for the sample collected on 840103 should be considered as part of the data set or whether it is anomalous.

Date On	pH	NO_3^-	SO_4^{2-}	Cl^-	NH_4^+	Ca^{2+}	Mg^{2+}
830920	4.60	0.82	1.30	0.17	0.10	0.13	0.038
830927	4.70	0.57	0.79	0.38	0.03	0.05	0.035
831108	4.17	1.79	1.65	0.19	0.23	0.04	0.17
831122	4.31	1.11	1.65	0.29	0.09	0.04	0.125
831129	4.43	1.21	1.17	0.23	0.10	0.04	0.018
831206	4.44	0.79	1.33	1.71	0.08	0.05	0.118
831313	4.60	0.94	1.36	1.46	0.06	0.05	0.106
831220	4.54	0.49	0.89	0.24	0.02	0.02	0.020
831227	3.87	0.52	1.30	0.26	0.07	0.05	0.019
840103	4.55	15.11	5.00	1.32	0.52	0.65	0.142
840110	4.53	2.79	2.04	0.52	0.16	0.13	0.110
840313	4.53	0.78	1.18	0.05	0.05	0.06	0.017
840320	4.13	2.38	3.23	1.12	0.22	0.64	0.112
840327	4.40	0.17	1.05	0.29	0.03	0.25	0.069
840403	4.41	0.48	1.57	0.39	0.09	0.03	0.028

840410	4.32	1.18	1.45	0.12	0.17	0.09	0.022
840417	4.00	4.30	3.05	0.23	0.28	0.31	0.080
840522	4.30	1.21	1.87	0.28	0.21	0.06	0.030
840626	4.07	2.32	2.30	0.10	0.18	0.04	0.015
840703	4.18	1.62	2.13	0.17	0.19	0.06	0.022
840710	3.99	1.66	3.59	0.11	0.22	0.07	0.016

41. Using the data from problem 40:

a. Calculate the covariance matrix (about the mean and about the origin) for the data.

b. Calculate the correlation matrix (about the mean and about the origin) for the data. Remember that the element in the correlation matrix is obtained by dividing the corresponding element in the covariance matrix by the product of the two standard deviations of the row and column.

c. Obtain the vector of eigenvalues using the equation:

$$|\rho - \lambda \cdot I| = 0$$

where ρ is the correlation matrix, λ is the vector of eigenvalues, and I is the identity matrix.

d. Once you have obtained the vector of eigenvalues, calculate the first and second principal components.

e. Plot the first and second principal components against each other to determine if there are some groupings that can be made of the analytes. Interpret those groupings by giving a possible rationale for each grouping.

f. Calculate the percentage contribution to the variability of each of the principal components.

42. When using the generalized standards addition method, it is generally assumed that volume changes are negligible on addition(s) of the standard. There are situations where this assumption is not valid. Derive an expression for the K matrix that can be used that is independent of volume. (Hint, consider that the R matrix can be written as

$$R_i = \sum_{s=1}^{r} \frac{N_s}{V} k_{si}$$

where N_s is the total number of moles of analyte s in the volume, V.)

43. In an example in Chapter 14, the generalized standard addition method was used to calculate the concentration of four metals in an aqueous solution. The results that were obtained were

	Cr	**Co**	**Ni**	**Cu**
N_o	0.006	1.014	1.018	1.002

Are the results acceptable at the 5% significance level if the accepted values for the concentration are

	Cr	**Co**	**Ni**	**Cu**
N_o	0.000	1.000	1.000	1.000

44. An alternative method of calculating concentrations with the generalized standard addition method is to record the cumulative **total** added concentration rather than the incremental added concentration. Such a data analysis is given in the following data matrices for the analysis of Ni, Cr, Co, and Cr in an aqueous solution. Determine the concentration of the metals in the original solution. The response matrix at four different wavelengths for the original solution is

Q_0			
Ni	**Cu**	**Co**	**Cr**
0.546	0.861	0.500	0.117

The response and cumulative concentration added matrices are given by

ΔQ				N Cumulative added			
Ni	Cu	Co	Cr	Ni	Cu	Co	Cr
0.491	0.058	-0.01	0.0422	1	0	0	0
0.979	0.093	-0.012	0.0822	2	0	0	0
1.48	0.198	0.004	0.0151	3	0	0	0
1.99	0.266	0.018	0.209	4	0	0	0
1.98	1.81	0.005	0.265	4	2	0	0
2.00	3.28	0.002	0.326	4	4	0	0
2.00	4.88	0.017	0.413	4	6	0	0
2.04	6.44	0.044	0.507	4	8	0	0
2.16	6.52	1.99	0.565	4	8	4	0
2.17	6.39	3.85	0.502	4	8	8	0
2.32	6.61	5.77	0.606	4	8	12	0
2.54	6.70	7.80	0.733	4	8	16	0
4.22	6.57	7.87	8.71	4	8	16	8
5.85	6.18	7.90	16.6	4	8	16	16
7.70	6.38	8.06	24.9	4	8	16	24
9.99	6.58	8.45	32.4	4	8	16	32

45. Give examples of enumerative, continuous, and discrete data. Defend your examples.

46. Why is it necessary to have an understanding of the following attributes of a data set?

a. Measure of central tendency (e.g., mean, median or mode).
b. Measure of variability.
c. Measure of the distribution.

47. Why is it necessary for a data set to be *representative*?

48. Why is it necessary for a data set to be *repeatable*?

49. Why is it necessary for a data set to be *accurate*?

50. Discuss the meaning of *data comparability*. Include in the discussion the need for comparable means, standard deviations, and distributions.

51. The following table was generated by a student who did not remember all of the concepts that related variance, standard deviation, standard deviation of the mean, and the coefficient of variation. Help complete the table and identify those entries that cannot be completed with the data given.

s_x	n	s_x^2	$\bar{x}$	Coefficient of Variation	$s_{\bar{x}}$
3.87	16	___	5.20	___	___
___	___	6.10	___	4.17	1.34
___	5	___	18.00	3.21	___
2.47	___	___	18.32	___	___

52. Discuss the concept of *statistical control* and its relation to overall quality of data.

53. Given the following data, use a normal probability plot to decide if the data are normally distributed.

18.35 18.70 18.20 18.60 18.50 18.65 18.35 18.45 18.55

54. Explain what is meant by a *normal distribution*. Give several examples of data sets that are normally distributed.

55. What is the minimum number of statistics that must be reported for a data set? What are they?

56. The following statement could be found in a technical report:

> "A sample of particulates in air gave a value of 3.84 fibers of asbestos per m^3 with a standard deviation of 0.03 asbestos fibers per m^3".
>
> Critically discuss the statement.

57. An analyst obtained the following values for the concentration (mg/L) of dioxin in a sample of river water:

18.25 18.67 18.14 18.34 18.62 18.25 18.50 18.56 18.45

What is the probability that the analyst will make a Type II error if the true concentration of dioxin is 18.85?

58. Two analysts have each analyzed a standard reference material to ascertain their relative precisions. If analyst A analyzed the SRM eight times with a standard deviation of 0.23 and analyst B analyzed the same SRM six times with a standard deviation of 0.43, are their precisions significantly different at the 95% confidence level?

59. A question that is often encountered in data analysis is the problem of *outliers* or data that appear to be out of the norm. Discuss the philosophy of discarding outliers in light of the consequences of retaining the suspect data and risking a false conclusion. Also discuss in light of rejecting the suspect data and risking a false conclusion. Discuss in terms of Type I and Type II errors.

60. Manufacturers are concerned about costs to replace automobile parts that fail before the expiration of the warranty. Warranty periods are often calculated from the average lifetime of the part. A manufacturer of a spark plug that has an improved electrode material wants to have a warranty period of 48 months. An acceptable failure rate is at most 5 failures out of 1000 parts during the warranty period and at most 50 of 1000 within 50 months of the purchase. What is the expected value for the lifetime of the spark plug?

61. You have visited your doctor who has given you the results of a medical test for which you tested below the normal range (95% confidence interval). A person who has the disease will always register low. What is the probability you have the disease if (a) 1 person in 1000 of the population actually has the disease? (b) 1 person in 100 has the disease? (c) 1 person in 10,000 has the disease?

62. The Karl Fischer titration is well known as a sensitive method to test for water in a sample. Assume that the standard deviation of the titration is always the same for a group of samples regardless of how many are analyzed. One set of samples ($n=4$) was analyzed with a 95% confidence interval of 82.3 ± 5.0. How many samples must be analyzed to reduce the confidence interval to less than 1.25?

63. An analytical technique for dieldrin in water is used to determine the amount of dieldrin in a standard reference material. The standard reference material is known to have an expected value (μ) of 18.42 ng/L and a σ_x of 0.09 ng/L. Seven determinations were made and gave an average of 18.49 ng/L. Is there reason to believe the method may be out of control?

64. The analytical technique of standardizing a base solution (such as NaOH) against the primary standard, potassium hydrogen phthalate (KHP), is a very good technique to demonstrate systematic and random errors. Identify whether the following are systematic or random. If the error that results is a systematic error, indicate whether the result in the calculated normality of the NaOH is high or low. The neutralization reaction for the standardization is

$$C_6H_4(COOH)(COO^-) + OH^- \rightleftharpoons C_6H_4(COO^-)_2 + H_2O$$

and the concentration of NaOH would be found using:

$$\text{Normality} = \frac{\text{mass KHP}}{(\text{molar mass KHP}) \cdot (\text{Volume NaOH})}$$

a. The balance is not properly calibrated and always reads 0.75 grams low.

b. The weighing bottle containing the KHP is not placed in the exact center of the balance pan.

c. The indicator chosen for the analysis changes color between pH 3 and 4 rather than between 7 and 8.

d. A bubble is lodged in the tip of the buret, but is dislodged during the titration.

e. The NaOH is not allowed to cool to ambient temperature before being titrated.

f. The volumetric flask used to prepare the NaOH contained 5 mL of H_2O before adding the NaOH pellets.

g. The KHP was taken directly from the reagent bottle without being dried.

h. The protocol called for 25 mL H_2O to be added to the 250.00 mL volumetric flask to dissolve the solid KHP, but 30.00 mL was added.

i. The analyst did not rinse down the sides of the titration flask near the equivalence point.

j. Rather than weighing a portion of KHP for each repetitive determination the analyst decides to prepare a stock solution. When using the pipet to transfer an aliquot of KHP for analysis, the analyst always blows out the last drop of KHP solution from the tip of the pipet.

k. The analyst always uses a fresh sheet of weighing paper to weigh out each portion of KHP solid. He tares only the first sheet.

References for Homework Problems

1. Table reprinted with permission of American Institute of Aeronautics and Astronautics, **Journal of Aircraft**, **1983**, 552-556.

2. Reprinted with the permission of the American Institute of Chemical Engineers, ***Chemical Engineering Progress***, **1968**, 79-84.

3. Data furnished courtesy of the National Atmospheric Deposition Program, Natural Resources Ecology Laboratory, Colorado State University, Fort Collins, Colorado.

4. **Environmental Studies, 1982, p. 62-66.**

5. Reprinted with permission of the Water Pollution Control Federation. **Journal of Water Pollution Control Federation**, **1981**, 99-112.

6. Reprinted with permission of the Soil Science Society of America, **Soil Science**, **1984**, 109.

7. **Analytical Chemistry**, 1980, **52**, 1045-1049. Reprinted with the permission of the American Chemical Society.

8. Reproduced from ***Journal of Chromatographic Science***, 1980, **18**, 381 by permission of Preston Publications, a division of Preston Industries, Inc.

9. ***Anal. Chem.***, 1980, **52,** 2344. Reprinted with the permission of the American Chemical Society.

10. Reprinted by permission of the ***J. Environ. Anal. Chem.*** 1979, **7**, 41-54,.

11. ***Anal. Chem.***, 1980, **52**, 2344. Reprinted with the permission of the American Chemical Society.

APPENDIX C -- Statistical Tables

Cumulative Probabilities of the Normal Distribution Values in the center of the table represent the cumulative probability, $P(Z \leq z)$. The values are given for the standardized normal distribution:

$$z = \frac{(x - \mu)}{\sigma}$$

	.09	.08	.07	.06	.05	.04	.03	.02	.01	.00
-3.7	0.0000	0.0000	0.0000	0.0000	0.0000	0.0000	0.0000	0.0001	0.0001	0.0001
-3.6	0.0001	0.0001	0.0001	0.0001	0.0001	0.0001	0.0001	0.0001	0.0001	0.0001
-3.5	0.0001	0.0001	0.0001	0.0001	0.0001	0.0002	0.0002	0.0002	0.0002	0.0002
-3.4	0.0002	0.0002	0.0002	0.0002	0.0002	0.0002	0.0003	0.0003	0.0003	0.0003
-3.3	0.0003	0.0003	0.0003	0.0003	0.0004	0.0004	0.0004	0.0004	0.0004	0.0004
-3.2	0.0005	0.0005	0.0005	0.0005	0.0005	0.0005	0.0006	0.0006	0.0006	0.0006
-3.1	0.0007	0.0007	0.0007	0.0007	0.0008	0.0008	0.0008	0.0009	0.0009	0.0009
-3.0	0.0010	0.0010	0.0010	0.0011	0.0011	0.0011	0.0012	0.0012	0.0013	0.0013
-2.9	0.0013	0.0014	0.0014	0.0015	0.0015	0.0016	0.0016	0.0017	0.0018	0.0018
-2.8	0.0019	0.0019	0.0020	0.0021	0.0021	0.0022	0.0023	0.0024	0.0024	0.0025
-2.7	0.0026	0.0027	0.0028	0.0028	0.0029	0.0030	0.0031	0.0032	0.0033	0.0034
-2.6	0.0035	0.0036	0.0037	0.0039	0.0040	0.0041	0.0042	0.0043	0.0045	0.0046
-2.5	0.0047	0.0049	0.0050	0.0052	0.0053	0.0055	0.0057	0.0058	0.0060	0.0062
-2.4	0.0063	0.0065	0.0067	0.0069	0.0071	0.0073	0.0075	0.0077	0.0079	0.0081
-2.3	0.0084	0.0086	0.0088	0.0091	0.0093	0.0096	0.0099	0.0101	0.0104	0.0107
-2.2	0.0110	0.0113	0.0116	0.0119	0.0122	0.0125	0.0128	0.0132	0.0135	0.0139
-2.1	0.0142	0.0146	0.0150	0.0153	0.0157	0.0161	0.0165	0.0170	0.0174	0.0178
-2.0	0.0183	0.0187	0.0192	0.0196	0.0201	0.0206	0.0211	0.0216	0.0222	0.0227
-1.9	0.0232	0.0238	0.0244	0.0249	0.0255	0.0261	0.0268	0.0274	0.0280	0.0287
-1.8	0.0293	0.0300	0.0307	0.0314	0.0321	0.0328	0.0336	0.0343	0.0351	0.0359
-1.7	0.0367	0.0375	0.0383	0.0392	0.0400	0.0409	0.0418	0.0427	0.0436	0.0445
-1.6	0.0455	0.0464	0.0474	0.0484	0.0494	0.0505	0.0515	0.0526	0.0536	0.0547
-1.5	0.0559	0.0570	0.0582	0.0593	0.0605	0.0617	0.0630	0.0642	0.0655	0.0668
-1.4	0.0681	0.0694	0.0707	0.0721	0.0735	0.0749	0.0763	0.0778	0.0792	0.0807
-1.3	0.0822	0.0837	0.0853	0.0869	0.0885	0.0901	0.0917	0.0934	0.0950	0.0968
-1.2	0.0985	0.1002	0.1020	0.1038	0.1056	0.1074	0.1093	0.1112	0.1131	0.1150
-1.1	0.1170	0.1190	0.1210	0.1230	0.1250	0.1271	0.1292	0.1313	0.1334	0.1356
-1.0	0.1378	0.1400	0.1423	0.1445	0.1468	0.1491	0.1515	0.1538	0.1562	0.1586
-0.9	0.1610	0.1635	0.1660	0.1685	0.1710	0.1736	0.1761	0.1787	0.1814	0.1840
-0.8	0.1867	0.1894	0.1921	0.1948	0.1976	0.2004	0.2032	0.2061	0.2089	0.2118
-0.7	0.2147	0.2176	0.2206	0.2236	0.2266	0.2296	0.2326	0.2357	0.2388	0.2419
-0.6	0.2450	0.2482	0.2514	0.2546	0.2578	0.2610	0.2643	0.2676	0.2709	0.2742
-0.5	0.2775	0.2809	0.2843	0.2877	0.2911	0.2945	0.2980	0.3015	0.3050	0.3085
-0.4	0.3120	0.3156	0.3191	0.3227	0.3263	0.3299	0.3335	0.3372	0.3409	0.3445
-0.3	0.3482	0.3519	0.3556	0.3594	0.3631	0.3669	0.3706	0.3744	0.3782	0.3820
-0.2	0.3859	0.3897	0.3935	0.3974	0.4012	0.4051	0.4090	0.4129	0.4168	0.4207
-0.1	0.4246	0.4285	0.4325	0.4364	0.4403	0.4443	0.4482	0.4522	0.4562	0.4601
-0.0	0.4641	0.4681	0.4720	0.4760	0.4800	0.4840	0.4880	0.4920	0.4960	0.4999

This table was generated using MINITAB.

Cumulative Probabilities of the Normal Distribution (continued)

	.00	.01	.02	.03	.04	.05	.06	.07	.08	.09
0.0	0.5000	0.5039	0.5079	0.5119	0.5159	0.5199	0.5239	0.5279	0.5318	0.5358
0.1	0.5398	0.5437	0.5477	0.5517	0.5556	0.5596	0.5635	0.5674	0.5714	0.5753
0.2	0.5792	0.5831	0.5870	0.5909	0.5948	0.5987	0.6025	0.6064	0.6102	0.6140
0.3	0.6179	0.6217	0.6255	0.6292	0.6330	0.6368	0.6405	0.6443	0.6480	0.6517
0.4	0.6554	0.6590	0.6627	0.6664	0.6700	0.6736	0.6772	0.6808	0.6843	0.6879
0.5	0.6914	0.6949	0.6984	0.7019	0.7054	0.7088	0.7122	0.7156	0.7190	0.7224
0.6	0.7257	0.7290	0.7323	0.7356	0.7389	0.7421	0.7453	0.7485	0.7517	0.7549
0.7	0.7580	0.7611	0.7642	0.7673	0.7703	0.7733	0.7763	0.7793	0.7823	0.7852
0.8	0.7881	0.7910	0.7938	0.7967	0.7995	0.8023	0.8051	0.8078	0.8105	0.8132
0.9	0.8159	0.8185	0.8212	0.8238	0.8263	0.8289	0.8314	0.8339	0.8364	0.8389
1.0	0.8413	0.8437	0.8461	0.8484	0.8508	0.8531	0.8554	0.8576	0.8599	0.8621
1.1	0.8643	0.8665	0.8686	0.8707	0.8728	0.8749	0.8769	0.8789	0.8809	0.8829
1.2	0.8849	0.8868	0.8887	0.8906	0.8925	0.8943	0.8961	0.8979	0.8997	0.9014
1.3	0.9031	0.9049	0.9065	0.9082	0.9098	0.9114	0.9130	0.9146	0.9162	0.9177
1.4	0.9192	0.9207	0.9221	0.9236	0.9250	0.9264	0.9278	0.9292	0.9305	0.9318
1.5	0.9331	0.9344	0.9357	0.9369	0.9382	0.9394	0.9406	0.9417	0.9429	0.9440
1.6	0.9452	0.9463	0.9473	0.9484	0.9494	0.9505	0.9515	0.9525	0.9535	0.9544
1.7	0.9554	0.9563	0.9572	0.9581	0.9590	0.9599	0.9607	0.9616	0.9624	0.9632
1.8	0.9640	0.9648	0.9656	0.9663	0.9671	0.9678	0.9685	0.9692	0.9699	0.9706
1.9	0.9712	0.9719	0.9725	0.9731	0.9738	0.9744	0.9750	0.9755	0.9761	0.9767
2.0	0.9772	0.9777	0.9783	0.9788	0.9793	0.9798	0.9803	0.9807	0.9812	0.9816
2.1	0.9821	0.9825	0.9829	0.9834	0.9838	0.9842	0.9846	0.9849	0.9853	0.9857
2.2	0.9860	0.9864	0.9867	0.9871	0.9874	0.9877	0.9880	0.9883	0.9886	0.9889
2.3	0.9892	0.9895	0.9898	0.9900	0.9903	0.9906	0.9908	0.9911	0.9913	0.9915
2.4	0.9918	0.9920	0.9922	0.9924	0.9926	0.9928	0.9930	0.9932	0.9934	0.9936
2.5	0.9937	0.9939	0.9941	0.9942	0.9944	0.9946	0.9947	0.9949	0.9950	0.9952
2.6	0.9953	0.9954	0.9956	0.9957	0.9958	0.9959	0.9960	0.9962	0.9963	0.9964
2.7	0.9965	0.9966	0.9967	0.9968	0.9969	0.9970	0.9971	0.9971	0.9972	0.9973
2.8	0.9974	0.9975	0.9975	0.9976	0.9977	0.9978	0.9978	0.9979	0.9980	0.9980
2.9	0.9981	0.9981	0.9982	0.9983	0.9983	0.9984	0.9984	0.9985	0.9985	0.9986
3.0	0.9986	0.9986	0.9987	0.9987	0.9988	0.9988	0.9988	0.9989	0.9989	0.9989
3.1	0.9990	0.9990	0.9990	0.9991	0.9991	0.9991	0.9992	0.9992	0.9992	0.9992
3.2	0.9993	0.9993	0.9993	0.9993	0.9994	0.9994	0.9994	0.9994	0.9994	0.9994
3.3	0.9995	0.9995	0.9995	0.9995	0.9995	0.9995	0.9996	0.9996	0.9996	0.9996
3.4	0.9996	0.9996	0.9996	0.9996	0.9997	0.9997	0.9997	0.9997	0.9997	0.9997
3.5	0.9997	0.9997	0.9997	0.9997	0.9998	0.9998	0.9998	0.9998	0.9998	0.9998
3.6	0.9998	0.9998	0.9998	0.9998	0.9998	0.9998	0.9998	0.9998	0.9998	0.9998
3.7	0.9998	0.9998	0.9999	0.9999	0.9999	0.9999	0.9999	0.9999	0.9999	0.9999

This table was generated using MINITAB.

Table of Values for the F-Distribution

F-table for 1 Degrees of freedom in the denominator

numerator degrees of freedom

	1	2	3	4	5
0.00010	*	*	*	*	*
0.001	0	0.00	*	*	*
0.005	0	0.00	0.000	0.000	0.000
0.010	0	0.00	0.000	0.000	0.000
0.020	0	0.00	0.001	0.001	0.001
0.030	0	0.00	0.002	0.002	0.002
0.040	0	0.00	0.003	0.003	0.003
0.050	0	0.01	0.005	0.004	0.004
0.060	0	0.01	0.007	0.006	0.006
0.070	0	0.01	0.009	0.009	0.009
0.080	0	0.01	0.012	0.011	0.011
0.090	0	0.02	0.015	0.014	0.014
0.100	0	0.02	0.019	0.018	0.017
0.200	0	0.08	0.077	0.073	0.071
0.250	0	0.13	0.122	0.117	0.113
0.300	0	0.20	0.180	0.172	0.167
0.350	0	0.28	0.252	0.240	0.233
0.400	1	0.38	0.342	0.323	0.313
0.450	1	0.51	0.451	0.425	0.410
0.500	1	0.67	0.585	0.549	0.528
0.550	1	0.87	0.751	0.700	0.671
0.600	2	1.13	0.957	0.885	0.846
0.650	3	1.46	1.220	1.118	1.062
0.700	4	1.92	1.562	1.415	1.336
0.750	6	2.57	2.024	1.807	1.693
0.800	9	3.56	2.682	2.351	2.178
0.850	17	5.21	3.703	3.162	2.888
0.900	40	8.53	5.539	4.545	4.060
0.950	161	19	10.128	7.709	6.608
0.960	253	24	12.124	8.991	7.598
0.970	450	32	15.179	10.874	9.017
0.980	1013	49	21	14.040	11.323
0.990	4052	99	34	21	16.258
0.995	1.62E4	199	56	31	23
0.99900	4.05E5	999	167	74	47
0.99990	4.05E7	9997	784	242	125

Table was generated using MINITAB

F-table for 1 Degrees of freedom in the denominator
numerator degrees of freedom

	6	7	8	9	10
0.00010	*	*	*	*	*
0.001	*	*	*	*	*
0.005	0.000	*	*	*	*
0.010	0.000	0.000	0.000	0.000	0.000
0.020	0.000	0.000	0.000	0.000	0.001
0.030	0.002	0.002	0.002	0.002	0.002
0.040	0.003	0.003	0.003	0.003	0.003
0.050	0.004	0.004	0.004	0.004	0.004
0.060	0.006	0.006	0.006	0.006	0.006
0.070	0.008	0.008	0.008	0.008	0.008
0.080	0.011	0.010	0.010	0.010	0.010
0.090	0.013	0.013	0.013	0.013	0.013
0.100	0.017	0.017	0.016	0.016	0.016
0.200	0.070	0.069	0.068	0.068	0.067
0.250	0.111	0.109	0.108	0.107	0.107
0.300	0.163	0.161	0.159	0.158	0.157
0.350	0.227	0.224	0.222	0.220	0.218
0.400	0.306	0.301	0.298	0.295	0.293
0.450	0.400	0.394	0.389	0.385	0.382
0.500	0.514	0.505	0.499	0.493	0.489
0.550	0.652	0.640	0.630	0.623	0.618
0.600	0.820	0.802	0.790	0.780	0.772
0.650	1.026	1.002	0.985	0.971	0.961
0.700	1.286	1.252	1.228	1.209	1.194
0.750	1.621	1.573	1.538	1.512	1.491
0.800	2.072	2.002	1.951	1.912	1.882
0.850	2.723	2.613	2.535	2.476	2.431
0.900	3.775	3.589	3.457	3.360	3.285
0.950	5.987	5.591	5.317	5.117	4.964
0.960	6.823	6.333	5.997	5.752	5.566
0.970	8.002	7.368	6.937	6.624	6.388
0.980	9.876	8.987	8.389	7.960	7.638
0.990	13.745	12.246	11.258	10.561	10.044
0.995	18	16.235	14.688	13.613	12.826
0.99900	36	29.244	25.414	22.857	21.039

Table was generated using MINITAB

F-table for 1 Degrees of freedom in the denominator
numerator degrees of freedom

α	11	12	13	14	15
0.00010	*	*	*	*	*
0.00100	*	*	*	*	*
0.00500	*	*	*	*	*
0.01000	0.0002	0.0002	0.0002	0.0002	0.0002
0.02000	0.0007	0.0007	0.0007	0.0007	0.0006
0.03000	0.0015	0.0015	0.0015	0.0015	0.0015
0.04000	0.0026	0.0026	0.0026	0.0026	0.0026
0.05000	0.0041	0.0041	0.0041	0.0041	0.0041
0.06000	0.0059	0.0059	0.0059	0.0059	0.0059
0.07000	0.0081	0.0080	0.0080	0.0080	0.0080
0.08000	0.0106	0.0105	0.0105	0.0105	0.0104
0.09000	0.0134	0.0133	0.0133	0.0133	0.0132
0.10000	0.0165	0.0165	0.0164	0.0164	0.0163
0.20000	0.0674	0.0671	0.0669	0.0667	0.0665
0.25000	0.1067	0.1063	0.1059	0.1056	0.1053
0.30000	0.1565	0.1558	0.1552	0.1547	0.1543
0.35000	0.2176	0.2166	0.2157	0.2150	0.2144
0.40000	0.2915	0.2901	0.2889	0.2879	0.2870
0.45000	0.3803	0.3783	0.3766	0.3752	0.3740
0.50000	0.4864	0.4837	0.4814	0.4794	0.4777
0.55000	0.6135	0.6098	0.6067	0.6039	0.6016
0.60000	0.7666	0.7615	0.7572	0.7535	0.7504
0.65000	0.9528	0.9458	0.9399	0.9349	0.9306
0.70000	1.1830	1.1733	1.1652	1.1583	1.1525
0.75000	1.4749	1.4613	1.4499	1.4403	1.4321
0.80000	1.8589	1.8393	1.8230	1.8091	1.7972
0.85000	2.3949	2.3654	2.3407	2.3199	2.3020
0.90000	3.2252	3.1765	3.1362	3.1023	3.0732
0.95000	4.8442	4.7472	4.6673	4.6001	4.5432
0.96000	5.4202	5.3025	5.2056	5.1246	5.0559
0.97000	6.2035	6.0550	5.9335	5.8317	5.7456
0.98000	7.3879	7.1878	7.0242	6.8880	6.7730
0.99000	9.6460	9.3302	9.0737	8.8616	8.6831
0.99500	12.2262	11.7543	11.3736	11.0602	10.7982
0.99900	19.6869	18.6434	17.8154	17.1435	16.5874
0.99990	35.0592	32.4255	30.3870	28.7651	27.4470

The Table was generated using MINITAB

F-table for 1 Degrees of freedom in the denominator
numerator degrees of freedom

α	16	17	18	19	20
0.00010	*	*	*	*	*
0.00100	*	*	*	*	*
0.00500	*	*	*	*	*
0.01000	0.0002	0.0002	0.0002	0.0002	0.0002
0.02000	0.0006	0.0006	0.0006	0.0006	0.0006
0.03000	0.0015	0.0015	0.0015	0.0015	0.0014
0.04000	0.0026	0.0026	0.0026	0.0026	0.0026
0.05000	0.0041	0.0041	0.0040	0.0040	0.0040
0.06000	0.0058	0.0058	0.0058	0.0058	0.0058
0.07000	0.0080	0.0079	0.0079	0.0079	0.0079
0.08000	0.0104	0.0104	0.0104	0.0104	0.0103
0.09000	0.0132	0.0132	0.0131	0.0131	0.0131
0.10000	0.0163	0.0163	0.0162	0.0162	0.0162
0.20000	0.0664	0.0662	0.0661	0.0660	0.0659
0.25000	0.1051	0.1049	0.1047	0.1045	0.1044
0.30000	0.1539	0.1536	0.1533	0.1530	0.1528
0.35000	0.2138	0.2134	0.2129	0.2126	0.2122
0.40000	0.2863	0.2855	0.2850	0.2844	0.2839
0.45000	0.3729	0.3720	0.3711	0.3704	0.3697
0.50000	0.4763	0.4750	0.4739	0.4728	0.4719
0.55000	0.5997	0.5979	0.5963	0.5949	0.5937
0.60000	0.7477	0.7452	0.7431	0.7412	0.7396
0.65000	0.9268	0.9236	0.9207	0.9182	0.9158
0.70000	1.1473	1.1428	1.1388	1.1353	1.1321
0.75000	1.4249	1.4186	1.4130	1.4081	1.4037
0.80000	1.7869	1.7779	1.7699	1.7629	1.7565
0.85000	2.2865	2.2730	2.2612	2.2505	2.2410
0.90000	3.0480	3.0263	3.0070	2.9898	2.9747
0.95000	4.4940	4.4512	4.4140	4.3807	4.3513
0.96000	4.9967	4.9456	4.9005	4.8607	4.8253
0.97000	5.6717	5.6078	5.5515	5.5021	5.4580
0.98000	6.6744	6.5891	6.5147	6.4491	6.3907
0.99000	8.5310	8.3998	8.2854	8.1850	8.0960
0.99500	10.5755	10.3842	10.2182	10.0725	9.9439
0.99900	16.1201	15.7223	15.3795	15.0810	14.8188
0.99990	16.3560	25.4387	24.6574	23.9844	23.3988

Table generated using MINITAB

F-table for 1 Degrees of freedom in the denominator

numerator degrees of freedom

α	25	50	75	100	200
0.00010	*	*	*	*	*
0.00100	*	*	*	*	*
0.00500	*	*	*	*	*
0.01000	*	*	*	*	*
0.02000	0.000	0.000	0.000	0.000	0.000
0.03000	0.001	0.000	0.000	0.000	0.000
0.04000	0.002	0.002	0.002	0.002	0.002
0.05000	0.004	0.004	0.004	0.004	0.004
0.06000	0.005	0.005	0.005	0.005	0.005
0.07000	0.007	0.007	0.007	0.007	0.007
0.08000	0.010	0.010	0.010	0.010	0.010
0.09000	0.013	0.012	0.012	0.012	0.012
0.10000	0.016	0.015	0.015	0.015	0.015
0.20000	0.065	0.064	0.064	0.064	0.645
0.25000	0.103	0.102	0.102	0.102	0.102
0.30000	0.151	0.150	0.149	0.149	0.149
0.35000	0.210	0.208	0.207	0.207	0.207
0.40000	0.282	0.278	0.277	0.276	0.275
0.45000	0.367	0.362	0.360	0.359	0.358
0.50000	0.468	0.461	0.459	0.458	0.457
0.55000	0.589	0.579	0.576	0.575	0.574
0.60000	0.733	0.720	0.716	0.714	0.713
0.65000	0.907	0.890	0.884	0.875	0.864
0.70000	1.120	1.096	1.089	1.075	1.069
0.75000	1.387	1.354	1.344	1.321	1.315
0.80000	1.732	1.686	1.671	1.664	1.654
0.85000	2.205	2.137	2.115	2.104	2.095
0.90000	2.917	2.808	2.773	2.756	2.744
0.95000	4.241	4.034	3.968	3.936	3.888
0.96000	4.694	4.446	4.368	4.330	4.273
0.97000	5.295	4.989	4.893	4.846	4.777
0.98000	6.175	5.775	5.650	5.589	5.499
0.99000	7.769	7.170	6.985	6.895	6.763
0.99500	9.475	8.625	8.366	8.240	8.057
0.99900	13.876	12.222	11.730	11.495	11.154
0.99990	21.335	17.878	16.893	16.429	15.764

Table was generated using MINITAB

F-table for 2 Degrees of freedom in the denominator
numerator degrees of freedom

α	1	2	3	4	5
0.00010	0	0.00	0.000	0.000	0.0001
0.00100	0	0.00	0.001	0.001	0.0010
0.00500	0	0.01	0.005	0.005	0.0050
0.01000	0	0.01	0.010	0.010	0.0101
0.02000	0	0.02	0.020	0.020	0.0203
0.03000	0	0.03	0.031	0.031	0.0306
0.04000	0	0.04	0.041	0.041	0.0412
0.05000	0	0.05	0.052	0.052	0.0518
0.06000	0	0.06	0.063	0.063	0.0626
0.07000	0	0.08	0.074	0.074	0.0736
0.08000	0	0.09	0.086	0.085	0.0848
0.09000	0	0.10	0.097	0.097	0.0961
0.10000	0	0.11	0.109	0.108	0.1076
0.20000	0	0.25	0.241	0.236	0.2334
0.25000	0	0.33	0.317	0.309	0.3049
0.30000	1	0.43	0.403	0.390	0.3834
0.35000	1	0.54	0.499	0.481	0.4701
0.40000	1	0.67	0.609	0.582	0.5668
0.45000	1	0.82	0.734	0.697	0.6754
0.50000	2	1.00	0.881	0.828	0.7988
0.55000	2	1.22	1.054	0.981	0.9408
0.60000	3	1.50	1.263	1.162	1.1067
0.65000	4	1.86	1.520	1.381	1.3047
0.70000	5	2.33	1.847	1.651	1.5466
0.75000	8	3.00	2.280	2.000	1.8528
0.80000	12	4.00	2.886	2.472	2.2592
0.85000	22	5.67	3.813	3.164	2.8396
0.90000	49	9.00	5.463	4.325	3.7798
0.95000	199	19.00	9.552	6.944	5.7860
0.96000	312	24.00	11.325	8.000	6.5597
0.97000	555	32.33	14.036	9.547	7.6647
0.98000	1249	49.00	18.858	12.142	9.4543
0.99000	4999	99.00	30.816	18.000	13.2739
0.99500	19999	199.00	49.800	26.284	18.3137
0.99900	499996	999.01	148.497	61.246	37.1227
0.99990	49986888	9997.34	694.650	197.983	97.0194

The table was generated using MINITAB

F-table for 2 Degrees of freedom in the denominator
numerator degrees of freedom

α	6	7	8	9	10
0.00010	0.0001	0.0001	0.0001	0.0001	0.0001
0.00100	0.0010	0.0010	0.0010	0.0010	0.0010
0.00500	0.0050	0.0050	0.0050	0.0050	0.0050
0.01000	0.0101	0.0101	0.0101	0.0101	0.0101
0.02000	0.0203	0.0203	0.0203	0.0202	0.0202
0.03000	0.0306	0.0306	0.0306	0.0306	0.0306
0.04000	0.0411	0.0411	0.0410	0.0410	0.0410
0.05000	0.0517	0.0517	0.0516	0.0516	0.0516
0.06000	0.0625	0.0624	0.0624	0.0623	0.0623
0.07000	0.0735	0.0733	0.0732	0.0732	0.0731
0.08000	0.0846	0.0844	0.0843	0.0841	0.0841
0.09000	0.0958	0.0956	0.0954	0.0953	0.0952
0.10000	0.1072	0.1070	0.1068	0.1066	0.1065
0.20000	0.2317	0.2304	0.2295	0.2288	0.2282
0.25000	0.3019	0.2998	0.2983	0.2971	0.2961
0.30000	0.3788	0.3755	0.3731	0.3712	0.3697
0.35000	0.4632	0.4584	0.4548	0.4521	0.4499
0.40000	0.5569	0.5500	0.5449	0.5410	0.5378
0.45000	0.6616	0.6519	0.6448	0.6394	0.6350
0.50000	0.7798	0.7665	0.7568	0.7494	0.7435
0.55000	0.9149	0.8970	0.8838	0.8738	0.8658
0.60000	1.0717	1.0474	1.0297	1.0162	1.0056
0.65000	1.2570	1.2242	1.2004	1.1824	1.1682
0.70000	1.4814	1.4370	1.4048	1.3804	1.3613
0.75000	1.7622	1.7010	1.6568	1.6236	1.5975
0.80000	2.1299	2.0434	1.9813	1.9349	1.8987
0.85000	2.6462	2.5183	2.4275	2.3597	2.3073
0.90000	3.4633	3.2575	3.1130	3.0064	2.9245
0.95000	5.1431	4.7373	4.4590	4.2566	4.1029
0.96000	5.7721	5.2796	4.9443	4.7016	4.5182
0.97000	6.6549	6.0319	5.6111	5.3091	5.0820
0.98000	8.0522	7.2025	6.6366	6.2340	5.9336
0.99000	10.9246	9.5464	8.6490	8.0215	7.5594
0.99500	14.5441	12.4040	11.0424	10.1068	9.4270
0.99900	27.0005	21.6889	18.4939	16.3872	14.9052
0.99990	61.6285	45.1301	35.9983	30.3405	26.5471

Table generated using MINITAB

F-table for 2 Degrees of freedom in the denominator
numerator degrees of freedom

α	11	12	13	14	15
0.00010	0.000	0.000	0.000	0.000	0.000
0.00100	0.001	0.001	0.001	0.001	0.001
0.00500	0.005	0.005	0.005	0.005	0.005
0.01000	0.010	0.010	0.010	0.010	0.010
0.02000	0.020	0.020	0.020	0.020	0.020
0.03000	0.030	0.030	0.030	0.030	0.030
0.04000	0.041	0.041	0.040	0.040	0.040
0.05000	0.051	0.051	0.051	0.051	0.051
0.06000	0.062	0.062	0.062	0.062	0.062
0.07000	0.073	0.073	0.073	0.073	0.072
0.08000	0.084	0.084	0.083	0.083	0.083
0.09000	0.095	0.095	0.095	0.094	0.094
0.10000	0.106	0.106	0.106	0.106	0.106
0.20000	0.227	0.227	0.227	0.226	0.226
0.25000	0.295	0.294	0.294	0.293	0.293
0.30000	0.368	0.367	0.366	0.365	0.365
0.35000	0.448	0.446	0.445	0.444	0.443
0.40000	0.535	0.533	0.531	0.529	0.528
0.45000	0.631	0.628	0.626	0.624	0.622
0.50000	0.738	0.734	0.731	0.728	0.726
0.55000	0.859	0.854	0.849	0.845	0.842
0.60000	0.997	0.989	0.984	0.979	0.974
0.65000	1.156	1.147	1.139	1.132	1.126
0.70000	1.345	1.333	1.322	1.313	1.306
0.75000	1.576	1.559	1.545	1.533	1.522
0.80000	1.869	1.846	1.826	1.809	1.795
0.85000	2.265	2.231	2.203	2.179	2.158
0.90000	2.859	2.806	2.763	2.726	2.695
0.95000	3.982	3.885	3.805	3.738	3.682
0.96000	4.374	4.259	4.165	4.086	4.020
0.97000	4.905	4.763	4.648	4.551	4.470
0.98000	5.701	5.516	5.365	5.240	5.135
0.99000	7.205	6.926	6.701	6.515	6.358
0.99500	8.912	8.509	8.186	7.921	7.700
0.99600	9.509	9.059	8.699	8.405	8.159
0.99700	10.315	9.799	9.387	9.051	8.772
0.99800	11.525	10.903	10.409	10.008	9.676
0.99900	13.811	12.973	12.312	11.779	11.339
0.99990	23.850	21.848	20.309	19.092	18.108

The table was generated using MINITAB

F-table for 2 Degrees of freedom in the denominator
numerator degrees of freedom

α	16	17	18	19	20
0.00010	0.000	0.000	0.000	0.000	0.000
0.00100	0.001	0.001	0.001	0.001	0.001
0.00500	0.005	0.005	0.005	0.005	0.005
0.01000	0.010	0.010	0.010	0.010	0.010
0.02000	0.020	0.020	0.020	0.020	0.020
0.03000	0.030	0.030	0.030	0.030	0.030
0.04000	0.040	0.040	0.040	0.040	0.040
0.05000	0.051	0.051	0.051	0.051	0.051
0.06000	0.062	0.062	0.062	0.062	0.062
0.07000	0.072	0.072	0.072	0.072	0.072
0.08000	0.083	0.083	0.083	0.083	0.083
0.09000	0.094	0.094	0.094	0.094	0.094
0.10000	0.106	0.106	0.106	0.105	0.105
0.20000	0.226	0.226	0.225	0.225	0.225
0.25000	0.292	0.292	0.292	0.292	0.291
0.30000	0.364	0.364	0.363	0.363	0.363
0.35000	0.442	0.441	0.441	0.440	0.440
0.40000	0.527	0.526	0.525	0.524	0.524
0.45000	0.620	0.619	0.618	0.617	0.616
0.50000	0.724	0.722	0.720	0.719	0.717
0.55000	0.839	0.837	0.835	0.833	0.831
0.60000	0.970	0.967	0.964	0.961	0.959
0.65000	1.121	1.117	1.113	1.110	1.106
0.70000	1.299	1.293	1.288	1.283	1.279
0.75000	1.513	1.505	1.498	1.492	1.487
0.80000	1.782	1.771	1.762	1.753	1.746
0.85000	2.141	2.125	2.111	2.099	2.089
0.90000	2.668	2.644	2.624	2.605	2.589
0.95000	3.633	3.591	3.554	3.521	3.492
0.96000	3.962	3.913	3.869	3.831	3.797
0.97000	4.400	4.340	4.287	4.241	4.200
0.98000	5.045	4.967	4.900	4.840	4.787
0.99000	6.226	6.112	6.012	5.925	5.848
0.99500	7.513	7.353	7.214	7.093	6.986
0.99600	7.952	7.775	7.621	7.487	7.369
0.99700	8.536	8.335	8.161	8.010	7.876
0.99800	9.396	9.158	8.952	8.773	8.616
0.99900	10.971	10.658	10.390	10.156	9.952
0.99990	17.297	16.618	16.042	15.547	15.118

Table was generated using MINITAB

F-table for 2 Degrees of freedom in the denominator
numerator degrees of freedom

α	25	50	75	100	200
0.00010	0.000	0.000	0.000	0.000	0.000
0.00100	0.001	0.001	0.001	0.001	0.001
0.00500	0.005	0.005	0.005	0.005	0.005
0.01000	0.010	0.010	0.010	0.010	0.010
0.02000	0.020	0.020	0.020	0.020	0.020
0.03000	0.030	0.030	0.030	0.030	0.030
0.04000	0.040	0.040	0.040	0.040	0.040
0.05000	0.051	0.051	0.051	0.051	0.051
0.06000	0.062	0.062	0.061	0.061	0.061
0.07000	0.072	0.072	0.072	0.072	0.072
0.08000	0.083	0.083	0.083	0.083	0.083
0.09000	0.094	0.094	0.094	0.094	0.094
0.10000	0.105	0.105	0.105	0.105	0.105
0.20000	0.225	0.224	0.223	0.223	0.223
0.25000	0.291	0.289	0.288	0.288	0.288
0.30000	0.361	0.359	0.358	0.358	0.357
0.35000	0.438	0.434	0.433	0.432	0.431
0.40000	0.521	0.516	0.514	0.513	0.512
0.45000	0.612	0.605	0.602	0.601	0.599
0.50000	0.712	0.702	0.699	0.698	0.695
0.55000	0.824	0.811	0.807	0.804	0.801
0.60000	0.950	0.933	0.927	0.924	0.920
0.65000	1.095	1.072	1.064	1.060	1.055
0.70000	1.263	1.233	1.223	1.218	1.211
0.75000	1.466	1.425	1.412	1.405	1.395
0.80000	1.717	1.662	1.644	1.635	1.622
0.85000	2.048	1.971	1.945	1.933	1.915
0.90000	2.528	2.412	2.374	2.356	2.329
0.95000	3.385	3.182	3.118	3.087	3.041
0.96000	3.671	3.435	3.361	3.324	3.271
0.97000	4.047	3.764	3.675	3.632	3.568
0.98000	4.593	4.234	4.123	4.069	3.989
0.99000	5.567	5.056	4.899	4.823	4.712
0.99500	6.598	5.901	5.690	5.589	5.441
0.99900	9.222	7.956	7.584	7.407	7.151
0.99990	13.615	11.135	10.439	10.113	9.647

The table was generated using MINITAB

F-table for 3 Degrees of freedom in the denominator
numerator degrees of freedom

α	1	2	3	4	5
0.00010	0	0.00	0.001	0.002	0.0016
0.00100	0	0.01	0.007	0.007	0.0075
0.00500	0	0.02	0.021	0.022	0.0221
0.01000	0	0.03	0.034	0.035	0.0354
0.02000	0	0.05	0.055	0.057	0.0574
0.03000	0	0.07	0.074	0.075	0.0765
0.04000	0	0.09	0.091	0.093	0.0942
0.05000	0	0.10	0.108	0.110	0.1109
0.06000	0	0.12	0.124	0.126	0.1271
0.07000	0	0.14	0.140	0.141	0.1428
0.08000	0	0.15	0.155	0.157	0.1582
0.09000	0	0.17	0.170	0.172	0.1733
0.10000	0	0.18	0.186	0.187	0.1883
0.20000	0	0.35	0.341	0.338	0.3372
0.25000	0	0.44	0.425	0.418	0.4150
0.30000	1	0.54	0.515	0.504	0.4974
0.35000	1	0.66	0.616	0.597	0.5858
0.40000	1	0.79	0.727	0.699	0.6821
0.45000	1	0.95	0.854	0.812	0.7884
0.50000	2	1.13	1.000	0.941	0.9071
0.55000	2	1.36	1.171	1.088	1.0420
0.60000	3	1.64	1.375	1.261	1.1978
0.65000	4	2.00	1.624	1.467	1.3817
0.70000	6	2.48	1.940	1.722	1.6046
0.75000	8	3.15	2.355	2.047	1.8843
0.80000	13	4.16	2.936	2.485	2.2530
0.85000	24	5.83	3.821	3.124	2.7763
0.90000	54	9.16	5.391	4.191	3.6194
0.95000	216	19.16	9.277	6.591	5.4094
0.96000	337	24.16	10.960	7.557	6.0981
0.97000	600	32.50	13.534	8.972	7.0802
0.98000	1350	49.17	18.109	11.343	8.6704
0.99000	5403	99.17	29.457	16.695	12.0600
0.99500	21613	199.18	47.469	24.259	16.5294
0.99900	540349	999.23	141.112	56.178	33.2022
0.99990	54024744	9997.49	659.260	181.002	86.2864

The table was generated using MINITAB

F-table for 3 Degrees of freedom in the denominator
numerator degrees of freedom

α	6	7	8	9	10
0.00010	0.0017	0.0017	0.0016	0.0016	0.0016
0.00100	0.0075	0.0076	0.0077	0.0077	0.0077
0.00500	0.0223	0.0225	0.0227	0.0228	0.0229
0.01000	0.0358	0.0362	0.0364	0.0366	0.0367
0.02000	0.0580	0.0584	0.0588	0.0591	0.0593
0.03000	0.0773	0.0778	0.0782	0.0786	0.0789
0.04000	0.0950	0.0957	0.0962	0.0965	0.0969
0.05000	0.1119	0.1125	0.1131	0.1135	0.1138
0.06000	0.1280	0.1287	0.1293	0.1297	0.1301
0.07000	0.1437	0.1445	0.1450	0.1455	0.1459
0.08000	0.1591	0.1599	0.1604	0.1608	0.1612
0.09000	0.1743	0.1750	0.1755	0.1760	0.1763
0.10000	0.1892	0.1899	0.1904	0.1908	0.1912
0.20000	0.3366	0.3362	0.3360	0.3358	0.3357
0.25000	0.4129	0.4115	0.4104	0.4097	0.4091
0.30000	0.4932	0.4903	0.4882	0.4865	0.4853
0.35000	0.5789	0.5741	0.5705	0.5678	0.5656
0.40000	0.6716	0.6642	0.6588	0.6546	0.6513
0.45000	0.7731	0.7625	0.7546	0.7486	0.7439
0.50000	0.8858	0.8709	0.8600	0.8517	0.8451
0.55000	1.0127	0.9924	0.9776	0.9662	0.9572
0.60000	1.1581	1.1307	1.1107	1.0954	1.0834
0.65000	1.3281	1.2912	1.2645	1.2441	1.2280
0.70000	1.5318	1.4823	1.4463	1.4191	1.3978
0.75000	1.7844	1.7169	1.6683	1.6315	1.6029
0.80000	2.1126	2.0186	1.9513	1.9007	1.8614
0.85000	2.5700	2.4334	2.3366	2.2644	2.2085
0.90000	3.2888	3.0741	2.9238	2.8128	2.7277
0.95000	4.7570	4.3469	4.0661	3.8625	3.7083
0.96000	5.3049	4.8113	4.4761	4.2344	4.0520
0.97000	6.0729	5.4545	5.0386	4.7406	4.5172
0.98000	7.2870	6.4540	5.9014	5.5097	5.2182
0.99000	9.7796	8.4514	7.5909	6.9920	6.5523
0.99500	12.9166	10.8825	9.5965	8.7172	8.0807
0.99900	23.7035	18.7725	15.8297	13.9020	12.5527
0.99990	53.6776	38.6743	30.4551	25.4023	22.0369

Table generated using MINITAB

F-table for 3 Degrees of freedom in the denominator
numerator degrees of freedom

α	11	12	13	14	15
0.00010	0.001	0.001	0.001	0.001	0.001
0.00100	0.007	0.007	0.007	0.007	0.007
0.00500	0.023	0.023	0.023	0.023	0.023
0.01000	0.036	0.037	0.037	0.037	0.037
0.02000	0.059	0.059	0.059	0.059	0.060
0.03000	0.079	0.079	0.079	0.079	0.079
0.04000	0.097	0.097	0.097	0.097	0.097
0.05000	0.114	0.114	0.114	0.114	0.114
0.06000	0.130	0.130	0.130	0.131	0.131
0.07000	0.146	0.146	0.146	0.146	0.147
0.08000	0.161	0.161	0.162	0.162	0.162
0.09000	0.176	0.176	0.177	0.177	0.177
0.10000	0.191	0.191	0.192	0.192	0.192
0.20000	0.335	0.335	0.335	0.335	0.335
0.25000	0.408	0.408	0.407	0.407	0.407
0.30000	0.484	0.483	0.482	0.482	0.481
0.35000	0.563	0.562	0.561	0.560	0.559
0.40000	0.648	0.646	0.644	0.643	0.641
0.45000	0.740	0.736	0.734	0.731	0.729
0.50000	0.839	0.835	0.831	0.828	0.825
0.55000	0.949	0.944	0.938	0.934	0.930
0.60000	1.073	1.065	1.058	1.053	1.048
0.65000	1.215	1.204	1.195	1.188	1.181
0.70000	1.380	1.366	1.354	1.344	1.336
0.75000	1.579	1.560	1.545	1.531	1.520
0.80000	1.829	1.804	1.782	1.764	1.749
0.85000	2.164	2.127	2.097	2.072	2.050
0.90000	2.660	2.605	2.560	2.522	2.489
0.95000	3.587	3.490	3.410	3.343	3.287
0.96000	3.909	3.795	3.702	3.624	3.558
0.97000	4.343	4.204	4.091	3.997	3.918
0.98000	4.993	4.814	4.669	4.548	4.447
0.99000	6.216	5.952	5.739	5.563	5.416
0.99500	7.600	7.225	6.925	6.680	6.476
0.99900	11.561	10.804	10.209	9.729	9.335
0.99990	19.657	17.898	16.551	15.490	14.634

The table was generated using MINITAB

F-table for 3 Degrees of freedom in the denominator
numerator degrees of freedom

α	16	17	18	19	20
0.00010	0.001	0.001	0.001	0.001	0.001
0.00100	0.007	0.007	0.007	0.007	0.007
0.00500	0.023	0.023	0.023	0.023	0.023
0.01000	0.037	0.037	0.037	0.037	0.037
0.02000	0.060	0.060	0.060	0.060	0.060
0.03000	0.079	0.080	0.080	0.080	0.080
0.04000	0.098	0.098	0.098	0.098	0.098
0.05000	0.115	0.115	0.115	0.115	0.115
0.06000	0.131	0.131	0.131	0.131	0.131
0.07000	0.147	0.147	0.147	0.147	0.147
0.08000	0.162	0.162	0.162	0.162	0.163
0.09000	0.177	0.177	0.177	0.178	0.178
0.10000	0.192	0.192	0.192	0.192	0.192
0.20000	0.335	0.335	0.335	0.335	0.335
0.25000	0.407	0.406	0.406	0.406	0.406
0.30000	0.481	0.480	0.480	0.480	0.479
0.35000	0.558	0.557	0.557	0.556	0.556
0.40000	0.640	0.639	0.638	0.637	0.636
0.45000	0.728	0.726	0.725	0.724	0.723
0.50000	0.823	0.821	0.819	0.817	0.816
0.55000	0.927	0.924	0.922	0.920	0.918
0.60000	1.044	1.040	1.036	1.033	1.031
0.65000	1.175	1.170	1.166	1.162	1.159
0.70000	1.328	1.321	1.316	1.310	1.306
0.75000	1.510	1.501	1.493	1.486	1.480
0.80000	1.735	1.723	1.713	1.704	1.695
0.85000	2.031	2.015	2.000	1.987	1.976
0.90000	2.461	2.437	2.416	2.397	2.380
0.95000	3.238	3.196	3.159	3.127	3.098
0.96000	3.501	3.452	3.409	3.371	3.338
0.97000	3.850	3.791	3.739	3.694	3.654
0.98000	4.360	4.286	4.221	4.164	4.113
0.99000	5.292	5.185	5.091	5.010	4.938
0.99500	6.303	6.155	6.027	5.916	5.817
0.99900	9.005	8.726	8.487	8.280	8.098
0.99990	13.931	13.343	12.845	12.419	12.049

Table was generated using MINITAB

F-table for 3 Degrees of freedom in the denominator
numerator degrees of freedom

α	25	50	75	100	200
0.00010	0.001	0.001	0.001	0.001	0.001
0.00100	0.008	0.008	0.008	0.008	0.008
0.00500	0.023	0.023	0.023	0.023	0.023
0.01000	0.037	0.037	0.038	0.038	0.038
0.02000	0.060	0.061	0.061	0.061	0.061
0.03000	0.080	0.081	0.081	0.081	0.081
0.04000	0.098	0.099	0.099	0.099	0.099
0.05000	0.115	0.116	0.116	0.116	0.117
0.06000	0.132	0.132	0.133	0.133	0.133
0.07000	0.148	0.148	0.149	0.149	0.149
0.08000	0.163	0.164	0.164	0.164	0.164
0.09000	0.178	0.179	0.179	0.179	0.179
0.10000	0.193	0.194	0.194	0.194	0.194
0.20000	0.335	0.335	0.335	0.335	0.335
0.25000	0.406	0.405	0.404	0.404	0.404
0.30000	0.478	0.476	0.475	0.475	0.475
0.35000	0.554	0.550	0.549	0.548	0.548
0.40000	0.634	0.628	0.626	0.625	0.624
0.45000	0.719	0.711	0.708	0.707	0.705
0.50000	0.810	0.799	0.795	0.794	0.791
0.55000	0.910	0.895	0.890	0.888	0.884
0.60000	1.021	1.001	0.994	0.991	0.986
0.65000	1.145	1.119	1.111	1.106	1.100
0.70000	1.288	1.254	1.243	1.238	1.229
0.75000	1.457	1.412	1.398	1.390	1.380
0.80000	1.664	1.604	1.585	1.575	1.561
0.85000	1.933	1.850	1.824	1.811	1.791
0.90000	2.317	2.196	2.158	2.139	2.111
0.95000	2.991	2.789	2.726	2.695	2.649
0.96000	3.214	2.982	2.909	2.873	2.821
0.97000	3.506	3.231	3.145	3.103	3.042
0.98000	3.927	3.585	3.479	3.428	3.352
0.99000	4.675	4.199	4.054	3.983	3.881
0.99500	5.461	4.825	4.634	4.542	4.408
0.99900	7.451	6.336	6.011	5.856	5.634
0.99990	10.760	8.652	8.065	7.791	7.401

The table was generated using MINITAB

F-table for 4 Degrees of freedom in the denominator
numerator degrees of freedom

α	1	2	3	4	5
0.00010	0	0.01	0.006	0.006	0.0061
0.00100	0	0.02	0.018	0.019	0.0194
0.00500	0	0.04	0.041	0.043	0.0446
0.01000	0	0.06	0.060	0.063	0.0644
0.02000	0	0.08	0.088	0.092	0.0942
0.03000	0	0.10	0.111	0.116	0.1185
0.04000	0	0.13	0.132	0.137	0.1400
0.05000	0	0.14	0.152	0.157	0.1598
0.06000	0	0.16	0.170	0.175	0.1785
0.07000	0	0.18	0.188	0.193	0.1964
0.08000	0	0.20	0.205	0.210	0.2137
0.09000	0	0.21	0.222	0.227	0.2305
0.10000	0	0.23	0.239	0.243	0.2469
0.20000	0	0.40	0.402	0.403	0.4035
0.25000	1	0.50	0.489	0.484	0.4825
0.30000	1	0.61	0.581	0.570	0.5649
0.35000	1	0.72	0.682	0.663	0.6524
0.40000	1	0.86	0.793	0.763	0.7467
0.45000	1	1.02	0.919	0.875	0.8499
0.50000	2	1.21	1.063	1.000	0.9646
0.55000	2	1.44	1.231	1.143	1.0940
0.60000	3	1.72	1.432	1.310	1.2428
0.65000	4	2.08	1.676	1.509	1.4175
0.70000	6	2.56	1.984	1.753	1.6286
0.75000	9	3.23	2.390	2.064	1.8927
0.80000	14	4.24	2.955	2.483	2.2397
0.85000	25	5.91	3.817	3.092	2.7310
0.90000	56	9.24	5.343	4.107	3.5203
0.95000	225	19.25	9.117	6.388	5.1923
0.96000	351	24.25	10.752	7.305	5.8347
0.97000	625	32.58	13.250	8.648	6.7508
0.98000	1406	49.25	17.694	10.899	8.2331
0.99000	5625	99.25	28.710	15.977	11.3920
0.99500	22501	199.25	46.196	23.154	15.5558
0.99900	562463	999.26	137.100	53.438	31.0847
0.99990	56235252	9997.59	640.124	171.854	80.5218

The table was generated using MINITAB

F-table for 4 Degrees of freedom in the denominator
numerator degrees of freedom

α	6	7	8	9	10
0.00010	0.0059	0.0060	0.0064	0.0066	0.0067
0.00100	0.0199	0.0201	0.0205	0.0206	0.0208
0.00500	0.0455	0.0463	0.0468	0.0473	0.0477
0.01000	0.0657	0.0668	0.0676	0.0682	0.0687
0.02000	0.0960	0.0974	0.0984	0.0993	0.1000
0.03000	0.1205	0.1221	0.1234	0.1244	0.1252
0.04000	0.1423	0.1440	0.1453	0.1465	0.1474
0.05000	0.1623	0.1641	0.1655	0.1667	0.1677
0.06000	0.1810	0.1830	0.1845	0.1857	0.1867
0.07000	0.1990	0.2009	0.2024	0.2037	0.2047
0.08000	0.2162	0.2182	0.2198	0.2210	0.2220
0.09000	0.2330	0.2350	0.2365	0.2378	0.2388
0.10000	0.2494	0.2513	0.2529	0.2541	0.2551
0.20000	0.4043	0.4050	0.4056	0.4061	0.4066
0.25000	0.4816	0.4810	0.4807	0.4805	0.4803
0.30000	0.5615	0.5593	0.5576	0.5564	0.5555
0.35000	0.6458	0.6413	0.6380	0.6355	0.6335
0.40000	0.7360	0.7286	0.7232	0.7190	0.7158
0.45000	0.8340	0.8229	0.8148	0.8086	0.8036
0.50000	0.9419	0.9262	0.9147	0.9058	0.8988
0.55000	1.0627	1.0410	1.0252	1.0130	1.0034
0.60000	1.2003	1.1709	1.1496	1.1332	1.1203
0.65000	1.3603	1.3210	1.2923	1.2705	1.2533
0.70000	1.5513	1.4985	1.4603	1.4312	1.4084
0.75000	1.7872	1.7157	1.6642	1.6253	1.5949
0.80000	2.0925	1.9937	1.9230	1.8699	1.8286
0.85000	2.5163	2.3746	2.2740	2.1990	2.1408
0.90000	3.1807	2.9606	2.8065	2.6927	2.6053
0.95000	4.5337	4.1203	3.8379	3.6331	3.4780
0.96000	5.0377	4.5427	4.2072	3.9653	3.7831
0.97000	5.7439	5.1272	4.7132	4.4171	4.1953
0.98000	6.8595	6.0347	5.4890	5.1026	4.8156
0.99000	9.1482	7.8465	7.0061	6.4221	5.9944
0.99500	12.0275	10.0504	8.8052	7.9560	7.3429
0.99900	21.9238	17.1981	14.3917	12.5604	11.2828
0.99990	49.4166	35.2200	27.4917	22.7653	19.6293

Table generated using MINITAB

F-table for 4 Degrees of freedom in the denominator
numerator degrees of freedom

α	11	12	13	14	15
0.00010	0.006	0.006	0.006	0.006	0.006
0.00100	0.021	0.021	0.021	0.021	0.021
0.00500	0.048	0.048	0.048	0.048	0.048
0.01000	0.069	0.069	0.069	0.070	0.070
0.02000	0.100	0.101	0.101	0.101	0.102
0.03000	0.125	0.126	0.127	0.127	0.127
0.04000	0.148	0.148	0.149	0.149	0.150
0.05000	0.168	0.169	0.169	0.170	0.170
0.06000	0.187	0.188	0.188	0.189	0.189
0.07000	0.205	0.206	0.206	0.207	0.208
0.08000	0.222	0.223	0.224	0.224	0.225
0.09000	0.239	0.240	0.241	0.241	0.242
0.10000	0.256	0.256	0.257	0.257	0.258
0.20000	0.407	0.407	0.407	0.407	0.408
0.25000	0.480	0.480	0.480	0.480	0.480
0.30000	0.554	0.554	0.553	0.553	0.552
0.35000	0.632	0.630	0.629	0.628	0.627
0.40000	0.713	0.711	0.709	0.707	0.706
0.45000	0.799	0.796	0.793	0.791	0.789
0.50000	0.893	0.888	0.884	0.881	0.878
0.55000	0.995	0.989	0.983	0.979	0.975
0.60000	1.109	1.101	1.094	1.087	1.082
0.65000	1.239	1.228	1.218	1.210	1.203
0.70000	1.390	1.375	1.362	1.351	1.342
0.75000	1.570	1.550	1.533	1.519	1.507
0.80000	1.795	1.768	1.745	1.726	1.710
0.85000	2.094	2.056	2.025	1.998	1.976
0.90000	2.536	2.480	2.433	2.394	2.361
0.95000	3.356	3.259	3.179	3.112	3.055
0.96000	3.640	3.526	3.433	3.355	3.290
0.97000	4.023	3.885	3.773	3.680	3.601
0.98000	4.594	4.418	4.276	4.157	4.058
0.99000	5.668	5.411	5.205	5.035	4.893
0.99500	6.881	6.521	6.233	5.998	5.802
0.99900	10.346	9.632	9.072	8.622	8.252
0.99990	17.420	15.792	14.547	13.569	12.782

The table was generated using MINITAB

F-table for 4 Degrees of freedom in the denominator
numerator degrees of freedom

α	16	17	18	19	20
0.00010	0.006	0.006	0.006	0.006	0.006
0.00100	0.021	0.021	0.021	0.021	0.022
0.00500	0.049	0.049	0.049	0.049	0.051
0.01000	0.070	0.070	0.071	0.071	0.074
0.02000	0.102	0.102	0.103	0.103	0.106
0.03000	0.128	0.128	0.128	0.129	0.133
0.04000	0.150	0.151	0.151	0.151	0.156
0.05000	0.171	0.171	0.171	0.172	0.177
0.06000	0.190	0.190	0.191	0.191	0.196
0.07000	0.208	0.208	0.209	0.209	0.214
0.08000	0.225	0.226	0.226	0.226	0.232
0.09000	0.242	0.242	0.243	0.243	0.249
0.10000	0.258	0.259	0.259	0.259	0.265
0.20000	0.408	0.408	0.408	0.408	0.411
0.25000	0.480	0.480	0.480	0.480	0.480
0.30000	0.552	0.552	0.552	0.551	0.548
0.35000	0.627	0.626	0.626	0.625	0.618
0.40000	0.705	0.704	0.703	0.702	0.689
0.45000	0.787	0.785	0.784	0.783	0.763
0.50000	0.875	0.873	0.871	0.869	0.841
0.55000	0.971	0.968	0.966	0.963	0.925
0.60000	1.077	1.073	1.070	1.067	1.015
0.65000	1.197	1.191	1.187	1.182	1.115
0.70000	1.334	1.327	1.321	1.315	1.227
0.75000	1.496	1.487	1.479	1.471	1.356
0.80000	1.696	1.683	1.672	1.663	1.510
0.85000	1.956	1.939	1.924	1.911	1.704
0.90000	2.332	2.307	2.285	2.266	1.970
0.95000	3.006	2.964	2.927	2.895	2.412
0.96000	3.233	3.184	3.142	3.104	2.552
0.97000	3.534	3.476	3.425	3.380	2.731
0.98000	3.973	3.900	3.836	3.780	2.981
0.99000	4.772	4.669	4.579	4.500	3.405
0.99500	5.637	5.496	5.374	5.268	3.825
0.99900	7.944	7.683	7.459	7.265	4.793
0.99990	12.136	11.598	11.142	10.752	6.174

Table generated using MINITAB

F-table for 4 Degrees of freedom in the denominator

numerator degrees of freedom

α	25	50	75	100	200
0.00010	0.006	0.006	0.006	0.006	0.006
0.00100	0.021	0.022	0.022	0.022	0.022
0.00500	0.049	0.050	0.051	0.051	0.051
0.01000	0.071	0.073	0.073	0.073	0.073
0.02000	0.104	0.105	0.106	0.106	0.106
0.03000	0.130	0.131	0.132	0.132	0.133
0.04000	0.152	0.154	0.155	0.155	0.156
0.05000	0.173	0.175	0.176	0.176	0.177
0.06000	0.192	0.194	0.195	0.195	0.196
0.07000	0.210	0.213	0.213	0.214	0.214
0.08000	0.228	0.230	0.231	0.231	0.232
0.09000	0.244	0.247	0.248	0.248	0.249
0.10000	0.261	0.263	0.264	0.264	0.265
0.20000	0.409	0.410	0.411	0.411	0.411
0.25000	0.480	0.480	0.480	0.480	0.480
0.30000	0.551	0.549	0.549	0.549	0.548
0.35000	0.623	0.620	0.619	0.618	0.618
0.40000	0.698	0.693	0.691	0.690	0.689
0.45000	0.778	0.769	0.767	0.765	0.763
0.50000	0.862	0.850	0.846	0.844	0.842
0.55000	0.953	0.937	0.932	0.929	0.925
0.60000	1.053	1.032	1.025	1.021	1.016
0.65000	1.164	1.136	1.127	1.123	1.116
0.70000	1.291	1.255	1.243	1.237	1.228
0.75000	1.440	1.392	1.377	1.369	1.357
0.80000	1.621	1.558	1.537	1.527	1.512
0.85000	1.854	1.768	1.740	1.726	1.706
0.90000	2.184	2.060	2.021	2.001	1.973
0.95000	2.758	2.557	2.493	2.462	2.416
0.96000	2.947	2.716	2.644	2.609	2.557
0.97000	3.194	2.923	2.838	2.797	2.736
0.98000	3.549	3.215	3.112	3.062	2.988
0.99000	4.177	3.719	3.580	3.512	3.414
0.99500	4.835	4.231	4.050	3.963	3.836
0.99900	6.493	5.459	5.159	5.016	4.811
0.99990	9.239	7.329	6.801	6.554	6.205

The table was generated using MINITAB

F-table for 5 Degrees of freedom in the denominator
numerator degrees of freedom

α	1	2	3	4	5
0.00010	0	0.01	0.012	0.013	0.0129
0.00100	0	0.03	0.030	0.032	0.0336
0.00500	0	0.05	0.060	0.064	0.0669
0.01000	0	0.08	0.083	0.088	0.0912
0.02000	0	0.11	0.115	0.121	0.1257
0.03000	0	0.13	0.141	0.148	0.1530
0.04000	0	0.15	0.164	0.171	0.1765
0.05000	0	0.17	0.185	0.193	0.1980
0.06000	0	0.19	0.204	0.212	0.2180
0.07000	0	0.21	0.223	0.231	0.2369
0.08000	0	0.23	0.241	0.249	0.2550
0.09000	0	0.25	0.259	0.267	0.2725
0.10000	0	0.26	0.276	0.284	0.2896
0.20000	0	0.44	0.444	0.446	0.4489
0.25000	1	0.54	0.531	0.528	0.5278
0.30000	1	0.65	0.623	0.614	0.6094
0.35000	1	0.77	0.724	0.705	0.6955
0.40000	1	0.90	0.835	0.805	0.7879
0.45000	1	1.06	0.960	0.914	0.8886
0.50000	2	1.25	1.102	1.037	1.0000
0.55000	2	1.48	1.268	1.177	1.1254
0.60000	3	1.76	1.466	1.339	1.2693
0.65000	4	2.13	1.707	1.533	1.4378
0.70000	6	2.61	2.011	1.770	1.6410
0.75000	9	3.28	2.409	2.072	1.8947
0.80000	14	4.28	2.965	2.478	2.2276
0.85000	25	5.96	3.811	3.068	2.6980
0.90000	57	9.29	5.309	4.050	3.4531
0.95000	230	19.30	9.013	6.256	5.0502
0.96000	360	24.30	10.617	7.143	5.6635
0.97000	640	32.63	13.069	8.441	6.5383
0.98000	1441	49.30	17.429	10.616	7.9530
0.99000	5764	99.30	28.237	15.522	10.9670
0.99500	23056	199.30	45.390	22.456	14.9397
0.99900	576409	999.29	134.584	51.711	29.7527
0.99990	57626396	9997.20	628.113	166.121	76.9060

The table was generated using MINITAB

F-table for 5 Degrees of freedom in the denominator
numerator degrees of freedom

α	6	7	8	9	10
0.00010	0.0130	0.0138	0.0142	0.0142	0.0148
0.00100	0.0347	0.0354	0.0362	0.0368	0.0373
0.00500	0.0689	0.0705	0.0716	0.0726	0.0734
0.01000	0.0937	0.0957	0.0972	0.0984	0.0995
0.02000	0.1289	0.1313	0.1333	0.1349	0.1362
0.03000	0.1565	0.1593	0.1614	0.1632	0.1647
0.04000	0.1804	0.1833	0.1857	0.1876	0.1892
0.05000	0.2020	0.2051	0.2075	0.2096	0.2112
0.06000	0.2221	0.2253	0.2278	0.2299	0.2315
0.07000	0.2411	0.2443	0.2469	0.2490	0.2507
0.08000	0.2592	0.2625	0.2650	0.2671	0.2689
0.09000	0.2767	0.2799	0.2825	0.2846	0.2864
0.10000	0.2937	0.2969	0.2995	0.3015	0.3033
0.20000	0.4510	0.4527	0.4541	0.4553	0.4563
0.25000	0.5279	0.5281	0.5285	0.5288	0.5292
0.30000	0.6067	0.6051	0.6040	0.6032	0.6026
0.35000	0.6893	0.6852	0.6822	0.6799	0.6782
0.40000	0.7773	0.7699	0.7646	0.7605	0.7573
0.45000	0.8723	0.8610	0.8527	0.8464	0.8414
0.50000	0.9766	0.9603	0.9483	0.9392	0.9319
0.55000	1.0928	1.0703	1.0537	1.0410	1.0310
0.60000	1.2249	1.1942	1.1718	1.1547	1.1412
0.65000	1.3780	1.3370	1.3070	1.2842	1.2662
0.70000	1.5605	1.5055	1.4656	1.4352	1.4114
0.75000	1.7852	1.7110	1.6575	1.6170	1.5853
0.80000	2.0756	1.9735	1.9005	1.8455	1.8027
0.85000	2.4780	2.3324	2.2290	2.1519	2.0922
0.90000	3.1076	2.8834	2.7264	2.6106	2.5216
0.95000	4.3874	3.9715	3.6876	3.4816	3.3259
0.96000	4.8637	4.3674	4.0311	3.7889	3.6064
0.97000	5.5311	4.9150	4.5019	4.2065	3.9853
0.98000	6.5847	5.7646	5.2227	4.8395	4.5549
0.99000	8.7460	7.4605	6.6319	6.0570	5.6363
0.99500	11.4637	9.5220	8.3018	7.4712	6.8724
0.99900	20.8027	16.2058	13.4847	11.7137	10.4807
0.99990	46.7440	33.0546	25.6340	21.1113	18.1197

Table generated using MINITAB

F-table for 5 Degrees of freedom in the denominator
numerator degrees of freedom

α	11	12	13	14	15
0.00010	0.015	0.014	0.015	0.015	0.015
0.00100	0.037	0.037	0.038	0.038	0.038
0.00500	0.074	0.074	0.075	0.075	0.076
0.01000	0.100	0.101	0.101	0.102	0.102
0.02000	0.137	0.138	0.139	0.139	0.140
0.03000	0.165	0.167	0.167	0.168	0.169
0.04000	0.190	0.191	0.192	0.193	0.194
0.05000	0.212	0.213	0.214	0.215	0.216
0.06000	0.233	0.234	0.235	0.236	0.237
0.07000	0.252	0.253	0.254	0.255	0.256
0.08000	0.270	0.271	0.272	0.273	0.274
0.09000	0.287	0.289	0.290	0.291	0.292
0.10000	0.304	0.306	0.307	0.308	0.308
0.20000	0.457	0.458	0.458	0.459	0.459
0.25000	0.529	0.529	0.530	0.530	0.530
0.30000	0.602	0.601	0.601	0.601	0.601
0.35000	0.676	0.675	0.674	0.674	0.673
0.40000	0.754	0.752	0.750	0.749	0.748
0.45000	0.837	0.834	0.831	0.828	0.826
0.50000	0.926	0.921	0.917	0.913	0.910
0.55000	1.022	1.016	1.010	1.005	1.001
0.60000	1.130	1.121	1.113	1.107	1.101
0.65000	1.251	1.239	1.229	1.221	1.213
0.70000	1.392	1.376	1.363	1.351	1.342
0.75000	1.559	1.539	1.521	1.506	1.493
0.80000	1.768	1.740	1.716	1.697	1.680
0.85000	2.044	2.005	1.973	1.946	1.922
0.90000	2.451	2.394	2.346	2.307	2.273
0.95000	3.203	3.105	3.025	2.958	2.901
0.96000	3.464	3.350	3.256	3.178	3.113
0.97000	3.813	3.676	3.565	3.472	3.393
0.98000	4.335	4.161	4.020	3.903	3.805
0.99000	5.316	5.064	4.861	4.695	4.555
0.99500	6.421	6.071	5.791	5.562	5.372
0.99900	9.578	8.892	8.354	7.921	7.567
0.99990	16.017	14.470	13.290	12.365	11.620

The table was generated using MINITAB

F-table for 5 Degrees of freedom in the denominator
numerator degrees of freedom

α	16	17	18	19	20
0.00010	0.015	0.015	0.015	0.015	0.015
0.00100	0.038	0.039	0.039	0.039	0.039
0.00500	0.076	0.076	0.077	0.077	0.077
0.01000	0.103	0.103	0.104	0.104	0.104
0.02000	0.141	0.141	0.141	0.142	0.142
0.03000	0.170	0.170	0.171	0.171	0.171
0.04000	0.194	0.195	0.196	0.196	0.196
0.05000	0.217	0.217	0.218	0.218	0.219
0.06000	0.237	0.238	0.239	0.239	0.239
0.07000	0.257	0.257	0.258	0.258	0.259
0.08000	0.275	0.275	0.276	0.277	0.277
0.09000	0.292	0.293	0.294	0.294	0.295
0.10000	0.309	0.310	0.310	0.311	0.311
0.20000	0.460	0.460	0.461	0.461	0.461
0.25000	0.530	0.530	0.531	0.531	0.531
0.30000	0.601	0.600	0.600	0.600	0.600
0.35000	0.672	0.672	0.671	0.671	0.671
0.40000	0.747	0.746	0.745	0.744	0.743
0.45000	0.825	0.823	0.822	0.820	0.819
0.50000	0.908	0.905	0.903	0.902	0.900
0.55000	0.998	0.994	0.992	0.989	0.987
0.60000	1.096	1.092	1.088	1.085	1.082
0.65000	1.207	1.201	1.196	1.192	1.188
0.70000	1.333	1.326	1.319	1.314	1.308
0.75000	1.482	1.473	1.464	1.456	1.449
0.80000	1.665	1.652	1.641	1.630	1.621
0.85000	1.902	1.885	1.869	1.855	1.843
0.90000	2.243	2.218	2.195	2.175	2.158
0.95000	2.852	2.810	2.772	2.740	2.710
0.96000	3.056	3.007	2.965	2.927	2.893
0.97000	3.326	3.268	3.218	3.173	3.134
0.98000	3.721	3.649	3.586	3.530	3.481
0.99000	4.437	4.335	4.247	4.170	4.102
0.99500	5.211	5.074	4.956	4.852	4.761
0.99900	7.271	7.021	6.807	6.622	6.460
0.99990	11.010	10.502	10.072	9.705	9.387

Table generated using MINITAB

F-table for 5 Degrees of freedom in the denominator

numerator degrees of freedom

α	25	50	75	100	200
0.00010	0.015	0.015	0.015	0.015	0.015
0.00100	0.039	0.041	0.041	0.041	0.041
0.00500	0.078	0.080	0.081	0.081	0.081
0.01000	0.105	0.108	0.109	0.109	0.110
0.02000	0.144	0.147	0.148	0.148	0.149
0.03000	0.173	0.176	0.178	0.178	0.179
0.04000	0.198	0.202	0.203	0.204	0.205
0.05000	0.221	0.224	0.226	0.227	0.228
0.06000	0.241	0.245	0.247	0.247	0.248
0.07000	0.261	0.265	0.266	0.267	0.268
0.08000	0.279	0.283	0.284	0.285	0.286
0.09000	0.296	0.300	0.302	0.303	0.304
0.10000	0.313	0.317	0.319	0.319	0.320
0.20000	0.462	0.465	0.466	0.467	0.467
0.25000	0.531	0.533	0.533	0.534	0.534
0.30000	0.600	0.600	0.600	0.600	0.599
0.35000	0.669	0.667	0.666	0.666	0.665
0.40000	0.741	0.735	0.734	0.733	0.732
0.45000	0.815	0.807	0.804	0.803	0.801
0.50000	0.894	0.882	0.878	0.876	0.873
0.55000	0.978	0.962	0.956	0.953	0.949
0.60000	1.070	1.048	1.041	1.037	1.031
0.65000	1.173	1.143	1.133	1.129	1.121
0.70000	1.289	1.250	1.237	1.231	1.222
0.75000	1.424	1.373	1.357	1.349	1.337
0.80000	1.587	1.521	1.500	1.489	1.473
0.85000	1.797	1.708	1.679	1.665	1.643
0.90000	2.092	1.966	1.925	1.905	1.876
0.95000	2.602	2.400	2.336	2.305	2.259
0.96000	2.770	2.539	2.467	2.431	2.379
0.97000	2.988	2.718	2.634	2.593	2.533
0.98000	3.302	2.972	2.870	2.820	2.748
0.99000	3.854	3.407	3.271	3.205	3.109
0.99500	4.432	3.848	3.673	3.589	3.467
0.99900	5.885	4.901	4.616	4.481	4.287
0.99990	8.284	6.498	6.006	5.776	5.451

The table was generated using MINITAB

F-table for 6 Degrees of freedom in the denominator
numerator degrees of freedom

α	1	2	3	4	5
0.00010	0	0.02	0.018	0.020	0.0217
0.00100	0	0.04	0.042	0.046	0.0482
0.00500	0	0.07	0.077	0.083	0.0872
0.01000	0	0.09	0.102	0.109	0.1144
0.02000	0	0.12	0.137	0.146	0.1519
0.03000	0	0.15	0.165	0.174	0.1808
0.04000	0	0.17	0.189	0.198	0.2056
0.05000	0	0.19	0.210	0.221	0.2279
0.06000	0	0.21	0.231	0.241	0.2486
0.07000	0	0.23	0.250	0.260	0.2681
0.08000	0	0.25	0.268	0.279	0.2866
0.09000	0	0.27	0.286	0.297	0.3045
0.10000	0	0.29	0.304	0.314	0.3218
0.20000	0	0.47	0.473	0.478	0.4818
0.25000	1	0.57	0.560	0.560	0.5601
0.30000	1	0.68	0.653	0.645	0.6408
0.35000	1	0.80	0.753	0.735	0.7257
0.40000	1	0.93	0.863	0.833	0.8164
0.45000	2	1.09	0.987	0.941	0.9151
0.50000	2	1.28	1.129	1.062	1.0240
0.55000	2	1.51	1.293	1.199	1.1465
0.60000	3	1.80	1.489	1.359	1.2866
0.65000	4	2.16	1.727	1.548	1.4507
0.70000	6	2.64	2.028	1.781	1.6482
0.75000	9	3.31	2.422	2.077	1.8945
0.80000	14	4.32	2.971	2.473	2.2174
0.85000	26	5.99	3.806	3.050	2.6734
0.90000	58	9.33	5.285	4.010	3.4045
0.95000	234	19.33	8.941	6.163	4.9503
0.96000	366	24.33	10.524	7.028	5.5437
0.97000	651	32.67	12.943	8.296	6.3897
0.98000	1465	49.33	17.245	10.419	7.7577
0.99000	5859	99.33	27.911	15.207	10.6721
0.99500	23436	199.34	44.838	21.975	14.5134
0.99900	585904	999.38	132.847	50.525	28.8347
0.99990	58575016	9997.26	619.822	162.173	74.4208

The table was generated using MINITAB

F-table for 6 Degrees of freedom in the denominator
numerator degrees of freedom

α	6	7	8	9	10
0.00010	0.0220	0.0232	0.0239	0.0238	0.0248
0.00100	0.0500	0.0514	0.0524	0.0535	0.0542
0.00500	0.0904	0.0927	0.0947	0.0963	0.0976
0.01000	0.1181	0.1210	0.1234	0.1254	0.1270
0.02000	0.1564	0.1600	0.1628	0.1652	0.1671
0.03000	0.1858	0.1897	0.1929	0.1954	0.1976
0.04000	0.2109	0.2150	0.2183	0.2211	0.2233
0.05000	0.2334	0.2377	0.2412	0.2440	0.2463
0.06000	0.2542	0.2586	0.2621	0.2649	0.2674
0.07000	0.2737	0.2781	0.2817	0.2846	0.2870
0.08000	0.2923	0.2967	0.3003	0.3032	0.3056
0.09000	0.3101	0.3145	0.3181	0.3210	0.3234
0.10000	0.3274	0.3317	0.3352	0.3381	0.3405
0.20000	0.4850	0.4876	0.4898	0.4916	0.4931
0.25000	0.5611	0.5621	0.5631	0.5639	0.5647
0.30000	0.6389	0.6378	0.6371	0.6367	0.6364
0.35000	0.7199	0.7161	0.7134	0.7114	0.7099
0.40000	0.8059	0.7987	0.7935	0.7895	0.7864
0.45000	0.8985	0.8871	0.8788	0.8724	0.8674
0.50000	1.0000	0.9833	0.9711	0.9618	0.9543
0.55000	1.1129	1.0896	1.0726	1.0595	1.0492
0.60000	1.2409	1.2093	1.1861	1.1684	1.1545
0.65000	1.3891	1.3467	1.3157	1.2921	1.2735
0.70000	1.5653	1.5087	1.4674	1.4361	1.4115
0.75000	1.7821	1.7059	1.6508	1.6091	1.5765
0.80000	2.0619	1.9575	1.8825	1.8263	1.7822
0.85000	2.4493	2.3008	2.1955	2.1168	2.0557
0.90000	3.0546	2.8274	2.6684	2.5508	2.4605
0.95000	4.2839	3.8660	3.5806	3.3738	3.2172
0.96000	4.7412	4.2436	3.9065	3.6637	3.4808
0.97000	5.3816	4.7657	4.3527	4.0577	3.8366
0.98000	6.3927	5.5756	5.0359	4.6545	4.3715
0.99000	8.4661	7.1913	6.3707	5.8018	5.3858
0.99500	11.0731	9.1555	7.9519	7.1338	6.5446
0.99900	20.0296	15.5210	12.8580	11.1282	9.9258
0.99990	44.9066	31.5648	24.3557	19.9731	17.0799

Table generated using MINITAB

F-table for 6 Degrees of freedom in the denominator
numerator degrees of freedom

α	11	12	13	14	15
0.00010	0.025	0.025	0.025	0.025	0.025
0.00100	0.054	0.055	0.056	0.056	0.056
0.00500	0.098	0.099	0.100	0.101	0.102
0.01000	0.128	0.129	0.130	0.131	0.132
0.02000	0.168	0.170	0.171	0.172	0.173
0.03000	0.199	0.200	0.202	0.203	0.204
0.04000	0.225	0.226	0.228	0.229	0.230
0.05000	0.248	0.250	0.251	0.252	0.253
0.06000	0.269	0.271	0.272	0.274	0.275
0.07000	0.289	0.290	0.292	0.293	0.294
0.08000	0.307	0.309	0.310	0.312	0.313
0.09000	0.325	0.327	0.328	0.330	0.331
0.10000	0.342	0.344	0.345	0.347	0.348
0.20000	0.494	0.495	0.496	0.497	0.498
0.25000	0.565	0.566	0.566	0.567	0.567
0.30000	0.636	0.636	0.636	0.636	0.636
0.35000	0.708	0.707	0.707	0.706	0.705
0.40000	0.784	0.781	0.780	0.778	0.777
0.45000	0.863	0.860	0.857	0.854	0.852
0.50000	0.948	0.943	0.939	0.935	0.932
0.55000	1.040	1.033	1.028	1.023	1.018
0.60000	1.143	1.133	1.125	1.119	1.113
0.65000	1.258	1.246	1.235	1.226	1.219
0.70000	1.391	1.375	1.361	1.349	1.339
0.75000	1.550	1.528	1.510	1.495	1.482
0.80000	1.747	1.718	1.694	1.673	1.656
0.85000	2.007	1.967	1.934	1.906	1.882
0.90000	2.389	2.331	2.283	2.242	2.208
0.95000	3.094	2.996	2.915	2.847	2.790
0.96000	3.338	3.223	3.130	3.052	2.986
0.97000	3.665	3.528	3.416	3.324	3.245
0.98000	4.153	3.980	3.839	3.723	3.625
0.99000	5.069	4.820	4.620	4.455	4.318
0.99500	6.101	5.757	5.481	5.257	5.070
0.99900	9.046	8.378	7.855	7.435	7.091
0.99990	15.050	13.559	12.424	11.533	10.819

The table was generated using MINITAB

F-table for 6 Degrees of freedom in the denominator
numerator degrees of freedom

α	16	17	18	19	20
0.00010	0.026	0.026	0.026	0.026	0.026
0.00100	0.057	0.057	0.057	0.058	0.058
0.00500	0.102	0.103	0.103	0.103	0.104
0.01000	0.133	0.133	0.134	0.134	0.135
0.02000	0.174	0.175	0.175	0.176	0.176
0.03000	0.205	0.206	0.207	0.207	0.208
0.04000	0.231	0.232	0.233	0.234	0.234
0.05000	0.255	0.255	0.256	0.257	0.258
0.06000	0.276	0.277	0.277	0.278	0.279
0.07000	0.295	0.296	0.297	0.298	0.299
0.08000	0.314	0.315	0.316	0.317	0.317
0.09000	0.332	0.333	0.334	0.334	0.335
0.10000	0.349	0.350	0.351	0.351	0.352
0.20000	0.499	0.499	0.500	0.500	0.501
0.25000	0.567	0.568	0.568	0.568	0.569
0.30000	0.636	0.636	0.636	0.636	0.636
0.35000	0.705	0.704	0.704	0.704	0.703
0.40000	0.776	0.775	0.774	0.773	0.773
0.45000	0.851	0.849	0.848	0.846	0.845
0.50000	0.930	0.927	0.925	0.923	0.922
0.55000	1.015	1.011	1.009	1.006	1.004
0.60000	1.108	1.104	1.100	1.096	1.093
0.65000	1.212	1.206	1.201	1.196	1.192
0.70000	1.331	1.323	1.316	1.310	1.305
0.75000	1.470	1.460	1.451	1.443	1.436
0.80000	1.640	1.627	1.615	1.605	1.596
0.85000	1.861	1.843	1.827	1.813	1.800
0.90000	2.178	2.152	2.129	2.109	2.091
0.95000	2.741	2.698	2.661	2.628	2.598
0.96000	2.929	2.880	2.838	2.800	2.766
0.97000	3.179	3.121	3.070	3.026	2.986
0.98000	3.542	3.470	3.408	3.353	3.304
0.99000	4.201	4.101	4.014	3.938	3.871
0.99500	4.913	4.778	4.662	4.561	4.472
0.99900	6.804	6.562	6.354	6.175	6.018
0.99990	10.233	9.746	9.334	8.982	8.678

Table generated using MINITAB

F-table for 6 Degrees of freedom in the denominator
numerator degrees of freedom

α	25	50	75	100	200
0.00010	0.026	0.028	0.027	0.028	0.028
0.00100	0.059	0.061	0.062	0.062	0.062
0.00500	0.105	0.108	0.110	0.110	0.111
0.01000	0.137	0.141	0.142	0.143	0.144
0.02000	0.179	0.183	0.185	0.186	0.187
0.03000	0.210	0.215	0.217	0.218	0.220
0.04000	0.237	0.242	0.244	0.245	0.247
0.05000	0.260	0.266	0.268	0.269	0.270
0.06000	0.282	0.287	0.289	0.290	0.292
0.07000	0.301	0.307	0.309	0.310	0.312
0.08000	0.320	0.326	0.328	0.329	0.331
0.09000	0.338	0.344	0.346	0.347	0.348
0.10000	0.355	0.361	0.363	0.364	0.365
0.20000	0.503	0.507	0.508	0.509	0.510
0.25000	0.570	0.572	0.573	0.574	0.574
0.30000	0.636	0.637	0.637	0.637	0.637
0.35000	0.702	0.701	0.700	0.700	0.699
0.40000	0.770	0.766	0.764	0.763	0.762
0.45000	0.841	0.833	0.830	0.829	0.827
0.50000	0.915	0.903	0.899	0.897	0.894
0.55000	0.995	0.977	0.972	0.969	0.965
0.60000	1.081	1.058	1.050	1.046	1.040
0.65000	1.176	1.146	1.135	1.130	1.123
0.70000	1.284	1.244	1.231	1.224	1.214
0.75000	1.409	1.357	1.340	1.332	1.319
0.80000	1.560	1.492	1.470	1.459	1.442
0.85000	1.753	1.661	1.632	1.617	1.595
0.90000	2.024	1.895	1.854	1.833	1.803
0.95000	2.490	2.286	2.222	2.190	2.144
0.96000	2.642	2.411	2.338	2.303	2.250
0.97000	2.841	2.571	2.487	2.446	2.386
0.98000	3.125	2.798	2.696	2.647	2.575
0.99000	3.627	3.186	3.052	2.987	2.893
0.99500	4.149	3.578	3.407	3.325	3.205
0.99900	5.461	4.511	4.237	4.107	3.920
0.99990	7.624	5.922	5.454	5.237	4.929

The table was generated using MINITAB

F-table for 7 Degrees of freedom in the denominator

numerator degrees of freedom

α	1	2	3	4	5
0.00010	0	0.02	0.026	0.028	0.0305
0.00100	0	0.05	0.053	0.058	0.0618
0.00500	0	0.08	0.092	0.100	0.1050
0.01000	0	0.10	0.118	0.127	0.1340
0.02000	0	0.14	0.155	0.166	0.1734
0.03000	0	0.17	0.183	0.195	0.2034
0.04000	0	0.19	0.208	0.220	0.2290
0.05000	0	0.21	0.230	0.243	0.2518
0.06000	0	0.23	0.251	0.264	0.2728
0.07000	0	0.25	0.270	0.283	0.2925
0.08000	0	0.27	0.289	0.302	0.3113
0.09000	0	0.29	0.307	0.320	0.3294
0.10000	0	0.31	0.325	0.338	0.3468
0.20000	0	0.49	0.495	0.502	0.5067
0.25000	1	0.59	0.582	0.583	0.5844
0.30000	1	0.70	0.675	0.667	0.6642
0.35000	1	0.82	0.774	0.757	0.7480
0.40000	1	0.95	0.884	0.854	0.8374
0.45000	2	1.11	1.008	0.961	0.9344
0.50000	2	1.30	1.148	1.080	1.0414
0.55000	3	1.53	1.312	1.215	1.1615
0.60000	3	1.82	1.506	1.372	1.2988
0.65000	4	2.18	1.742	1.559	1.4595
0.70000	6	2.66	2.040	1.788	1.6527
0.75000	9	3.34	2.430	2.079	1.8935
0.80000	14	4.34	2.974	2.469	2.2091
0.85000	26	6.01	3.801	3.036	2.6543
0.90000	59	9.35	5.266	3.979	3.3680
0.95000	237	19.35	8.887	6.094	4.8758
0.96000	370	24.35	10.454	6.944	5.4544
0.97000	658	32.69	12.850	8.189	6.2795
0.98000	1482	49.36	17.111	10.274	7.6137
0.99000	5929	99.35	27.672	14.976	10.4556
0.99500	23715	199.36	44.434	21.621	14.2003
0.99900	592890	999.40	131.580	49.658	28.1627
0.99990	59272864	9997.87	613.806	159.297	72.6058

The table was generated using MINITAB

F-table for 7 Degrees of freedom in the denominator
numerator degrees of freedom

α	6	7	8	9	10
0.00010	0.0312	0.0323	0.0345	0.0335	0.0343
0.00100	0.0643	0.0667	0.0683	0.0698	0.0710
0.00500	0.1092	0.1126	0.1152	0.1175	0.1193
0.01000	0.1391	0.1430	0.1462	0.1489	0.1511
0.02000	0.1793	0.1840	0.1877	0.1908	0.1934
0.03000	0.2098	0.2148	0.2189	0.2222	0.2250
0.04000	0.2356	0.2409	0.2451	0.2486	0.2515
0.05000	0.2587	0.2641	0.2684	0.2720	0.2750
0.06000	0.2798	0.2853	0.2897	0.2933	0.2964
0.07000	0.2996	0.3051	0.3095	0.3132	0.3163
0.08000	0.3183	0.3239	0.3283	0.3319	0.3350
0.09000	0.3363	0.3418	0.3462	0.3498	0.3529
0.10000	0.3537	0.3591	0.3634	0.3670	0.3700
0.20000	0.5109	0.5143	0.5171	0.5195	0.5215
0.25000	0.5862	0.5878	0.5893	0.5906	0.5918
0.30000	0.6628	0.6622	0.6620	0.6619	0.6620
0.35000	0.7426	0.7390	0.7366	0.7349	0.7336
0.40000	0.8269	0.8199	0.8148	0.8109	0.8079
0.45000	0.9177	0.9062	0.8979	0.8914	0.8864
0.50000	1.0169	1.0000	0.9876	0.9780	0.9705
0.55000	1.1272	1.1035	1.0860	1.0726	1.0621
0.60000	1.2520	1.2197	1.1960	1.1778	1.1635
0.65000	1.3965	1.3531	1.3213	1.2971	1.2780
0.70000	1.5680	1.5101	1.4679	1.4358	1.4105
0.75000	1.7789	1.7012	1.6448	1.6022	1.5688
0.80000	2.0509	1.9445	1.8681	1.8107	1.7658
0.85000	2.4270	2.2765	2.1694	2.0894	2.0273
0.90000	3.0145	2.7849	2.6241	2.5053	2.4139
0.95000	4.2067	3.7870	3.5004	3.2928	3.1355
0.96000	4.6499	4.1512	3.8134	3.5700	3.3866
0.97000	5.2708	4.6545	4.2416	3.9466	3.7256
0.98000	6.2508	5.4355	4.8972	4.5170	4.2347
0.99000	8.2601	6.9929	6.1777	5.6128	5.2001
0.99500	10.7860	8.8855	7.6942	6.8850	6.3025
0.99900	19.4634	15.0186	12.3981	10.6980	9.5175
0.99990	43.5643	30.4755	23.4200	19.1395	16.3182

Table generated using MINITAB

F-table for 7 Degrees of freedom in the denominator
numerator degrees of freedom

α	11	12	13	14	15
0.00010	0.036	0.036	0.037	0.037	0.038
0.00100	0.072	0.072	0.073	0.074	0.075
0.00500	0.120	0.122	0.123	0.124	0.125
0.01000	0.152	0.154	0.156	0.157	0.158
0.02000	0.195	0.197	0.199	0.200	0.201
0.03000	0.227	0.229	0.231	0.232	0.234
0.04000	0.254	0.256	0.258	0.259	0.261
0.05000	0.277	0.279	0.281	0.283	0.284
0.06000	0.299	0.301	0.303	0.304	0.306
0.07000	0.318	0.321	0.323	0.324	0.326
0.08000	0.337	0.339	0.341	0.343	0.345
0.09000	0.355	0.357	0.359	0.361	0.362
0.10000	0.372	0.374	0.376	0.378	0.379
0.20000	0.523	0.524	0.526	0.527	0.528
0.25000	0.592	0.593	0.594	0.595	0.595
0.30000	0.662	0.662	0.662	0.662	0.662
0.35000	0.732	0.731	0.731	0.730	0.730
0.40000	0.805	0.803	0.802	0.800	0.799
0.45000	0.882	0.879	0.876	0.873	0.871
0.50000	0.964	0.959	0.955	0.951	0.948
0.55000	1.053	1.046	1.040	1.035	1.031
0.60000	1.151	1.142	1.134	1.127	1.121
0.65000	1.262	1.249	1.239	1.229	1.222
0.70000	1.390	1.373	1.359	1.347	1.336
0.75000	1.541	1.519	1.501	1.485	1.471
0.80000	1.729	1.700	1.675	1.654	1.636
0.85000	1.977	1.937	1.903	1.875	1.850
0.90000	2.341	2.282	2.234	2.193	2.158
0.95000	3.012	2.913	2.832	2.764	2.706
0.96000	3.243	3.129	3.035	2.957	2.890
0.97000	3.554	3.417	3.305	3.213	3.134
0.98000	4.017	3.845	3.705	3.589	3.491
0.99000	4.886	4.639	4.441	4.277	4.141
0.99500	5.864	5.524	5.252	5.031	4.847
0.99900	8.655	8.000	7.488	7.077	6.740
0.99990	14.341	12.891	11.788	10.924	10.230

The table was generated using MINITAB

F-table for 7 Degrees of freedom in the denominator
numerator degrees of freedom

α	16	17	18	19	20
0.00010	0.038	0.038	0.038	0.037	0.038
0.00100	0.075	0.076	0.076	0.077	0.077
0.00500	0.126	0.127	0.127	0.128	0.129
0.01000	0.159	0.160	0.161	0.161	0.162
0.02000	0.203	0.204	0.205	0.205	0.206
0.03000	0.235	0.236	0.237	0.238	0.239
0.04000	0.262	0.263	0.264	0.265	0.266
0.05000	0.286	0.287	0.288	0.289	0.290
0.06000	0.307	0.308	0.310	0.311	0.311
0.07000	0.327	0.328	0.330	0.330	0.331
0.08000	0.346	0.347	0.348	0.349	0.350
0.09000	0.364	0.365	0.366	0.367	0.368
0.10000	0.381	0.382	0.383	0.384	0.385
0.20000	0.529	0.530	0.530	0.531	0.532
0.25000	0.596	0.597	0.597	0.597	0.598
0.30000	0.662	0.663	0.663	0.663	0.663
0.35000	0.729	0.729	0.729	0.728	0.728
0.40000	0.798	0.797	0.796	0.796	0.795
0.45000	0.870	0.868	0.867	0.865	0.864
0.50000	0.945	0.943	0.941	0.939	0.937
0.55000	1.027	1.023	1.020	1.018	1.015
0.60000	1.116	1.111	1.107	1.103	1.100
0.65000	1.215	1.209	1.203	1.199	1.194
0.70000	1.327	1.320	1.313	1.306	1.301
0.75000	1.460	1.449	1.440	1.432	1.425
0.80000	1.621	1.607	1.595	1.584	1.575
0.85000	1.829	1.811	1.795	1.780	1.767
0.90000	2.127	2.101	2.078	2.058	2.039
0.95000	2.657	2.614	2.576	2.543	2.513
0.96000	2.834	2.785	2.742	2.704	2.670
0.97000	3.067	3.009	2.959	2.914	2.875
0.98000	3.408	3.337	3.274	3.219	3.171
0.99000	4.025	3.926	3.840	3.765	3.698
0.99500	4.692	4.559	4.444	4.344	4.256
0.99900	6.460	6.223	6.020	5.845	5.692
0.99990	9.663	9.190	8.792	8.451	8.157

Table generated using MINITAB

F-table for 7 Degrees of freedom in the denominator

numerator degrees of freedom

α	25	50	75	100	200
0.00010	0.038	0.042	0.042	0.042	0.042
0.00100	0.078	0.082	0.083	0.083	0.084
0.00500	0.131	0.135	0.137	0.138	0.139
0.01000	0.165	0.170	0.172	0.173	0.175
0.02000	0.209	0.216	0.218	0.219	0.221
0.03000	0.242	0.249	0.252	0.253	0.255
0.04000	0.269	0.277	0.279	0.281	0.283
0.05000	0.293	0.301	0.303	0.305	0.307
0.06000	0.315	0.323	0.325	0.327	0.329
0.07000	0.335	0.343	0.345	0.347	0.349
0.08000	0.354	0.361	0.364	0.365	0.368
0.09000	0.371	0.379	0.382	0.383	0.385
0.10000	0.388	0.396	0.399	0.400	0.402
0.20000	0.534	0.540	0.541	0.542	0.544
0.25000	0.599	0.603	0.604	0.605	0.606
0.30000	0.663	0.665	0.665	0.666	0.666
0.35000	0.727	0.726	0.726	0.726	0.726
0.40000	0.793	0.788	0.787	0.786	0.785
0.45000	0.860	0.852	0.849	0.848	0.846
0.50000	0.931	0.918	0.914	0.912	0.909
0.55000	1.006	0.989	0.983	0.980	0.975
0.60000	1.088	1.064	1.056	1.052	1.046
0.65000	1.178	1.146	1.136	1.130	1.122
0.70000	1.280	1.238	1.224	1.217	1.207
0.75000	1.397	1.343	1.326	1.317	1.304
0.80000	1.539	1.468	1.445	1.434	1.417
0.85000	1.719	1.625	1.595	1.579	1.557
0.90000	1.971	1.840	1.798	1.777	1.747
0.95000	2.404	2.199	2.134	2.102	2.055
0.96000	2.546	2.313	2.240	2.204	2.151
0.97000	2.730	2.460	2.375	2.334	2.274
0.98000	2.993	2.666	2.565	2.516	2.444
0.99000	3.456	3.020	2.887	2.823	2.729
0.99500	3.939	3.376	3.208	3.127	3.009
0.99900	5.148	4.222	3.955	3.828	3.646
0.99990	7.138	5.497	5.047	4.838	4.543

The table was generated using MINITAB

F-table for 8 Degrees of freedom in the denominator
numerator degrees of freedom

α	1	2	3	4	5
0.00010	0	0.03	0.034	0.036	0.0389
0.00100	0	0.05	0.063	0.070	0.0741
0.00500	0	0.09	0.104	0.114	0.1205
0.01000	0	0.12	0.132	0.143	0.1508
0.02000	0	0.15	0.169	0.182	0.1915
0.03000	0	0.18	0.198	0.212	0.2221
0.04000	0	0.20	0.223	0.238	0.2480
0.05000	0	0.22	0.246	0.261	0.2712
0.06000	0	0.24	0.267	0.282	0.2924
0.07000	0	0.26	0.287	0.302	0.3123
0.08000	0	0.28	0.306	0.320	0.3312
0.09000	0	0.30	0.324	0.339	0.3493
0.10000	0	0.32	0.342	0.356	0.3668
0.20000	1	0.50	0.512	0.520	0.5262
0.25000	1	0.60	0.599	0.601	0.6033
0.30000	1	0.71	0.691	0.685	0.6823
0.35000	1	0.83	0.791	0.774	0.7651
0.40000	1	0.97	0.900	0.870	0.8534
0.45000	2	1.13	1.023	0.975	0.9491
0.50000	2	1.32	1.163	1.093	1.0545
0.55000	3	1.55	1.325	1.227	1.1728
0.60000	3	1.84	1.518	1.383	1.3079
0.65000	5	2.20	1.753	1.567	1.4659
0.70000	6	2.68	2.048	1.793	1.6558
0.75000	9	3.35	2.436	2.080	1.8923
0.80000	15	4.36	2.976	2.465	2.2021
0.85000	26	6.03	3.797	3.025	2.6391
0.90000	59	9.37	5.252	3.955	3.3392
0.95000	239	19.37	8.845	6.041	4.8184
0.96000	373	24.37	10.401	6.879	5.3857
0.97000	664	32.71	12.780	8.106	6.1948
0.98000	1495	49.37	17.007	10.162	7.5028
0.99000	5981	99.38	27.489	14.799	10.2893
0.99500	23925	199.38	44.127	21.352	13.9612
0.99900	598185	999.35	130.621	48.996	27.6501
0.99990	59795064	9998.09	609.203	157.098	71.2209

The table was generated using MINITAB

F-table for 8 Degrees of freedom in the denominator
numerator degrees of freedom

α	6	7	8	9	10
0.00010	0.0401	0.0430	0.0449	0.0457	0.0455
0.00100	0.0776	0.0807	0.0830	0.0851	0.0868
0.00500	0.1257	0.1300	0.1334	0.1363	0.1387
0.01000	0.1570	0.1619	0.1659	0.1692	0.1720
0.02000	0.1986	0.2042	0.2088	0.2126	0.2158
0.03000	0.2297	0.2358	0.2406	0.2447	0.2481
0.04000	0.2560	0.2622	0.2673	0.2715	0.2751
0.05000	0.2793	0.2857	0.2908	0.2951	0.2988
0.06000	0.3006	0.3071	0.3123	0.3166	0.3203
0.07000	0.3205	0.3270	0.3322	0.3366	0.3403
0.08000	0.3394	0.3458	0.3511	0.3554	0.3591
0.09000	0.3574	0.3638	0.3690	0.3733	0.3769
0.10000	0.3748	0.3811	0.3862	0.3904	0.3940
0.20000	0.5312	0.5353	0.5387	0.5415	0.5440
0.25000	0.6057	0.6080	0.6099	0.6116	0.6131
0.30000	0.6815	0.6813	0.6814	0.6816	0.6819
0.35000	0.7600	0.7568	0.7546	0.7531	0.7520
0.40000	0.8431	0.8361	0.8312	0.8274	0.8245
0.45000	0.9323	0.9208	0.9124	0.9060	0.9010
0.50000	1.0297	1.0126	1.0000	0.9904	0.9828
0.55000	1.1379	1.1138	1.0960	1.0824	1.0717
0.60000	1.2603	1.2274	1.2031	1.1846	1.1700
0.65000	1.4017	1.3575	1.3252	1.3004	1.2809
0.70000	1.5696	1.5107	1.4677	1.4349	1.4091
0.75000	1.7759	1.6969	1.6396	1.5961	1.5620
0.80000	2.0418	1.9339	1.8563	1.7979	1.7523
0.85000	2.4094	2.2570	2.1486	2.0675	2.0046
0.90000	2.9830	2.7516	2.5893	2.4694	2.3772
0.95000	4.1468	3.7258	3.4382	3.2296	3.0717
0.96000	4.5794	4.0796	3.7410	3.4971	3.3133
0.97000	5.1853	4.5687	4.1555	3.8603	3.6393
0.98000	6.1415	5.3274	4.7900	4.4104	4.1289
0.99000	8.1016	6.8400	6.0289	5.4671	5.0566
0.99500	10.5659	8.6782	7.4960	6.6932	6.1159
0.99900	19.0305	14.6339	12.0456	10.3681	9.2041
0.99990	42.5390	29.6434	22.7047	18.5018	15.7353

Table generated using MINITAB

F-table for 8 Degrees of freedom in the denominator
numerator degrees of freedom

α	11	12	13	14	15
0.00010	0.047	0.048	0.049	0.049	0.049
0.00100	0.087	0.089	0.090	0.091	0.092
0.00500	0.140	0.142	0.144	0.145	0.146
0.01000	0.174	0.176	0.178	0.179	0.181
0.02000	0.218	0.220	0.222	0.224	0.226
0.03000	0.251	0.253	0.255	0.257	0.259
0.04000	0.278	0.280	0.283	0.285	0.286
0.05000	0.301	0.304	0.306	0.308	0.310
0.06000	0.323	0.326	0.328	0.330	0.332
0.07000	0.343	0.346	0.348	0.350	0.352
0.08000	0.362	0.364	0.367	0.369	0.371
0.09000	0.380	0.382	0.385	0.387	0.388
0.10000	0.397	0.399	0.402	0.404	0.405
0.20000	0.546	0.547	0.549	0.550	0.552
0.25000	0.614	0.615	0.616	0.617	0.618
0.30000	0.682	0.682	0.682	0.683	0.683
0.35000	0.751	0.750	0.749	0.749	0.749
0.40000	0.822	0.820	0.818	0.817	0.816
0.45000	0.896	0.893	0.890	0.888	0.886
0.50000	0.976	0.971	0.967	0.963	0.960
0.55000	1.063	1.055	1.049	1.044	1.040
0.60000	1.158	1.148	1.140	1.133	1.126
0.65000	1.265	1.252	1.241	1.231	1.223
0.70000	1.388	1.371	1.356	1.344	1.333
0.75000	1.534	1.512	1.493	1.477	1.463
0.80000	1.715	1.685	1.660	1.639	1.620
0.85000	1.954	1.913	1.879	1.850	1.825
0.90000	2.304	2.244	2.195	2.153	2.118
0.95000	2.948	2.848	2.766	2.698	2.640
0.96000	3.169	3.055	2.960	2.882	2.816
0.97000	3.467	3.330	3.219	3.126	3.048
0.98000	3.911	3.739	3.600	3.484	3.387
0.99000	4.744	4.499	4.302	4.140	4.004
0.99500	5.682	5.345	5.076	4.856	4.674
0.99900	8.354	7.710	7.206	6.801	6.470
0.99990	13.799	12.380	11.301	10.456	9.779

The table was generated using MINITAB

F-table for 8 Degrees of freedom in the denominator
numerator degrees of freedom

α	16	17	18	19	20
0.00010	0.050	0.051	0.052	0.052	0.052
0.00100	0.093	0.093	0.094	0.094	0.095
0.00500	0.147	0.148	0.149	0.150	0.151
0.01000	0.182	0.183	0.184	0.185	0.186
0.02000	0.227	0.229	0.230	0.231	0.232
0.03000	0.260	0.262	0.263	0.264	0.265
0.04000	0.288	0.289	0.291	0.292	0.293
0.05000	0.312	0.313	0.315	0.316	0.317
0.06000	0.334	0.335	0.336	0.338	0.339
0.07000	0.354	0.355	0.356	0.358	0.359
0.08000	0.372	0.374	0.375	0.376	0.377
0.09000	0.390	0.391	0.393	0.394	0.395
0.10000	0.407	0.408	0.410	0.411	0.412
0.20000	0.553	0.554	0.555	0.556	0.556
0.25000	0.619	0.619	0.620	0.621	0.621
0.30000	0.683	0.684	0.684	0.684	0.684
0.35000	0.748	0.748	0.748	0.748	0.748
0.40000	0.815	0.814	0.813	0.813	0.812
0.45000	0.884	0.883	0.881	0.880	0.879
0.50000	0.957	0.955	0.953	0.951	0.949
0.55000	1.036	1.032	1.029	1.026	1.024
0.60000	1.121	1.116	1.112	1.109	1.105
0.65000	1.216	1.210	1.204	1.200	1.195
0.70000	1.324	1.316	1.309	1.303	1.297
0.75000	1.451	1.440	1.431	1.422	1.415
0.80000	1.605	1.591	1.578	1.567	1.558
0.85000	1.803	1.785	1.768	1.753	1.740
0.90000	2.087	2.061	2.037	2.017	1.998
0.95000	2.591	2.547	2.510	2.476	2.447
0.96000	2.759	2.709	2.666	2.628	2.594
0.97000	2.981	2.923	2.872	2.827	2.788
0.98000	3.304	3.232	3.170	3.115	3.067
0.99000	3.889	3.790	3.705	3.630	3.564
0.99500	4.520	4.389	4.275	4.177	4.089
0.99900	6.194	5.962	5.762	5.590	5.440
0.99990	9.225	8.764	8.375	8.043	7.757

Table generated using MINITAB

F-table for 8 Degrees of freedom in the denominator
numerator degrees of freedom

α	25	50	75	100	200
0.00010	0.052	0.055	0.055	0.055	0.058
0.00100	0.097	0.101	0.103	0.104	0.105
0.00500	0.154	0.160	0.163	0.164	0.166
0.01000	0.189	0.197	0.200	0.201	0.203
0.02000	0.236	0.244	0.247	0.249	0.251
0.03000	0.269	0.278	0.281	0.283	0.286
0.04000	0.297	0.306	0.310	0.311	0.314
0.05000	0.321	0.331	0.334	0.336	0.338
0.06000	0.343	0.352	0.356	0.358	0.360
0.07000	0.363	0.372	0.376	0.378	0.380
0.08000	0.382	0.391	0.394	0.396	0.399
0.09000	0.399	0.409	0.412	0.414	0.416
0.10000	0.416	0.425	0.429	0.430	0.433
0.20000	0.559	0.566	0.569	0.570	0.572
0.25000	0.623	0.628	0.630	0.630	0.632
0.30000	0.685	0.688	0.688	0.689	0.690
0.35000	0.747	0.747	0.746	0.746	0.746
0.40000	0.810	0.806	0.805	0.804	0.803
0.45000	0.875	0.867	0.864	0.863	0.861
0.50000	0.943	0.930	0.926	0.924	0.921
0.55000	1.015	0.997	0.991	0.988	0.983
0.60000	1.093	1.068	1.060	1.056	1.049
0.65000	1.179	1.146	1.135	1.129	1.121
0.70000	1.275	1.232	1.218	1.211	1.201
0.75000	1.387	1.331	1.313	1.304	1.290
0.80000	1.521	1.448	1.425	1.413	1.396
0.85000	1.691	1.595	1.564	1.549	1.526
0.90000	1.929	1.796	1.753	1.732	1.701
0.95000	2.337	2.129	2.064	2.032	1.984
0.96000	2.469	2.235	2.162	2.126	2.073
0.97000	2.642	2.371	2.287	2.246	2.185
0.98000	2.889	2.563	2.462	2.413	2.341
0.99000	3.323	2.890	2.758	2.694	2.601
0.99500	3.775	3.218	3.052	2.972	2.856
0.99900	4.906	3.998	3.736	3.612	3.434
0.99990	6.764	5.169	4.733	4.531	4.244

The table was generated using MINITAB

F-table for 9 Degrees of freedom in the denominator
numerator degrees of freedom

α	1	2	3	4	5
0.00010	0	0.03	0.039	0.044	0.0471
0.00100	0	0.06	0.072	0.080	0.0854
0.00500	0	0.10	0.115	0.126	0.1338
0.01000	0	0.12	0.143	0.156	0.1651
0.02000	0	0.16	0.181	0.196	0.2066
0.03000	0	0.19	0.211	0.226	0.2377
0.04000	0	0.21	0.236	0.252	0.2639
0.05000	0	0.23	0.259	0.275	0.2872
0.06000	0	0.26	0.280	0.296	0.3086
0.07000	0	0.28	0.300	0.316	0.3285
0.08000	0	0.30	0.319	0.335	0.3474
0.09000	0	0.31	0.338	0.354	0.3656
0.10000	0	0.33	0.356	0.371	0.3831
0.20000	1	0.52	0.526	0.535	0.5419
0.25000	1	0.62	0.613	0.615	0.6184
0.30000	1	0.72	0.705	0.699	0.6968
0.35000	1	0.85	0.804	0.787	0.7787
0.40000	1	0.98	0.913	0.882	0.8660
0.45000	2	1.14	1.035	0.987	0.9606
0.50000	2	1.33	1.174	1.104	1.0648
0.55000	3	1.56	1.336	1.237	1.1815
0.60000	3	1.85	1.528	1.391	1.3149
0.65000	5	2.21	1.761	1.574	1.4707
0.70000	6	2.69	2.055	1.797	1.6579
0.75000	9	3.37	2.441	2.081	1.8910
0.80000	15	4.37	2.978	2.462	2.1962
0.85000	26	6.04	3.794	3.015	2.6268
0.90000	60	9.38	5.240	3.936	3.3163
0.95000	241	19.39	8.812	5.999	4.7724
0.96000	376	24.39	10.359	6.828	5.3311
0.97000	669	32.72	12.723	8.041	6.1274
0.98000	1505	49.39	16.925	10.074	7.4153
0.99000	6023	99.39	27.345	14.659	10.1577
0.99500	24092	199.39	43.882	21.139	13.7716
0.99900	602359	999.45	129.862	48.475	27.2441
0.99990	60210920	9997.65	605.627	155.368	70.1284

The table was generated using MINITAB

F-table for 9 Degrees of freedom in the denominator
numerator degrees of freedom

α	6	7	8	9	10
0.00010	0.0504	0.0519	0.0534	0.0557	0.0570
0.00100	0.0898	0.0936	0.0962	0.0987	0.1012
0.00500	0.1401	0.1452	0.1494	0.1528	0.1559
0.01000	0.1724	0.1782	0.1829	0.1869	0.1902
0.02000	0.2148	0.2214	0.2268	0.2312	0.2350
0.03000	0.2464	0.2534	0.2590	0.2638	0.2678
0.04000	0.2730	0.2801	0.2860	0.2908	0.2949
0.05000	0.2964	0.3037	0.3097	0.3146	0.3188
0.06000	0.3178	0.3252	0.3311	0.3361	0.3403
0.07000	0.3378	0.3452	0.3511	0.3561	0.3603
0.08000	0.3567	0.3640	0.3699	0.3749	0.3790
0.09000	0.3747	0.3819	0.3878	0.3927	0.3968
0.10000	0.3920	0.3991	0.4050	0.4098	0.4139
0.20000	0.5476	0.5523	0.5562	0.5595	0.5623
0.25000	0.6215	0.6241	0.6265	0.6286	0.6304
0.30000	0.6963	0.6965	0.6969	0.6974	0.6980
0.35000	0.7739	0.7709	0.7690	0.7676	0.7666
0.40000	0.8559	0.8490	0.8441	0.8405	0.8377
0.45000	0.9438	0.9323	0.9239	0.9175	0.9124
0.50000	1.0398	1.0224	1.0097	1.0000	0.9923
0.55000	1.1462	1.1218	1.1038	1.0900	1.0791
0.60000	1.2666	1.2331	1.2086	1.1898	1.1749
0.65000	1.4057	1.3608	1.3279	1.3027	1.2829
0.70000	1.5707	1.5108	1.4671	1.4338	1.4077
0.75000	1.7733	1.6931	1.6350	1.5909	1.5563
0.80000	2.0342	1.9250	1.8466	1.7873	1.7411
0.85000	2.3949	2.2411	2.1316	2.0496	1.9859
0.90000	2.9578	2.7246	2.5613	2.4403	2.3473
0.95000	4.0990	3.6767	3.3881	3.1789	3.0204
0.96000	4.5232	4.0223	3.6831	3.4387	3.2545
0.97000	5.1171	4.5002	4.0868	3.7914	3.5702
0.98000	6.0547	5.2413	4.7045	4.3255	4.0442
0.99000	7.9762	6.7187	5.9106	5.3512	4.9424
0.99500	10.3915	8.5138	7.3385	6.5411	5.9675
0.99900	18.6880	14.3299	11.7665	10.1066	8.9558
0.99990	41.7296	28.9861	22.1393	17.9980	15.2743

Table generated using MINITAB

F-table for 9 Degrees of freedom in the denominator
numerator degrees of freedom

α	11	12	13	14	15
0.00010	0.057	0.059	0.059	0.060	0.060
0.00100	0.102	0.104	0.106	0.107	0.108
0.00500	0.158	0.160	0.162	0.164	0.165
0.01000	0.193	0.195	0.197	0.199	0.201
0.02000	0.238	0.241	0.243	0.245	0.247
0.03000	0.271	0.274	0.276	0.279	0.281
0.04000	0.298	0.301	0.304	0.306	0.308
0.05000	0.322	0.325	0.328	0.330	0.332
0.06000	0.343	0.347	0.349	0.352	0.354
0.07000	0.363	0.367	0.369	0.372	0.374
0.08000	0.382	0.385	0.388	0.390	0.392
0.09000	0.400	0.403	0.406	0.408	0.410
0.10000	0.417	0.420	0.423	0.425	0.427
0.20000	0.564	0.566	0.568	0.570	0.571
0.25000	0.632	0.633	0.634	0.635	0.636
0.30000	0.698	0.699	0.699	0.700	0.700
0.35000	0.765	0.765	0.765	0.764	0.764
0.40000	0.835	0.833	0.832	0.830	0.829
0.45000	0.908	0.905	0.902	0.899	0.897
0.50000	0.986	0.981	0.976	0.973	0.969
0.55000	1.070	1.062	1.056	1.051	1.046
0.60000	1.162	1.152	1.144	1.137	1.131
0.65000	1.266	1.253	1.242	1.232	1.224
0.70000	1.386	1.369	1.354	1.341	1.330
0.75000	1.528	1.505	1.486	1.469	1.455
0.80000	1.703	1.673	1.647	1.626	1.607
0.85000	1.935	1.893	1.858	1.829	1.804
0.90000	2.273	2.213	2.163	2.122	2.086
0.95000	2.896	2.796	2.714	2.645	2.587
0.96000	3.110	2.995	2.901	2.822	2.755
0.97000	3.398	3.261	3.149	3.056	2.978
0.98000	3.827	3.655	3.516	3.400	3.303
0.99000	4.631	4.387	4.191	4.029	3.894
0.99500	5.536	5.202	4.935	4.717	4.536
0.99900	8.116	7.479	6.981	6.582	6.255
0.99990	13.370	11.975	10.915	10.086	9.421

The table was generated using MINITAB

F-table for 9 Degrees of freedom in the denominator

numerator degrees of freedom

α	16	17	18	19	20
0.00010	0.061	0.061	0.062	0.063	0.062
0.00100	0.109	0.110	0.110	0.111	0.112
0.00500	0.167	0.168	0.169	0.170	0.171
0.01000	0.203	0.204	0.205	0.206	0.207
0.02000	0.249	0.250	0.252	0.253	0.254
0.03000	0.282	0.284	0.286	0.287	0.288
0.04000	0.310	0.312	0.313	0.315	0.316
0.05000	0.334	0.336	0.337	0.339	0.340
0.06000	0.356	0.357	0.359	0.360	0.362
0.07000	0.376	0.377	0.379	0.380	0.382
0.08000	0.394	0.396	0.398	0.399	0.400
0.09000	0.412	0.414	0.415	0.417	0.418
0.10000	0.429	0.430	0.432	0.433	0.435
0.20000	0.573	0.574	0.575	0.576	0.577
0.25000	0.637	0.638	0.639	0.639	0.640
0.30000	0.700	0.701	0.701	0.701	0.702
0.35000	0.764	0.764	0.763	0.763	0.763
0.40000	0.828	0.828	0.827	0.826	0.826
0.45000	0.896	0.894	0.893	0.892	0.890
0.50000	0.967	0.964	0.962	0.960	0.958
0.55000	1.042	1.039	1.036	1.033	1.031
0.60000	1.125	1.120	1.116	1.112	1.109
0.65000	1.217	1.211	1.205	1.200	1.195
0.70000	1.321	1.313	1.305	1.299	1.293
0.75000	1.443	1.432	1.423	1.414	1.406
0.80000	1.591	1.577	1.564	1.553	1.543
0.85000	1.782	1.763	1.746	1.731	1.718
0.90000	2.055	2.028	2.004	1.983	1.964
0.95000	2.537	2.494	2.456	2.422	2.392
0.96000	2.698	2.649	2.605	2.567	2.533
0.97000	2.911	2.853	2.802	2.757	2.717
0.98000	3.220	3.149	3.086	3.032	2.983
0.99000	3.780	3.682	3.597	3.522	3.456
0.99500	4.383	4.253	4.141	4.042	3.956
0.99900	5.983	5.754	5.557	5.387	5.239
0.99990	8.878	8.426	8.045	7.720	7.439

Table generated using MINITAB

F-table for 9 Degrees of freedom in the denominator
numerator degrees of freedom

α	25	50	75	100	200
0.00010	0.066	0.068	0.069	0.070	0.070
0.00100	0.115	0.121	0.123	0.124	0.126
0.00500	0.175	0.183	0.186	0.187	0.190
0.01000	0.212	0.221	0.224	0.226	0.229
0.02000	0.259	0.269	0.273	0.275	0.278
0.03000	0.293	0.304	0.308	0.310	0.313
0.04000	0.321	0.332	0.336	0.338	0.341
0.05000	0.345	0.356	0.360	0.362	0.366
0.06000	0.367	0.378	0.382	0.384	0.387
0.07000	0.387	0.398	0.402	0.404	0.407
0.08000	0.405	0.416	0.421	0.423	0.426
0.09000	0.423	0.434	0.438	0.440	0.443
0.10000	0.440	0.450	0.454	0.456	0.459
0.20000	0.580	0.588	0.591	0.593	0.595
0.25000	0.643	0.648	0.650	0.651	0.653
0.30000	0.703	0.706	0.707	0.708	0.709
0.35000	0.763	0.763	0.763	0.763	0.763
0.40000	0.824	0.820	0.819	0.818	0.818
0.45000	0.886	0.879	0.876	0.875	0.873
0.50000	0.952	0.939	0.935	0.933	0.930
0.55000	1.021	1.003	0.997	0.993	0.989
0.60000	1.096	1.071	1.062	1.058	1.052
0.65000	1.178	1.145	1.134	1.128	1.120
0.70000	1.271	1.227	1.212	1.205	1.194
0.75000	1.378	1.321	1.302	1.293	1.279
0.80000	1.505	1.432	1.408	1.396	1.378
0.85000	1.668	1.570	1.538	1.523	1.499
0.90000	1.894	1.759	1.716	1.694	1.663
0.95000	2.282	2.073	2.007	1.974	1.926
0.96000	2.408	2.172	2.098	2.062	2.008
0.97000	2.571	2.300	2.215	2.173	2.112
0.98000	2.806	2.479	2.378	2.328	2.256
0.99000	3.217	2.784	2.653	2.589	2.497
0.99500	3.644	3.092	2.926	2.847	2.731
0.99900	4.713	3.818	3.560	3.438	3.263
0.99990	6.467	4.908	4.483	4.285	4.006

The table was generated using MINITAB

F-table for 10 Degrees of freedom in the denominator
numerator degrees of freedom

α	1	2	3	4	5
0.00010	0	0.04	0.046	0.051	0.054
0.00100	0	0.07	0.080	0.089	0.095
0.00500	0	0.11	0.124	0.136	0.145
0.01000	0	0.13	0.153	0.167	0.177
0.02000	0	0.17	0.192	0.208	0.219
0.03000	0	0.20	0.221	0.238	0.250
0.04000	0	0.22	0.247	0.264	0.277
0.05000	0	0.24	0.270	0.288	0.300
0.06000	0	0.26	0.291	0.309	0.322
0.07000	0	0.28	0.311	0.329	0.342
0.08000	0	0.30	0.330	0.348	0.361
0.09000	0	0.32	0.349	0.366	0.379
0.10000	0	0.34	0.367	0.384	0.396
0.20000	1	0.53	0.537	0.547	0.554
0.25000	1	0.63	0.624	0.627	0.630
0.30000	1	0.73	0.715	0.710	0.708
0.35000	1	0.86	0.814	0.798	0.789
0.40000	1	0.99	0.923	0.893	0.876
0.45000	2	1.16	1.045	0.997	0.969
0.50000	2	1.35	1.183	1.113	1.073
0.55000	3	1.57	1.344	1.244	1.188
0.60000	3	1.86	1.535	1.397	1.320
0.65000	5	2.22	1.768	1.578	1.474
0.70000	6	2.70	2.061	1.800	1.659
0.75000	9	3.38	2.445	2.082	1.889
0.80000	15	4.38	2.979	2.460	2.191
0.85000	27	6.05	3.792	3.008	2.616
0.90000	60	9.39	5.230	3.920	3.297
0.95000	242	19.40	8.785	5.964	4.735
0.96000	378	24.40	10.325	6.786	5.286
0.97000	673	32.73	12.677	7.988	6.072
0.98000	1514	49.40	16.860	10.003	7.343
0.99000	6055	99.40	27.229	14.546	10.051
0.99500	24227	199.40	43.687	20.967	13.618
0.99900	605671	999.41	129.247	48.053	26.916
0.99990	60545796	9997.40	602.691	153.973	69.244

The table was generated using MINITAB

F-table for 10 Degrees of freedom in the denominator
numerator degrees of freedom

α	6	7	8	9	10
0.00010	0.059	0.062	0.064	0.064	0.066
0.00100	0.100	0.104	0.108	0.111	0.114
0.00500	0.152	0.158	0.163	0.167	0.171
0.01000	0.185	0.192	0.197	0.202	0.206
0.02000	0.228	0.236	0.242	0.247	0.251
0.03000	0.260	0.268	0.274	0.280	0.284
0.04000	0.287	0.295	0.301	0.307	0.311
0.05000	0.310	0.319	0.325	0.331	0.335
0.06000	0.332	0.340	0.347	0.352	0.357
0.07000	0.352	0.360	0.367	0.372	0.377
0.08000	0.371	0.379	0.385	0.391	0.395
0.09000	0.389	0.397	0.403	0.409	0.413
0.10000	0.406	0.414	0.420	0.426	0.430
0.20000	0.561	0.566	0.570	0.574	0.577
0.25000	0.634	0.637	0.640	0.642	0.644
0.30000	0.708	0.708	0.709	0.710	0.711
0.35000	0.785	0.782	0.780	0.779	0.778
0.40000	0.866	0.859	0.854	0.851	0.848
0.45000	0.953	0.941	0.933	0.926	0.921
0.50000	1.047	1.030	1.017	1.007	1.000
0.55000	1.152	1.128	1.109	1.096	1.084
0.60000	1.271	1.237	1.212	1.193	1.178
0.65000	1.408	1.363	1.329	1.304	1.284
0.70000	1.571	1.510	1.466	1.432	1.406
0.75000	1.770	1.689	1.631	1.586	1.551
0.80000	2.027	1.917	1.838	1.778	1.731
0.85000	2.383	2.227	2.117	2.034	1.970
0.90000	2.936	2.702	2.538	2.416	2.322
0.95000	4.059	3.636	3.347	3.137	2.978
0.96000	4.477	3.975	3.635	3.390	3.206
0.97000	5.061	4.444	4.030	3.735	3.513
0.98000	5.983	5.171	4.634	4.256	3.975
0.99000	7.874	6.620	5.814	5.256	4.849
0.99500	10.250	8.380	7.210	6.417	5.846
0.99900	18.410	14.083	11.540	9.894	8.753
0.99990	41.074	28.453	21.681	17.589	14.900

Table generated using MINITAB

F-table for 10 Degrees of freedom in the denominator

numerator degrees of freedom

α	11	12	13	14	15
0.00010	0.068	0.069	0.069	0.071	0.073
0.00100	0.116	0.118	0.119	0.121	0.122
0.00500	0.174	0.176	0.178	0.181	0.182
0.01000	0.209	0.212	0.215	0.217	0.219
0.02000	0.255	0.258	0.261	0.263	0.266
0.03000	0.288	0.291	0.294	0.297	0.299
0.04000	0.315	0.319	0.322	0.325	0.327
0.05000	0.339	0.343	0.346	0.349	0.351
0.06000	0.361	0.364	0.368	0.370	0.373
0.07000	0.381	0.384	0.387	0.390	0.392
0.08000	0.399	0.403	0.406	0.409	0.411
0.09000	0.417	0.421	0.424	0.426	0.429
0.10000	0.434	0.437	0.440	0.443	0.445
0.20000	0.580	0.582	0.584	0.586	0.588
0.25000	0.646	0.648	0.649	0.650	0.651
0.30000	0.711	0.712	0.713	0.713	0.714
0.35000	0.778	0.777	0.777	0.777	0.776
0.40000	0.846	0.844	0.843	0.841	0.840
0.45000	0.917	0.914	0.911	0.909	0.907
0.50000	0.993	0.988	0.984	0.980	0.977
0.55000	1.076	1.068	1.062	1.056	1.052
0.60000	1.166	1.156	1.147	1.140	1.134
0.65000	1.267	1.254	1.243	1.233	1.224
0.70000	1.384	1.366	1.351	1.339	1.328
0.75000	1.523	1.499	1.480	1.463	1.449
0.80000	1.694	1.663	1.637	1.615	1.596
0.85000	1.919	1.876	1.841	1.812	1.786
0.90000	2.248	2.187	2.137	2.095	2.059
0.95000	2.853	2.753	2.671	2.602	2.543
0.96000	3.062	2.946	2.851	2.772	2.706
0.97000	3.341	3.204	3.092	2.999	2.920
0.98000	3.758	3.586	3.447	3.331	3.234
0.99000	4.539	4.296	4.100	3.939	3.804
0.99500	5.418	5.085	4.819	4.603	4.423
0.99900	7.922	7.292	6.799	6.404	6.080
0.99990	13.021	11.646	10.601	9.784	9.130

The table was generated using MINITAB

F-table for 10 Degrees of freedom in the denominator
numerator degrees of freedom

α	16	17	18	19	20
0.00010	0.073	0.072	0.075	0.075	0.077
0.00100	0.124	0.125	0.126	0.127	0.128
0.00500	0.184	0.185	0.187	0.188	0.189
0.01000	0.221	0.222	0.224	0.225	0.226
0.02000	0.268	0.269	0.271	0.273	0.274
0.03000	0.301	0.303	0.305	0.307	0.308
0.04000	0.329	0.331	0.333	0.334	0.336
0.05000	0.353	0.355	0.357	0.359	0.360
0.06000	0.375	0.377	0.379	0.380	0.382
0.07000	0.395	0.397	0.398	0.400	0.401
0.08000	0.413	0.415	0.417	0.418	0.420
0.09000	0.431	0.433	0.434	0.436	0.437
0.10000	0.447	0.449	0.451	0.452	0.454
0.20000	0.589	0.591	0.592	0.593	0.594
0.25000	0.653	0.654	0.654	0.655	0.656
0.30000	0.714	0.715	0.715	0.716	0.716
0.35000	0.776	0.776	0.776	0.776	0.776
0.40000	0.840	0.839	0.838	0.838	0.837
0.45000	0.905	0.904	0.902	0.901	0.900
0.50000	0.974	0.972	0.969	0.968	0.966
0.55000	1.048	1.044	1.041	1.038	1.036
0.60000	1.128	1.123	1.119	1.115	1.112
0.65000	1.217	1.211	1.205	1.200	1.195
0.70000	1.318	1.310	1.302	1.296	1.290
0.75000	1.436	1.425	1.415	1.407	1.399
0.80000	1.579	1.565	1.552	1.541	1.531
0.85000	1.764	1.745	1.728	1.713	1.699
0.90000	2.028	2.000	1.976	1.955	1.936
0.95000	2.493	2.449	2.411	2.377	2.347
0.96000	2.648	2.599	2.555	2.517	2.483
0.97000	2.853	2.795	2.744	2.699	2.659
0.98000	3.151	3.080	3.018	2.963	2.914
0.99000	3.690	3.593	3.508	3.433	3.368
0.99500	4.271	4.142	4.030	3.932	3.846
0.99900	5.811	5.584	5.390	5.221	5.075
0.99990	8.595	8.151	7.776	7.456	7.180

Table generated using MINITAB

F-table for 10 Degrees of freedom in the denominator
numerator degrees of freedom

α	25	50	75	100	200
0.00010	0.078	0.084	0.085	0.086	0.088
0.00100	0.131	0.139	0.141	0.143	0.145
0.00500	0.194	0.203	0.207	0.209	0.212
0.01000	0.231	0.243	0.247	0.249	0.252
0.02000	0.279	0.291	0.296	0.298	0.302
0.03000	0.314	0.326	0.331	0.333	0.337
0.04000	0.342	0.354	0.359	0.362	0.365
0.05000	0.366	0.379	0.383	0.386	0.390
0.06000	0.387	0.400	0.405	0.407	0.411
0.07000	0.407	0.420	0.425	0.427	0.431
0.08000	0.426	0.438	0.443	0.445	0.449
0.09000	0.443	0.456	0.460	0.463	0.466
0.10000	0.460	0.472	0.476	0.479	0.482
0.20000	0.598	0.607	0.610	0.612	0.615
0.25000	0.659	0.665	0.668	0.669	0.671
0.30000	0.718	0.721	0.723	0.724	0.725
0.35000	0.776	0.777	0.777	0.777	0.777
0.40000	0.835	0.832	0.831	0.830	0.830
0.45000	0.896	0.888	0.886	0.884	0.883
0.50000	0.959	0.946	0.942	0.940	0.937
0.55000	1.026	1.007	1.001	0.998	0.993
0.60000	1.099	1.073	1.064	1.060	1.053
0.65000	1.178	1.144	1.132	1.126	1.118
0.70000	1.267	1.222	1.207	1.200	1.189
0.75000	1.370	1.312	1.293	1.283	1.269
0.80000	1.493	1.417	1.393	1.380	1.362
0.85000	1.648	1.549	1.517	1.501	1.477
0.90000	1.865	1.729	1.685	1.663	1.630
0.95000	2.236	2.026	1.959	1.926	1.878
0.96000	2.356	2.120	2.045	2.008	1.954
0.97000	2.513	2.240	2.155	2.113	2.052
0.98000	2.737	2.409	2.308	2.259	2.186
0.99000	3.129	2.698	2.566	2.503	2.410
0.99500	3.537	2.987	2.823	2.743	2.629
0.99900	4.555	3.671	3.416	3.295	3.122
0.99990	6.225	4.695	4.277	4.084	3.810

The table was generated using MINITAB

F-table for 11 Degrees of freedom in the denominator
numerator degrees of freedom

α	1	2	3	4	5
0.00010	0	0.04	0.051	0.058	0.0626
0.00100	0	0.07	0.087	0.097	0.1042
0.00500	0	0.11	0.132	0.145	0.1557
0.01000	0	0.14	0.161	0.176	0.1881
0.02000	0	0.18	0.200	0.218	0.2306
0.03000	0	0.20	0.230	0.249	0.2622
0.04000	0	0.23	0.256	0.275	0.2887
0.05000	0	0.25	0.279	0.298	0.3121
0.06000	0	0.27	0.300	0.319	0.3335
0.07000	0	0.29	0.320	0.339	0.3535
0.08000	0	0.31	0.339	0.358	0.3724
0.09000	0	0.33	0.358	0.377	0.3905
0.10000	0	0.35	0.376	0.394	0.4080
0.20000	1	0.53	0.546	0.557	0.5655
0.25000	1	0.63	0.633	0.637	0.6411
0.30000	1	0.74	0.724	0.719	0.7183
0.35000	1	0.86	0.823	0.807	0.7989
0.40000	1	1.00	0.931	0.901	0.8848
0.45000	2	1.16	1.053	1.004	0.9776
0.50000	2	1.35	1.191	1.120	1.0798
0.55000	3	1.58	1.351	1.251	1.1943
0.60000	3	1.87	1.542	1.402	1.3249
0.65000	5	2.23	1.773	1.582	1.4774
0.70000	6	2.71	2.065	1.803	1.6606
0.75000	9	3.39	2.448	2.082	1.8887
0.80000	15	4.39	2.980	2.457	2.1872
0.85000	27	6.06	3.789	3.001	2.6080
0.90000	60	9.40	5.223	3.907	3.2817
0.95000	243	19.40	8.763	5.936	4.7040
0.96000	380	24.41	10.296	6.751	5.2494
0.97000	676	32.74	12.639	7.944	6.0273
0.98000	1521	49.41	16.805	9.943	7.2846
0.99000	6083	99.41	27.133	14.452	9.9626
0.99500	24336	199.40	43.525	20.824	13.4913
0.99900	608417	999.46	128.741	47.704	26.6457
0.99990	60818740	9998.07	600.279	152.815	68.5167

The table was generated using MINITAB

F-table for 11 Degrees of freedom in the denominator
numerator degrees of freedom

α	6	7	8	9	10
0.00010	0.0670	0.0689	0.0718	0.0750	0.0770
0.00100	0.1105	0.1157	0.1195	0.1232	0.1264
0.00500	0.1638	0.1705	0.1760	0.1806	0.1845
0.01000	0.1973	0.2047	0.2108	0.2159	0.2203
0.02000	0.2407	0.2489	0.2556	0.2613	0.2661
0.03000	0.2729	0.2814	0.2884	0.2943	0.2992
0.04000	0.2995	0.3083	0.3155	0.3215	0.3266
0.05000	0.3232	0.3320	0.3392	0.3453	0.3504
0.06000	0.3446	0.3534	0.3607	0.3668	0.3719
0.07000	0.3646	0.3734	0.3806	0.3867	0.3918
0.08000	0.3834	0.3921	0.3993	0.4053	0.4104
0.09000	0.4013	0.4100	0.4170	0.4230	0.4280
0.10000	0.4186	0.4271	0.4340	0.4399	0.4448
0.20000	0.5724	0.5781	0.5829	0.5869	0.5903
0.25000	0.6451	0.6486	0.6516	0.6543	0.6566
0.30000	0.7186	0.7193	0.7203	0.7213	0.7222
0.35000	0.7946	0.7921	0.7904	0.7894	0.7887
0.40000	0.8748	0.8682	0.8635	0.8600	0.8573
0.45000	0.9608	0.9492	0.9408	0.9344	0.9294
0.50000	1.0544	1.0368	1.0240	1.0141	1.0063
0.55000	1.1583	1.1333	1.1149	1.1008	1.0897
0.60000	1.2756	1.2414	1.2162	1.1969	1.1817
0.65000	1.4110	1.3651	1.3314	1.3056	1.2852
0.70000	1.5716	1.5104	1.4657	1.4316	1.4047
0.75000	1.7687	1.6869	1.6275	1.5823	1.5469
0.80000	2.0224	1.9113	1.8312	1.7708	1.7235
0.85000	2.3729	2.2167	2.1054	2.0220	1.9572
0.90000	2.9196	2.6839	2.5186	2.3961	2.3018
0.95000	4.0275	3.6030	3.3130	3.1024	2.9430
0.96000	4.4391	3.9368	3.5963	3.3509	3.1657
0.97000	5.0157	4.3979	3.9839	3.6879	3.4663
0.98000	5.9252	5.1128	4.5768	4.1982	3.9172
0.99000	7.7896	6.5382	5.7344	5.1779	4.7715
0.99500	10.1328	8.2697	7.1046	6.3143	5.7462
0.99900	18.1818	13.8791	11.3526	9.7183	8.5864
0.99990	40.5334	28.0131	21.3030	17.2506	14.5904

Table generated using MINITAB

F-table for 11 Degrees of freedom in the denominator
numerator degrees of freedom

α	11	12	13	14	15
0.00010	0.077	0.080	0.081	0.082	0.084
0.00100	0.129	0.131	0.132	0.134	0.136
0.00500	0.187	0.190	0.193	0.195	0.198
0.01000	0.224	0.227	0.230	0.232	0.235
0.02000	0.270	0.273	0.277	0.279	0.282
0.03000	0.303	0.307	0.310	0.313	0.316
0.04000	0.331	0.334	0.338	0.341	0.343
0.05000	0.354	0.358	0.362	0.365	0.367
0.06000	0.376	0.380	0.383	0.386	0.389
0.07000	0.396	0.400	0.403	0.406	0.409
0.08000	0.414	0.418	0.421	0.424	0.427
0.09000	0.432	0.436	0.439	0.442	0.444
0.10000	0.449	0.452	0.456	0.458	0.461
0.20000	0.593	0.595	0.598	0.600	0.602
0.25000	0.658	0.660	0.662	0.663	0.664
0.30000	0.723	0.724	0.724	0.725	0.726
0.35000	0.788	0.787	0.787	0.787	0.787
0.40000	0.855	0.853	0.852	0.851	0.850
0.45000	0.925	0.922	0.919	0.917	0.915
0.50000	1.000	0.994	0.990	0.986	0.983
0.55000	1.080	1.073	1.066	1.061	1.056
0.60000	1.169	1.159	1.150	1.142	1.136
0.65000	1.268	1.255	1.243	1.233	1.225
0.70000	1.382	1.364	1.349	1.336	1.325
0.75000	1.518	1.494	1.474	1.457	1.443
0.80000	1.685	1.654	1.628	1.605	1.586
0.85000	1.905	1.862	1.827	1.797	1.771
0.90000	2.226	2.166	2.115	2.072	2.036
0.95000	2.817	2.717	2.634	2.565	2.506
0.96000	3.021	2.905	2.810	2.731	2.664
0.97000	3.294	3.156	3.044	2.951	2.872
0.98000	3.700	3.529	3.389	3.274	3.176
0.99000	4.462	4.219	4.024	3.864	3.729
0.99500	5.319	4.988	4.724	4.508	4.329
0.99900	7.761	7.136	6.647	6.255	5.935
0.99990	12.732	11.373	10.342	9.535	8.889

The table was generated using MINITAB

F-table for 11 Degrees of freedom in the denominator
numerator degrees of freedom

α	16	17	18	19	20
0.00010	0.084	0.086	0.087	0.088	0.089
0.00100	0.138	0.139	0.140	0.141	0.142
0.00500	0.199	0.201	0.203	0.204	0.205
0.01000	0.237	0.239	0.240	0.242	0.243
0.02000	0.284	0.286	0.288	0.290	0.291
0.03000	0.318	0.320	0.322	0.324	0.326
0.04000	0.346	0.348	0.350	0.352	0.353
0.05000	0.370	0.372	0.374	0.376	0.377
0.06000	0.391	0.393	0.395	0.397	0.399
0.07000	0.411	0.413	0.415	0.417	0.419
0.08000	0.429	0.432	0.434	0.435	0.437
0.09000	0.447	0.449	0.451	0.453	0.454
0.10000	0.463	0.465	0.467	0.469	0.471
0.20000	0.603	0.605	0.606	0.607	0.609
0.25000	0.666	0.667	0.668	0.668	0.669
0.30000	0.726	0.727	0.727	0.728	0.728
0.35000	0.787	0.787	0.787	0.787	0.787
0.40000	0.849	0.848	0.847	0.847	0.846
0.45000	0.913	0.911	0.910	0.909	0.908
0.50000	0.980	0.978	0.976	0.974	0.972
0.55000	1.052	1.049	1.045	1.043	1.040
0.60000	1.130	1.126	1.121	1.117	1.114
0.65000	1.217	1.211	1.205	1.200	1.195
0.70000	1.315	1.307	1.299	1.293	1.286
0.75000	1.430	1.419	1.409	1.400	1.393
0.80000	1.570	1.555	1.542	1.531	1.520
0.85000	1.749	1.730	1.712	1.697	1.683
0.90000	2.005	1.977	1.953	1.932	1.912
0.95000	2.456	2.412	2.374	2.340	2.310
0.96000	2.606	2.556	2.513	2.474	2.440
0.97000	2.805	2.746	2.695	2.651	2.611
0.98000	3.094	3.022	2.960	2.905	2.857
0.99000	3.616	3.518	3.433	3.359	3.294
0.99500	4.178	4.049	3.938	3.841	3.755
0.99900	5.668	5.443	5.250	5.083	4.938
0.99990	8.361	7.922	7.553	7.237	6.965

Table generated using MINITAB

F-table for 11 Degrees of freedom in the denominator
numerator degrees of freedom

α	25	50	75	100	200
0.00010	0.089	0.095	0.097	0.098	0.102
0.00100	0.146	0.156	0.159	0.160	0.163
0.00500	0.211	0.222	0.227	0.229	0.232
0.01000	0.249	0.262	0.267	0.269	0.273
0.02000	0.298	0.311	0.316	0.319	0.323
0.03000	0.332	0.346	0.351	0.354	0.358
0.04000	0.360	0.374	0.380	0.382	0.387
0.05000	0.384	0.398	0.404	0.407	0.411
0.06000	0.405	0.420	0.425	0.428	0.432
0.07000	0.425	0.440	0.445	0.448	0.452
0.08000	0.443	0.458	0.463	0.466	0.470
0.09000	0.460	0.475	0.480	0.482	0.487
0.10000	0.477	0.491	0.496	0.498	0.502
0.20000	0.613	0.623	0.627	0.629	0.632
0.25000	0.673	0.680	0.683	0.684	0.687
0.30000	0.730	0.735	0.736	0.737	0.739
0.35000	0.787	0.788	0.789	0.789	0.789
0.40000	0.845	0.842	0.841	0.840	0.840
0.45000	0.904	0.896	0.894	0.892	0.891
0.50000	0.965	0.952	0.948	0.946	0.943
0.55000	1.030	1.011	1.005	1.002	0.997
0.60000	1.101	1.074	1.065	1.061	1.054
0.65000	1.177	1.142	1.131	1.125	1.116
0.70000	1.263	1.218	1.203	1.195	1.184
0.75000	1.363	1.304	1.284	1.274	1.260
0.80000	1.481	1.405	1.380	1.367	1.348
0.85000	1.632	1.531	1.498	1.482	1.457
0.90000	1.841	1.702	1.658	1.636	1.603
0.95000	2.197	1.986	1.918	1.885	1.836
0.96000	2.313	2.075	2.000	1.963	1.909
0.97000	2.464	2.190	2.104	2.062	2.000
0.98000	2.679	2.351	2.249	2.199	2.127
0.99000	3.055	2.625	2.493	2.430	2.337
0.99500	3.446	2.899	2.735	2.656	2.542
0.99900	4.423	3.547	3.295	3.176	3.004
0.99990	6.024	4.517	4.106	3.915	3.646

The table was generated using MINITAB

F-table for 12 Degrees of freedom in the denominator
numerator degrees of freedom

α	1	2	3	4	5
0.00010	0	0.05	0.055	0.064	0.068
0.00100	0	0.08	0.093	0.104	0.112
0.00500	0	0.12	0.138	0.153	0.164
0.01000	0	0.14	0.168	0.185	0.197
0.02000	0	0.18	0.208	0.226	0.240
0.03000	0	0.21	0.238	0.257	0.272
0.04000	0	0.23	0.263	0.284	0.298
0.05000	0	0.26	0.287	0.307	0.322
0.06000	0	0.28	0.308	0.328	0.343
0.07000	0	0.30	0.328	0.348	0.363
0.08000	0	0.32	0.347	0.367	0.382
0.09000	0	0.34	0.366	0.386	0.400
0.10000	0	0.36	0.384	0.403	0.417
0.20000	1	0.54	0.554	0.565	0.574
0.25000	1	0.64	0.641	0.645	0.649
0.30000	1	0.75	0.732	0.727	0.726
0.35000	1	0.87	0.830	0.814	0.806
0.40000	1	1.01	0.938	0.908	0.891
0.45000	2	1.17	1.059	1.011	0.984
0.50000	2	1.36	1.197	1.126	1.085
0.55000	3	1.59	1.357	1.256	1.199
0.60000	3	1.88	1.547	1.407	1.328
0.65000	5	2.24	1.778	1.586	1.479
0.70000	6	2.72	2.069	1.804	1.661
0.75000	9	3.39	2.450	2.083	1.887
0.80000	15	4.40	2.981	2.455	2.183
0.85000	27	6.07	3.787	2.996	2.600
0.90000	61	9.41	5.215	3.896	3.268
0.95000	244	19.41	8.745	5.912	4.677
0.96000	381	24.41	10.272	6.722	5.218
0.97000	678	32.75	12.608	7.907	5.988
0.98000	1526	49.41	16.759	9.894	7.234
0.99000	6107	99.41	27.051	14.374	9.888
0.99500	24428	199.42	43.386	20.705	13.384
0.99900	610644	999.46	128.315	47.413	26.417
0.99990	61043284	9997.43	598.252	151.849	67.903

The table was generated using MINITAB

F-table for 12 Degrees of freedom in the denominator

numerator degrees of freedom

α	6	7	8	9	10
0.00010	0.074	0.077	0.081	0.084	0.086
0.00100	0.119	0.125	0.129	0.133	0.136
0.00500	0.173	0.181	0.187	0.192	0.196
0.01000	0.207	0.215	0.222	0.227	0.232
0.02000	0.251	0.260	0.267	0.273	0.278
0.03000	0.283	0.292	0.300	0.306	0.312
0.04000	0.310	0.319	0.327	0.333	0.339
0.05000	0.333	0.343	0.351	0.357	0.363
0.06000	0.355	0.364	0.372	0.379	0.384
0.07000	0.375	0.384	0.392	0.398	0.404
0.08000	0.393	0.403	0.410	0.417	0.422
0.09000	0.411	0.421	0.428	0.435	0.440
0.10000	0.429	0.438	0.445	0.451	0.457
0.20000	0.582	0.588	0.593	0.597	0.601
0.25000	0.654	0.658	0.661	0.664	0.666
0.30000	0.727	0.728	0.729	0.730	0.731
0.35000	0.802	0.800	0.798	0.797	0.797
0.40000	0.882	0.875	0.870	0.867	0.864
0.45000	0.967	0.955	0.947	0.940	0.935
0.50000	1.060	1.042	1.029	1.019	1.011
0.55000	1.162	1.137	1.119	1.104	1.093
0.60000	1.278	1.244	1.219	1.199	1.184
0.65000	1.413	1.366	1.332	1.306	1.285
0.70000	1.571	1.510	1.465	1.430	1.403
0.75000	1.766	1.684	1.624	1.578	1.543
0.80000	2.017	1.905	1.825	1.764	1.716
0.85000	2.364	2.207	2.095	2.011	1.945
0.90000	2.904	2.668	2.502	2.378	2.284
0.95000	3.999	3.574	3.283	3.073	2.913
0.96000	4.406	3.903	3.562	3.316	3.131
0.97000	4.976	4.358	3.944	3.648	3.426
0.98000	5.875	5.063	4.527	4.149	3.868
0.99000	7.718	6.469	5.666	5.111	4.705
0.99500	10.034	8.176	7.014	6.227	5.661
0.99900	17.988	13.707	11.194	9.570	8.445
0.99990	40.078	27.643	20.983	16.966	14.329

Table generated using MINITAB

F-table for 12 Degrees of freedom in the denominator
numerator degrees of freedom

α	11	12	13	14	15
0.00010	0.086	0.089	0.091	0.093	0.094
0.00100	0.140	0.142	0.145	0.146	0.148
0.00500	0.200	0.203	0.206	0.209	0.211
0.01000	0.237	0.240	0.243	0.246	0.249
0.02000	0.283	0.287	0.290	0.293	0.296
0.03000	0.316	0.320	0.324	0.327	0.330
0.04000	0.344	0.348	0.352	0.355	0.358
0.05000	0.368	0.372	0.375	0.379	0.382
0.06000	0.389	0.393	0.397	0.400	0.403
0.07000	0.409	0.413	0.417	0.420	0.423
0.08000	0.427	0.431	0.435	0.438	0.441
0.09000	0.445	0.449	0.452	0.455	0.458
0.10000	0.461	0.465	0.469	0.472	0.475
0.20000	0.604	0.607	0.609	0.612	0.614
0.25000	0.669	0.671	0.672	0.674	0.675
0.30000	0.732	0.733	0.734	0.735	0.736
0.35000	0.796	0.796	0.796	0.796	0.796
0.40000	0.862	0.861	0.859	0.858	0.857
0.45000	0.931	0.928	0.925	0.923	0.921
0.50000	1.005	1.000	0.995	0.991	0.988
0.55000	1.084	1.076	1.070	1.065	1.060
0.60000	1.171	1.161	1.152	1.144	1.138
0.65000	1.269	1.255	1.243	1.233	1.225
0.70000	1.381	1.363	1.347	1.334	1.323
0.75000	1.514	1.490	1.470	1.453	1.438
0.80000	1.678	1.646	1.620	1.597	1.578
0.85000	1.893	1.850	1.815	1.784	1.758
0.90000	2.208	2.147	2.096	2.053	2.017
0.95000	2.787	2.686	2.603	2.534	2.475
0.96000	2.986	2.870	2.775	2.696	2.628
0.97000	3.253	3.116	3.004	2.910	2.831
0.98000	3.651	3.480	3.340	3.225	3.127
0.99000	4.397	4.155	3.960	3.800	3.666
0.99500	5.236	4.906	4.642	4.428	4.249
0.99900	7.625	7.004	6.519	6.130	5.812
0.99990	12.489	11.143	10.122	9.324	8.685

The table was generated using MINITAB

F-table for 12 Degrees of freedom in the denominator

numerator degrees of freedom

α	16	17	18	19	20
0.00010	0.094	0.094	0.096	0.098	0.100
0.00100	0.150	0.152	0.153	0.154	0.156
0.00500	0.214	0.215	0.217	0.219	0.220
0.01000	0.251	0.253	0.255	0.257	0.259
0.02000	0.299	0.301	0.303	0.305	0.307
0.03000	0.333	0.335	0.337	0.339	0.341
0.04000	0.360	0.363	0.365	0.367	0.369
0.05000	0.384	0.387	0.389	0.391	0.393
0.06000	0.406	0.408	0.410	0.412	0.414
0.07000	0.425	0.428	0.430	0.432	0.434
0.08000	0.444	0.446	0.448	0.450	0.452
0.09000	0.461	0.463	0.465	0.467	0.469
0.10000	0.477	0.479	0.481	0.483	0.485
0.20000	0.615	0.617	0.618	0.620	0.621
0.25000	0.677	0.678	0.679	0.680	0.681
0.30000	0.736	0.737	0.738	0.738	0.739
0.35000	0.796	0.796	0.796	0.796	0.796
0.40000	0.857	0.856	0.855	0.855	0.854
0.45000	0.919	0.918	0.917	0.915	0.914
0.50000	0.985	0.983	0.981	0.979	0.977
0.55000	1.056	1.052	1.049	1.046	1.044
0.60000	1.132	1.127	1.123	1.119	1.115
0.65000	1.217	1.210	1.205	1.199	1.195
0.70000	1.313	1.304	1.297	1.290	1.284
0.75000	1.425	1.414	1.404	1.395	1.387
0.80000	1.561	1.546	1.533	1.521	1.511
0.85000	1.736	1.716	1.699	1.683	1.669
0.90000	1.985	1.957	1.933	1.911	1.892
0.95000	2.424	2.380	2.342	2.307	2.277
0.96000	2.571	2.521	2.477	2.438	2.404
0.97000	2.764	2.705	2.654	2.609	2.569
0.98000	3.045	2.973	2.911	2.856	2.807
0.99000	3.552	3.455	3.370	3.296	3.231
0.99500	4.099	3.970	3.859	3.763	3.677
0.99900	5.547	5.323	5.132	4.967	4.822
0.99990	8.163	7.730	7.364	7.052	6.783

Table generated using MINITAB

F-table for 12 Degrees of freedom in the denominator
numerator degrees of freedom

α	25	50	75	100	200
0.00010	0.102	0.108	0.111	0.115	0.114
0.00100	0.161	0.171	0.175	0.177	0.180
0.00500	0.226	0.240	0.244	0.247	0.251
0.01000	0.265	0.280	0.285	0.288	0.292
0.02000	0.314	0.329	0.335	0.338	0.343
0.03000	0.348	0.364	0.370	0.373	0.378
0.04000	0.376	0.392	0.398	0.401	0.406
0.05000	0.400	0.416	0.422	0.425	0.430
0.06000	0.421	0.437	0.443	0.446	0.451
0.07000	0.441	0.457	0.462	0.466	0.470
0.08000	0.459	0.475	0.480	0.483	0.488
0.09000	0.476	0.491	0.497	0.500	0.505
0.10000	0.492	0.507	0.513	0.516	0.520
0.20000	0.626	0.637	0.641	0.643	0.647
0.25000	0.684	0.693	0.696	0.697	0.700
0.30000	0.741	0.746	0.748	0.749	0.751
0.35000	0.797	0.798	0.799	0.799	0.800
0.40000	0.853	0.850	0.849	0.849	0.848
0.45000	0.910	0.903	0.900	0.899	0.897
0.50000	0.970	0.957	0.953	0.951	0.948
0.55000	1.034	1.014	1.008	1.005	1.000
0.60000	1.102	1.075	1.066	1.062	1.055
0.65000	1.177	1.141	1.129	1.123	1.114
0.70000	1.260	1.214	1.198	1.190	1.179
0.75000	1.357	1.296	1.277	1.267	1.252
0.80000	1.472	1.394	1.368	1.355	1.336
0.85000	1.617	1.515	1.482	1.465	1.440
0.90000	1.820	1.680	1.634	1.612	1.578
0.95000	2.164	1.951	1.883	1.850	1.800
0.96000	2.276	2.037	1.961	1.924	1.869
0.97000	2.422	2.146	2.060	2.018	1.956
0.98000	2.629	2.300	2.198	2.148	2.075
0.99000	2.993	2.562	2.431	2.367	2.274
0.99500	3.370	2.824	2.661	2.582	2.468
0.99900	4.311	3.442	3.192	3.073	2.903
0.99990	5.853	4.365	3.960	3.772	3.507

The table was generated using MINITAB

F-table for 13 Degrees of freedom in the denominator

numerator degrees of freedom

α	1	2	3	4	5
0.00010	0	0.05	0.060	0.067	0.075
0.00100	0	0.08	0.098	0.110	0.119
0.00500	0	0.12	0.144	0.160	0.172
0.01000	0	0.15	0.174	0.192	0.205
0.02000	0	0.19	0.214	0.234	0.248
0.03000	0	0.22	0.244	0.265	0.280
0.04000	0	0.24	0.270	0.291	0.307
0.05000	0	0.26	0.293	0.315	0.330
0.06000	0	0.28	0.315	0.336	0.351
0.07000	0	0.30	0.335	0.356	0.371
0.08000	0	0.32	0.354	0.375	0.390
0.09000	0	0.34	0.373	0.393	0.408
0.10000	0	0.36	0.391	0.411	0.426
0.20000	1	0.55	0.561	0.573	0.582
0.25000	1	0.65	0.647	0.652	0.657
0.30000	1	0.76	0.738	0.734	0.733
0.35000	1	0.88	0.836	0.821	0.813
0.40000	1	1.02	0.944	0.914	0.897
0.45000	2	1.18	1.065	1.016	0.989
0.50000	2	1.37	1.203	1.131	1.090
0.55000	3	1.60	1.362	1.260	1.203
0.60000	3	1.88	1.551	1.410	1.331
0.65000	5	2.25	1.782	1.588	1.482
0.70000	6	2.73	2.072	1.806	1.662
0.75000	9	3.40	2.452	2.083	1.886
0.80000	15	4.40	2.981	2.453	2.180
0.85000	27	6.08	3.786	2.991	2.594
0.90000	61	9.41	5.210	3.886	3.256
0.95000	245	19.42	8.729	5.891	4.655
0.96000	383	24.42	10.252	6.697	5.191
0.97000	680	32.76	12.580	7.876	5.956
0.98000	1531	49.42	16.720	9.852	7.192
0.99000	6125	99.43	26.982	14.307	9.825
0.99500	24505	199.42	43.271	20.603	13.293
0.99900	612670	999.40	127.953	47.164	26.224
0.99990	61243892	9998.09	596.558	151.024	67.382

The table was generated using MINITAB

F-table for 13 Degrees of freedom in the denominator
numerator degrees of freedom

α	6	7	8	9	10
0.00010	0.080	0.085	0.087	0.091	0.094
0.00100	0.127	0.133	0.138	0.143	0.147
0.00500	0.182	0.190	0.196	0.202	0.207
0.01000	0.216	0.225	0.232	0.238	0.243
0.02000	0.260	0.269	0.277	0.284	0.290
0.03000	0.292	0.302	0.310	0.317	0.323
0.04000	0.319	0.329	0.337	0.344	0.350
0.05000	0.343	0.353	0.361	0.368	0.374
0.06000	0.364	0.374	0.382	0.389	0.395
0.07000	0.384	0.394	0.402	0.409	0.415
0.08000	0.403	0.413	0.421	0.428	0.433
0.09000	0.420	0.430	0.438	0.445	0.451
0.10000	0.438	0.447	0.455	0.462	0.467
0.20000	0.590	0.596	0.602	0.606	0.610
0.25000	0.662	0.666	0.669	0.672	0.675
0.30000	0.734	0.735	0.737	0.738	0.739
0.35000	0.809	0.807	0.805	0.804	0.804
0.40000	0.888	0.881	0.877	0.873	0.871
0.45000	0.972	0.961	0.952	0.946	0.941
0.50000	1.064	1.046	1.033	1.023	1.016
0.55000	1.166	1.141	1.122	1.108	1.096
0.60000	1.281	1.246	1.221	1.201	1.186
0.65000	1.414	1.367	1.333	1.307	1.286
0.70000	1.571	1.509	1.464	1.429	1.402
0.75000	1.765	1.682	1.621	1.575	1.539
0.80000	2.013	1.901	1.819	1.758	1.710
0.85000	2.356	2.199	2.086	2.001	1.936
0.90000	2.892	2.654	2.487	2.364	2.268
0.95000	3.976	3.550	3.259	3.047	2.887
0.96000	4.379	3.875	3.534	3.287	3.102
0.97000	4.943	4.324	3.910	3.613	3.391
0.98000	5.833	5.021	4.485	4.107	3.826
0.99000	7.657	6.410	5.608	5.054	4.649
0.99500	9.950	8.096	6.938	6.153	5.588
0.99900	17.824	13.560	11.059	9.443	8.324
0.99990	39.691	27.327	20.712	16.722	14.106

Table generated using MINITAB

F-table for 13 Degrees of freedom in the denominator

numerator degrees of freedom

α	11	12	13	14	15
0.00010	0.095	0.099	0.101	0.101	0.103
0.00100	0.150	0.153	0.155	0.158	0.160
0.00500	0.211	0.215	0.218	0.221	0.224
0.01000	0.248	0.252	0.256	0.259	0.262
0.02000	0.295	0.299	0.303	0.306	0.309
0.03000	0.328	0.332	0.336	0.340	0.343
0.04000	0.355	0.360	0.364	0.367	0.370
0.05000	0.379	0.384	0.388	0.391	0.394
0.06000	0.400	0.405	0.409	0.412	0.416
0.07000	0.420	0.425	0.428	0.432	0.435
0.08000	0.438	0.443	0.447	0.450	0.453
0.09000	0.456	0.460	0.464	0.467	0.470
0.10000	0.472	0.477	0.480	0.484	0.487
0.20000	0.614	0.617	0.619	0.622	0.624
0.25000	0.678	0.680	0.682	0.683	0.685
0.30000	0.740	0.742	0.743	0.743	0.744
0.35000	0.804	0.804	0.804	0.804	0.804
0.40000	0.869	0.867	0.866	0.865	0.864
0.45000	0.937	0.934	0.931	0.929	0.927
0.50000	1.009	1.004	1.000	0.996	0.993
0.55000	1.087	1.080	1.073	1.068	1.063
0.60000	1.173	1.162	1.154	1.146	1.140
0.65000	1.269	1.255	1.243	1.233	1.224
0.70000	1.379	1.361	1.345	1.332	1.321
0.75000	1.510	1.486	1.465	1.448	1.433
0.80000	1.671	1.639	1.613	1.590	1.570
0.85000	1.883	1.840	1.804	1.773	1.747
0.90000	2.193	2.131	2.080	2.037	2.000
0.95000	2.761	2.660	2.576	2.507	2.448
0.96000	2.957	2.840	2.745	2.665	2.598
0.97000	3.219	3.081	2.968	2.875	2.796
0.98000	3.609	3.438	3.298	3.183	3.085
0.99000	4.341	4.099	3.905	3.745	3.611
0.99500	5.164	4.835	4.573	4.359	4.181
0.99900	7.509	6.892	6.409	6.022	5.706
0.99990	12.281	10.947	9.935	9.144	8.511

The table was generated using MINITAB

F-table for 13 Degrees of freedom in the denominator
numerator degrees of freedom

α	16	17	18	19	20
0.00010	0.104	0.104	0.107	0.107	0.109
0.00100	0.162	0.163	0.165	0.166	0.168
0.00500	0.226	0.228	0.230	0.232	0.234
0.01000	0.264	0.267	0.269	0.271	0.272
0.02000	0.312	0.314	0.317	0.319	0.321
0.03000	0.346	0.348	0.351	0.353	0.355
0.04000	0.373	0.376	0.378	0.380	0.382
0.05000	0.397	0.400	0.402	0.404	0.406
0.06000	0.418	0.421	0.423	0.425	0.427
0.07000	0.438	0.440	0.443	0.445	0.447
0.08000	0.456	0.459	0.461	0.463	0.465
0.09000	0.473	0.476	0.478	0.480	0.482
0.10000	0.489	0.492	0.494	0.496	0.498
0.20000	0.626	0.628	0.629	0.631	0.632
0.25000	0.686	0.688	0.689	0.690	0.691
0.30000	0.745	0.746	0.746	0.747	0.748
0.35000	0.804	0.804	0.804	0.804	0.804
0.40000	0.863	0.863	0.862	0.862	0.861
0.45000	0.925	0.923	0.922	0.921	0.920
0.50000	0.990	0.987	0.985	0.983	0.981
0.55000	1.059	1.055	1.052	1.049	1.046
0.60000	1.134	1.129	1.124	1.120	1.117
0.65000	1.217	1.210	1.204	1.199	1.194
0.70000	1.311	1.302	1.294	1.287	1.281
0.75000	1.420	1.409	1.399	1.390	1.382
0.80000	1.553	1.538	1.525	1.513	1.503
0.85000	1.724	1.705	1.687	1.671	1.657
0.90000	1.968	1.940	1.915	1.893	1.874
0.95000	2.397	2.353	2.314	2.280	2.249
0.96000	2.540	2.489	2.445	2.407	2.372
0.97000	2.728	2.669	2.618	2.573	2.533
0.98000	3.002	2.931	2.868	2.813	2.765
0.99000	3.498	3.400	3.316	3.242	3.176
0.99500	4.031	3.903	3.792	3.696	3.611
0.99900	5.443	5.221	5.031	4.866	4.723
0.99990	7.994	7.564	7.203	6.894	6.627

Table generated using MINITAB

F-table for 13 Degrees of freedom in the denominator

numerator degrees of freedom

α	25	50	75	100	200
0.00010	0.113	0.124	0.126	0.126	0.128
0.00100	0.174	0.186	0.190	0.193	0.196
0.00500	0.240	0.255	0.261	0.264	0.269
0.01000	0.280	0.296	0.302	0.305	0.310
0.02000	0.328	0.345	0.352	0.355	0.360
0.03000	0.363	0.380	0.387	0.390	0.395
0.04000	0.390	0.408	0.415	0.418	0.423
0.05000	0.414	0.432	0.438	0.442	0.447
0.06000	0.435	0.453	0.459	0.463	0.468
0.07000	0.455	0.472	0.478	0.482	0.487
0.08000	0.472	0.490	0.496	0.499	0.504
0.09000	0.489	0.506	0.512	0.516	0.521
0.10000	0.505	0.522	0.528	0.531	0.536
0.20000	0.637	0.649	0.654	0.656	0.660
0.25000	0.695	0.704	0.707	0.709	0.712
0.30000	0.750	0.756	0.758	0.759	0.761
0.35000	0.805	0.807	0.807	0.808	0.809
0.40000	0.860	0.857	0.857	0.856	0.856
0.45000	0.916	0.909	0.906	0.905	0.903
0.50000	0.975	0.962	0.957	0.955	0.952
0.55000	1.037	1.017	1.010	1.007	1.002
0.60000	1.103	1.076	1.067	1.062	1.055
0.65000	1.176	1.140	1.127	1.121	1.112
0.70000	1.257	1.210	1.194	1.186	1.174
0.75000	1.351	1.290	1.270	1.260	1.244
0.80000	1.463	1.384	1.358	1.345	1.326
0.85000	1.604	1.501	1.467	1.450	1.425
0.90000	1.801	1.660	1.614	1.591	1.557
0.95000	2.136	1.921	1.852	1.819	1.769
0.96000	2.244	2.003	1.927	1.890	1.834
0.97000	2.385	2.109	2.022	1.979	1.917
0.98000	2.586	2.257	2.154	2.104	2.031
0.99000	2.938	2.508	2.376	2.313	2.220
0.99500	3.304	2.759	2.596	2.517	2.403
0.99900	4.215	3.352	3.103	2.985	2.816
0.99990	5.707	4.236	3.835	3.649	3.387

The table was generated using MINITAB

F-table for 14 Degrees of freedom in the denominator
numerator degrees of freedom

α	1	2	3	4	5
0.00010	0	0.05	0.066	0.073	0.080
0.00100	0	0.09	0.103	0.116	0.126
0.00500	0	0.13	0.150	0.167	0.179
0.01000	0	0.15	0.180	0.199	0.212
0.02000	0	0.19	0.220	0.240	0.256
0.03000	0	0.22	0.250	0.272	0.288
0.04000	0	0.24	0.276	0.298	0.314
0.05000	0	0.27	0.299	0.321	0.338
0.06000	0	0.29	0.320	0.343	0.359
0.07000	0	0.31	0.341	0.363	0.379
0.08000	0	0.33	0.360	0.382	0.398
0.09000	0	0.35	0.378	0.400	0.416
0.10000	0	0.37	0.396	0.418	0.433
0.20000	1	0.55	0.567	0.579	0.589
0.25000	1	0.65	0.653	0.658	0.663
0.30000	1	0.76	0.744	0.740	0.739
0.35000	1	0.88	0.842	0.826	0.819
0.40000	1	1.02	0.949	0.919	0.903
0.45000	2	1.18	1.070	1.021	0.994
0.50000	2	1.37	1.207	1.135	1.094
0.55000	3	1.60	1.366	1.264	1.206
0.60000	3	1.89	1.555	1.413	1.334
0.65000	5	2.25	1.785	1.591	1.483
0.70000	6	2.73	2.074	1.807	1.662
0.75000	9	3.41	2.454	2.083	1.885
0.80000	15	4.41	2.982	2.452	2.177
0.85000	27	6.08	3.784	2.987	2.588
0.90000	61	9.42	5.205	3.878	3.246
0.95000	245	19.42	8.715	5.873	4.635
0.96000	384	24.43	10.234	6.675	5.168
0.97000	682	32.76	12.557	7.848	5.928
0.98000	1535	49.43	16.686	9.815	7.155
0.99000	6142	99.43	26.924	14.248	9.770
0.99500	24574	199.42	43.171	20.515	13.214
0.99900	614335	999.44	127.644	46.948	26.056
0.99990	61404648	9998.01	595.053	150.316	66.930

The table was generated using MINITAB

F-table for 14 Degrees of freedom in the denominator
numerator degrees of freedom

α	6	7	8	9	10
0.00010	0.086	0.092	0.096	0.098	0.102
0.00100	0.134	0.141	0.147	0.152	0.155
0.00500	0.190	0.198	0.205	0.212	0.217
0.01000	0.224	0.233	0.241	0.248	0.253
0.02000	0.268	0.278	0.286	0.294	0.300
0.03000	0.300	0.311	0.319	0.327	0.333
0.04000	0.327	0.338	0.346	0.354	0.360
0.05000	0.351	0.361	0.370	0.377	0.384
0.06000	0.372	0.383	0.391	0.399	0.405
0.07000	0.392	0.402	0.411	0.418	0.425
0.08000	0.411	0.421	0.430	0.437	0.443
0.09000	0.428	0.439	0.447	0.454	0.460
0.10000	0.445	0.456	0.464	0.471	0.477
0.20000	0.597	0.604	0.610	0.614	0.619
0.25000	0.668	0.673	0.677	0.680	0.683
0.30000	0.740	0.742	0.743	0.745	0.746
0.35000	0.815	0.813	0.811	0.811	0.810
0.40000	0.893	0.887	0.882	0.879	0.876
0.45000	0.977	0.965	0.957	0.951	0.946
0.50000	1.068	1.050	1.037	1.027	1.019
0.55000	1.169	1.144	1.125	1.111	1.099
0.60000	1.284	1.249	1.223	1.203	1.187
0.65000	1.415	1.368	1.334	1.307	1.286
0.70000	1.572	1.509	1.463	1.428	1.400
0.75000	1.763	1.680	1.619	1.573	1.536
0.80000	2.009	1.896	1.815	1.753	1.704
0.85000	2.350	2.191	2.078	1.993	1.927
0.90000	2.880	2.642	2.475	2.351	2.255
0.95000	3.955	3.529	3.237	3.025	2.864
0.96000	4.355	3.851	3.509	3.262	3.076
0.97000	4.914	4.295	3.880	3.584	3.361
0.98000	5.796	4.985	4.449	4.070	3.790
0.99000	7.604	6.359	5.558	5.005	4.600
0.99500	9.877	8.027	6.872	6.088	5.525
0.99900	17.682	13.434	10.943	9.333	8.220
0.99990	39.356	27.054	20.477	16.512	13.914

Table generated using MINITAB

F-table for 14 Degrees of freedom in the denominator
numerator degrees of freedom

α	11	12	13	14	15
0.00010	0.104	0.107	0.109	0.113	0.113
0.00100	0.159	0.162	0.166	0.168	0.171
0.00500	0.221	0.225	0.229	0.232	0.235
0.01000	0.258	0.263	0.267	0.270	0.273
0.02000	0.305	0.310	0.314	0.317	0.321
0.03000	0.338	0.343	0.347	0.351	0.354
0.04000	0.366	0.370	0.375	0.378	0.382
0.05000	0.389	0.394	0.398	0.402	0.406
0.06000	0.411	0.415	0.420	0.423	0.427
0.07000	0.430	0.435	0.439	0.443	0.446
0.08000	0.448	0.453	0.457	0.461	0.464
0.09000	0.466	0.470	0.474	0.478	0.481
0.10000	0.482	0.486	0.490	0.494	0.497
0.20000	0.622	0.626	0.628	0.631	0.633
0.25000	0.685	0.688	0.690	0.692	0.693
0.30000	0.748	0.749	0.750	0.751	0.752
0.35000	0.810	0.810	0.810	0.810	0.810
0.40000	0.875	0.873	0.872	0.871	0.870
0.45000	0.942	0.938	0.936	0.933	0.931
0.50000	1.013	1.008	1.003	1.000	0.996
0.55000	1.090	1.082	1.076	1.070	1.066
0.60000	1.175	1.164	1.155	1.147	1.141
0.65000	1.269	1.255	1.243	1.233	1.224
0.70000	1.378	1.359	1.344	1.330	1.319
0.75000	1.507	1.482	1.462	1.444	1.429
0.80000	1.665	1.633	1.606	1.583	1.564
0.85000	1.874	1.831	1.794	1.764	1.737
0.90000	2.179	2.117	2.065	2.022	1.985
0.95000	2.738	2.637	2.553	2.483	2.424
0.96000	2.931	2.814	2.718	2.638	2.571
0.97000	3.189	3.051	2.938	2.844	2.765
0.98000	3.573	3.401	3.262	3.146	3.049
0.99000	4.293	4.051	3.857	3.697	3.563
0.99500	5.103	4.774	4.512	4.299	4.121
0.99900	7.408	6.794	6.314	5.929	5.615
0.99990	12.102	10.777	9.773	8.988	8.360

The table was generated using MINITAB

F-table for 14 Degrees of freedom in the denominator
numerator degrees of freedom

α	16	17	18	19	20
0.00010	0.113	0.114	0.117	0.118	0.121
0.00100	0.172	0.175	0.177	0.178	0.179
0.00500	0.238	0.240	0.242	0.244	0.246
0.01000	0.276	0.278	0.281	0.283	0.285
0.02000	0.324	0.326	0.329	0.331	0.333
0.03000	0.357	0.360	0.363	0.365	0.367
0.04000	0.385	0.388	0.390	0.392	0.395
0.05000	0.409	0.411	0.414	0.416	0.418
0.06000	0.430	0.432	0.435	0.437	0.439
0.07000	0.449	0.452	0.454	0.457	0.459
0.08000	0.467	0.470	0.472	0.474	0.476
0.09000	0.484	0.487	0.489	0.491	0.493
0.10000	0.500	0.503	0.505	0.507	0.509
0.20000	0.635	0.637	0.639	0.640	0.642
0.25000	0.695	0.696	0.697	0.699	0.700
0.30000	0.753	0.754	0.754	0.755	0.756
0.35000	0.810	0.811	0.811	0.811	0.811
0.40000	0.869	0.869	0.868	0.868	0.867
0.45000	0.930	0.928	0.927	0.926	0.925
0.50000	0.993	0.991	0.989	0.987	0.985
0.55000	1.061	1.058	1.054	1.051	1.049
0.60000	1.135	1.130	1.125	1.121	1.118
0.65000	1.217	1.210	1.204	1.198	1.193
0.70000	1.309	1.300	1.292	1.285	1.279
0.75000	1.416	1.405	1.395	1.385	1.377
0.80000	1.547	1.532	1.518	1.506	1.496
0.85000	1.714	1.694	1.677	1.661	1.647
0.90000	1.953	1.925	1.900	1.878	1.858
0.95000	2.373	2.328	2.290	2.255	2.224
0.96000	2.513	2.462	2.418	2.379	2.345
0.97000	2.697	2.638	2.587	2.542	2.502
0.98000	2.966	2.894	2.831	2.776	2.728
0.99000	3.450	3.353	3.268	3.194	3.129
0.99500	3.972	3.844	3.734	3.637	3.553
0.99900	5.353	5.132	4.943	4.779	4.637
0.99990	7.847	7.421	7.063	6.756	6.492

Table generated using MINITAB

F-table for 14 Degrees of freedom in the denominator

numerator degrees of freedom

α	25	50	75	100	200
0.00010	0.123	0.133	0.139	0.142	0.144
0.00100	0.185	0.199	0.205	0.208	0.212
0.00500	0.253	0.270	0.276	0.280	0.285
0.01000	0.293	0.311	0.317	0.321	0.326
0.02000	0.341	0.360	0.367	0.371	0.377
0.03000	0.376	0.395	0.402	0.406	0.411
0.04000	0.403	0.422	0.429	0.433	0.439
0.05000	0.427	0.446	0.453	0.457	0.463
0.06000	0.448	0.467	0.474	0.477	0.483
0.07000	0.467	0.486	0.493	0.496	0.502
0.08000	0.485	0.503	0.510	0.514	0.519
0.09000	0.501	0.519	0.526	0.530	0.535
0.10000	0.517	0.535	0.541	0.545	0.550
0.20000	0.647	0.660	0.665	0.668	0.672
0.25000	0.704	0.714	0.717	0.719	0.722
0.30000	0.758	0.764	0.767	0.768	0.770
0.35000	0.812	0.814	0.815	0.816	0.817
0.40000	0.866	0.864	0.863	0.863	0.862
0.45000	0.921	0.914	0.911	0.910	0.909
0.50000	0.978	0.965	0.961	0.959	0.956
0.55000	1.039	1.019	1.012	1.009	1.004
0.60000	1.104	1.076	1.067	1.062	1.055
0.65000	1.175	1.138	1.126	1.120	1.110
0.70000	1.255	1.207	1.191	1.183	1.170
0.75000	1.346	1.284	1.264	1.253	1.238
0.80000	1.455	1.375	1.349	1.336	1.316
0.85000	1.593	1.489	1.454	1.437	1.411
0.90000	1.785	1.642	1.596	1.573	1.538
0.95000	2.111	1.894	1.825	1.791	1.741
0.96000	2.216	1.974	1.897	1.859	1.803
0.97000	2.353	2.076	1.988	1.946	1.882
0.98000	2.549	2.218	2.115	2.065	1.991
0.99000	2.891	2.460	2.329	2.265	2.172
0.99500	3.246	2.703	2.539	2.461	2.347
0.99900	4.131	3.273	3.025	2.908	2.739
0.99990	5.580	4.122	3.726	3.542	3.282

The table was generated using MINITAB

F-table for 15 Degrees of freedom in the denominator
numerator degrees of freedom

α	1	2	3	4	5
0.00010	0	0.05	0.069	0.078	0.085
0.00100	0	0.09	0.107	0.121	0.132
0.00500	0	0.13	0.154	0.172	0.186
0.01000	0	0.16	0.185	0.204	0.219
0.02000	0	0.19	0.225	0.246	0.262
0.03000	0	0.22	0.255	0.278	0.294
0.04000	0	0.25	0.281	0.304	0.321
0.05000	0	0.27	0.304	0.327	0.344
0.06000	0	0.29	0.326	0.349	0.366
0.07000	0	0.31	0.346	0.369	0.385
0.08000	0	0.33	0.365	0.388	0.404
0.09000	0	0.35	0.384	0.406	0.422
0.10000	0	0.37	0.402	0.423	0.440
0.20000	1	0.56	0.572	0.585	0.595
0.25000	1	0.66	0.658	0.664	0.669
0.30000	1	0.77	0.749	0.745	0.745
0.35000	1	0.89	0.846	0.831	0.823
0.40000	1	1.03	0.954	0.924	0.907
0.45000	2	1.19	1.074	1.025	0.998
0.50000	2	1.38	1.211	1.139	1.098
0.55000	3	1.61	1.370	1.267	1.209
0.60000	3	1.89	1.559	1.416	1.336
0.65000	5	2.26	1.788	1.593	1.485
0.70000	6	2.74	2.077	1.809	1.663
0.75000	9	3.41	2.455	2.083	1.885
0.80000	15	4.42	2.982	2.450	2.175
0.85000	27	6.09	3.783	2.984	2.583
0.90000	61	9.42	5.200	3.870	3.238
0.95000	246	19.43	8.703	5.858	4.618
0.96000	384	24.43	10.219	6.656	5.148
0.97000	684	32.76	12.537	7.825	5.903
0.98000	1539	49.43	16.657	9.783	7.123
0.99000	6157	99.43	26.872	14.198	9.722
0.99500	24633	199.44	43.085	20.438	13.146
0.99900	615766	999.40	127.374	46.762	25.911
0.99990	61558492	9997.46	593.789	149.694	66.539

The table was generated using MINITAB

F-table for 15 Degrees of freedom in the denominator
numerator degrees of freedom

α	6	7	8	9	10
0.00010	0.092	0.096	0.101	0.104	0.109
0.00100	0.141	0.148	0.154	0.159	0.164
0.00500	0.197	0.206	0.213	0.220	0.226
0.01000	0.231	0.241	0.249	0.256	0.262
0.02000	0.275	0.286	0.295	0.302	0.309
0.03000	0.308	0.319	0.328	0.335	0.342
0.04000	0.334	0.345	0.355	0.362	0.369
0.05000	0.358	0.369	0.378	0.386	0.393
0.06000	0.379	0.390	0.399	0.407	0.414
0.07000	0.399	0.410	0.419	0.427	0.433
0.08000	0.418	0.429	0.437	0.445	0.452
0.09000	0.435	0.446	0.455	0.462	0.469
0.10000	0.452	0.463	0.472	0.479	0.485
0.20000	0.603	0.610	0.616	0.622	0.626
0.25000	0.674	0.679	0.683	0.687	0.690
0.30000	0.746	0.748	0.749	0.751	0.752
0.35000	0.820	0.818	0.817	0.816	0.816
0.40000	0.898	0.891	0.887	0.884	0.881
0.45000	0.981	0.969	0.961	0.955	0.950
0.50000	1.072	1.054	1.041	1.031	1.023
0.55000	1.172	1.147	1.128	1.113	1.102
0.60000	1.286	1.250	1.224	1.205	1.189
0.65000	1.416	1.369	1.334	1.308	1.287
0.70000	1.572	1.509	1.462	1.427	1.399
0.75000	1.762	1.678	1.617	1.570	1.533
0.80000	2.006	1.893	1.810	1.748	1.700
0.85000	2.344	2.185	2.071	1.986	1.919
0.90000	2.871	2.632	2.464	2.339	2.243
0.95000	3.938	3.510	3.218	3.006	2.845
0.96000	4.334	3.829	3.487	3.240	3.053
0.97000	4.889	4.270	3.854	3.558	3.335
0.98000	5.764	4.953	4.417	4.039	3.758
0.99000	7.559	6.314	5.515	4.962	4.558
0.99500	9.814	7.967	6.814	6.032	5.470
0.99900	17.558	13.323	10.841	9.238	8.128
0.99990	39.065	26.817	20.272	16.329	13.746

Table generated using MINITAB

F-table for 15 Degrees of freedom in the denominator
numerator degrees of freedom

α	11	12	13	14	15
0.00010	0.112	0.114	0.117	0.119	0.122
0.00100	0.168	0.171	0.174	0.177	0.180
0.00500	0.230	0.235	0.239	0.242	0.245
0.01000	0.268	0.272	0.276	0.280	0.283
0.02000	0.314	0.319	0.324	0.327	0.331
0.03000	0.348	0.353	0.357	0.361	0.365
0.04000	0.375	0.380	0.384	0.388	0.392
0.05000	0.398	0.404	0.408	0.412	0.416
0.06000	0.420	0.425	0.429	0.433	0.437
0.07000	0.439	0.444	0.448	0.452	0.456
0.08000	0.457	0.462	0.466	0.470	0.474
0.09000	0.474	0.479	0.483	0.487	0.491
0.10000	0.491	0.495	0.499	0.503	0.507
0.20000	0.630	0.633	0.636	0.639	0.641
0.25000	0.692	0.695	0.697	0.699	0.701
0.30000	0.754	0.755	0.756	0.757	0.758
0.35000	0.816	0.816	0.816	0.816	0.816
0.40000	0.879	0.878	0.877	0.876	0.875
0.45000	0.946	0.943	0.940	0.938	0.936
0.50000	1.016	1.011	1.007	1.003	1.000
0.55000	1.092	1.085	1.078	1.073	1.068
0.60000	1.176	1.165	1.156	1.148	1.142
0.65000	1.269	1.255	1.243	1.233	1.224
0.70000	1.377	1.358	1.342	1.329	1.317
0.75000	1.504	1.479	1.459	1.441	1.426
0.80000	1.660	1.628	1.601	1.578	1.558
0.85000	1.866	1.822	1.786	1.755	1.728
0.90000	2.167	2.104	2.053	2.009	1.972
0.95000	2.718	2.616	2.533	2.463	2.403
0.96000	2.908	2.791	2.695	2.615	2.547
0.97000	3.162	3.024	2.911	2.817	2.738
0.98000	3.541	3.369	3.229	3.114	3.016
0.99000	4.250	4.009	3.815	3.655	3.522
0.99500	5.048	4.721	4.459	4.246	4.069
0.99900	7.321	6.709	6.231	5.848	5.535
0.99990	11.945	10.629	9.631	8.852	8.228

The table was generated using MINITAB

F-table for 15 Degrees of freedom in the denominator
numerator degrees of freedom

α	16	17	18	19	20
0.00010	0.123	0.125	0.127	0.128	0.128
0.00100	0.183	0.185	0.186	0.189	0.190
0.00500	0.248	0.251	0.253	0.255	0.257
0.01000	0.286	0.289	0.292	0.294	0.296
0.02000	0.334	0.337	0.340	0.342	0.344
0.03000	0.368	0.371	0.374	0.376	0.378
0.04000	0.395	0.398	0.401	0.403	0.406
0.05000	0.419	0.422	0.424	0.427	0.429
0.06000	0.440	0.443	0.445	0.448	0.450
0.07000	0.459	0.462	0.465	0.467	0.469
0.08000	0.477	0.480	0.482	0.485	0.487
0.09000	0.494	0.496	0.499	0.501	0.503
0.10000	0.510	0.512	0.515	0.517	0.519
0.20000	0.643	0.645	0.647	0.649	0.650
0.25000	0.702	0.704	0.705	0.706	0.707
0.30000	0.759	0.760	0.761	0.762	0.763
0.35000	0.816	0.817	0.817	0.817	0.817
0.40000	0.874	0.874	0.873	0.873	0.872
0.45000	0.934	0.932	0.931	0.930	0.929
0.50000	0.997	0.994	0.992	0.990	0.988
0.55000	1.064	1.060	1.057	1.054	1.051
0.60000	1.136	1.131	1.126	1.122	1.119
0.65000	1.216	1.209	1.203	1.198	1.193
0.70000	1.307	1.298	1.290	1.283	1.276
0.75000	1.413	1.401	1.391	1.381	1.373
0.80000	1.541	1.525	1.512	1.500	1.489
0.85000	1.705	1.685	1.667	1.651	1.637
0.90000	1.939	1.911	1.886	1.864	1.844
0.95000	2.352	2.307	2.268	2.234	2.203
0.96000	2.489	2.438	2.394	2.355	2.320
0.97000	2.670	2.611	2.560	2.514	2.474
0.98000	2.933	2.861	2.799	2.744	2.695
0.99000	3.408	3.311	3.227	3.153	3.088
0.99500	3.920	3.792	3.682	3.586	3.501
0.99900	5.274	5.054	4.866	4.703	4.561
0.99990	7.719	7.296	6.940	6.636	6.374

Table generated using MINITAB

F-table for 15 Degrees of freedom in the denominator

numerator degrees of freedom

α	25	50	75	100	200
0.00010	0.133	0.146	0.150	0.153	0.158
0.00100	0.197	0.212	0.218	0.222	0.226
0.00500	0.265	0.283	0.290	0.294	0.300
0.01000	0.305	0.324	0.331	0.335	0.342
0.02000	0.353	0.374	0.381	0.385	0.392
0.03000	0.387	0.408	0.416	0.420	0.426
0.04000	0.415	0.435	0.443	0.447	0.454
0.05000	0.438	0.459	0.466	0.470	0.477
0.06000	0.459	0.479	0.487	0.491	0.497
0.07000	0.478	0.498	0.506	0.509	0.516
0.08000	0.496	0.515	0.523	0.527	0.533
0.09000	0.512	0.531	0.539	0.542	0.548
0.10000	0.528	0.546	0.554	0.557	0.563
0.20000	0.656	0.670	0.675	0.678	0.682
0.25000	0.712	0.722	0.726	0.728	0.732
0.30000	0.765	0.772	0.775	0.776	0.779
0.35000	0.818	0.821	0.822	0.823	0.824
0.40000	0.871	0.869	0.869	0.868	0.868
0.45000	0.925	0.918	0.916	0.915	0.913
0.50000	0.982	0.968	0.964	0.962	0.959
0.55000	1.041	1.021	1.014	1.011	1.006
0.60000	1.105	1.077	1.067	1.063	1.056
0.65000	1.174	1.137	1.124	1.118	1.109
0.70000	1.252	1.204	1.187	1.179	1.167
0.75000	1.342	1.279	1.258	1.248	1.232
0.80000	1.448	1.368	1.341	1.327	1.307
0.85000	1.583	1.478	1.443	1.425	1.399
0.90000	1.770	1.626	1.579	1.556	1.521
0.95000	2.088	1.871	1.801	1.767	1.716
0.96000	2.191	1.948	1.871	1.832	1.776
0.97000	2.325	2.046	1.959	1.915	1.852
0.98000	2.516	2.184	2.081	2.030	1.956
0.99000	2.850	2.418	2.287	2.223	2.129
0.99500	3.196	2.653	2.489	2.411	2.296
0.99900	4.058	3.203	2.957	2.840	2.672
0.99990	5.469	4.023	3.630	3.447	3.190

The table was generated using MINITAB

F-table for 16 Degrees of freedom in the denominator
numerator degrees of freedom

α	1	2	3	4	5
0.00010	0	0.06	0.071	0.083	0.091
0.00100	0	0.09	0.111	0.126	0.137
0.00500	0	0.13	0.159	0.177	0.191
0.01000	0	0.16	0.189	0.210	0.225
0.02000	0	0.20	0.229	0.252	0.268
0.03000	0	0.23	0.260	0.283	0.300
0.04000	0	0.25	0.286	0.309	0.327
0.05000	0	0.28	0.309	0.333	0.350
0.06000	0	0.30	0.330	0.354	0.371
0.07000	0	0.32	0.350	0.374	0.391
0.08000	0	0.34	0.370	0.393	0.410
0.09000	0	0.36	0.388	0.411	0.428
0.10000	0	0.37	0.406	0.429	0.445
0.20000	1	0.56	0.576	0.590	0.600
0.25000	1	0.66	0.662	0.668	0.674
0.30000	1	0.77	0.753	0.749	0.749
0.35000	1	0.89	0.850	0.835	0.828
0.40000	1	1.03	0.958	0.928	0.911
0.45000	2	1.19	1.078	1.029	1.002
0.50000	2	1.38	1.215	1.142	1.101
0.55000	3	1.61	1.373	1.270	1.212
0.60000	3	1.90	1.561	1.418	1.338
0.65000	5	2.26	1.791	1.594	1.486
0.70000	6	2.74	2.079	1.810	1.663
0.75000	10	3.41	2.457	2.083	1.884
0.80000	15	4.42	2.982	2.449	2.172
0.85000	27	6.09	3.781	2.980	2.579
0.90000	61	9.43	5.196	3.864	3.230
0.95000	246	19.43	8.692	5.844	4.603
0.96000	385	24.43	10.205	6.639	5.130
0.97000	685	32.77	12.519	7.804	5.881
0.98000	1542	49.43	16.631	9.755	7.095
0.99000	6170	99.44	26.827	14.154	9.680
0.99500	24681	199.43	43.009	20.371	13.086
0.99900	617093	999.47	127.139	46.597	25.782
0.99990	61687144	9998.22	592.647	149.152	66.194

The table was generated using MINITAB

F-table for 16 Degrees of freedom in the denominator

numerator degrees of freedom

α	6	7	8	9	10
0.00010	0.097	0.103	0.109	0.113	0.115
0.00100	0.146	0.154	0.161	0.167	0.172
0.00500	0.203	0.213	0.221	0.228	0.234
0.01000	0.238	0.248	0.257	0.264	0.270
0.02000	0.282	0.293	0.302	0.310	0.317
0.03000	0.314	0.325	0.335	0.343	0.350
0.04000	0.341	0.352	0.362	0.370	0.377
0.05000	0.364	0.376	0.386	0.394	0.401
0.06000	0.386	0.397	0.407	0.415	0.422
0.07000	0.405	0.417	0.426	0.434	0.441
0.08000	0.424	0.435	0.445	0.452	0.459
0.09000	0.442	0.453	0.462	0.470	0.476
0.10000	0.459	0.469	0.478	0.486	0.493
0.20000	0.609	0.616	0.623	0.628	0.632
0.25000	0.680	0.684	0.689	0.692	0.696
0.30000	0.751	0.753	0.754	0.756	0.758
0.35000	0.824	0.822	0.821	0.821	0.821
0.40000	0.902	0.895	0.891	0.888	0.886
0.45000	0.985	0.973	0.965	0.958	0.953
0.50000	1.075	1.057	1.044	1.034	1.026
0.55000	1.175	1.149	1.130	1.115	1.104
0.60000	1.287	1.252	1.226	1.206	1.190
0.65000	1.417	1.370	1.335	1.308	1.287
0.70000	1.571	1.508	1.462	1.426	1.398
0.75000	1.760	1.676	1.615	1.568	1.531
0.80000	2.003	1.889	1.807	1.744	1.695
0.85000	2.339	2.179	2.065	1.979	1.913
0.90000	2.862	2.623	2.454	2.329	2.233
0.95000	3.922	3.494	3.201	2.988	2.827
0.96000	4.316	3.810	3.467	3.220	3.034
0.97000	4.867	4.247	3.832	3.535	3.312
0.98000	5.736	4.925	4.389	4.010	3.729
0.99000	7.518	6.275	5.476	4.924	4.520
0.99500	9.758	7.914	6.763	5.982	5.422
0.99900	17.449	13.226	10.751	9.153	8.048
0.99990	38.809	26.609	20.092	16.168	13.598

Table generated using MINITAB

F-table for 16 Degrees of freedom in the denominator
numerator degrees of freedom

α	11	12	13	14	15
0.00010	0.119	0.122	0.124	0.127	0.129
0.00100	0.176	0.180	0.183	0.186	0.189
0.00500	0.239	0.243	0.248	0.251	0.255
0.01000	0.276	0.281	0.285	0.289	0.293
0.02000	0.323	0.328	0.333	0.337	0.340
0.03000	0.356	0.361	0.366	0.370	0.374
0.04000	0.383	0.388	0.393	0.397	0.401
0.05000	0.407	0.412	0.417	0.421	0.425
0.06000	0.428	0.433	0.438	0.442	0.446
0.07000	0.447	0.452	0.457	0.461	0.465
0.08000	0.465	0.470	0.475	0.479	0.482
0.09000	0.482	0.487	0.492	0.496	0.499
0.10000	0.498	0.503	0.508	0.511	0.515
0.20000	0.636	0.640	0.643	0.646	0.648
0.25000	0.699	0.701	0.703	0.705	0.707
0.30000	0.759	0.761	0.762	0.763	0.764
0.35000	0.821	0.821	0.821	0.821	0.821
0.40000	0.884	0.882	0.881	0.880	0.879
0.45000	0.949	0.946	0.944	0.941	0.939
0.50000	1.019	1.014	1.009	1.006	1.002
0.55000	1.094	1.087	1.080	1.074	1.070
0.60000	1.177	1.166	1.157	1.149	1.143
0.65000	1.269	1.255	1.243	1.233	1.224
0.70000	1.376	1.357	1.341	1.327	1.315
0.75000	1.501	1.476	1.456	1.438	1.423
0.80000	1.656	1.623	1.596	1.573	1.553
0.85000	1.859	1.815	1.778	1.747	1.721
0.90000	2.156	2.093	2.041	1.998	1.960
0.95000	2.700	2.598	2.514	2.444	2.384
0.96000	2.888	2.771	2.675	2.594	2.526
0.97000	3.139	3.001	2.887	2.793	2.714
0.98000	3.513	3.341	3.201	3.085	2.988
0.99000	4.213	3.972	3.778	3.618	3.485
0.99500	5.001	4.674	4.413	4.200	4.023
0.99900	7.243	6.634	6.157	5.776	5.464
0.99990	11.807	10.498	9.506	8.732	8.112

The table was generated using MINITAB

F-table for 16 Degrees of freedom in the denominator

numerator degrees of freedom

α	16	17	18	19	20
0.00010	0.132	0.133	0.135	0.134	0.138
0.00100	0.192	0.194	0.196	0.198	0.200
0.00500	0.258	0.260	0.263	0.265	0.267
0.01000	0.296	0.299	0.302	0.304	0.306
0.02000	0.344	0.347	0.350	0.352	0.355
0.03000	0.377	0.381	0.383	0.386	0.388
0.04000	0.405	0.408	0.411	0.413	0.416
0.05000	0.428	0.431	0.434	0.437	0.439
0.06000	0.449	0.452	0.455	0.457	0.460
0.07000	0.468	0.471	0.474	0.476	0.479
0.08000	0.486	0.489	0.491	0.494	0.496
0.09000	0.502	0.505	0.508	0.510	0.513
0.10000	0.518	0.521	0.524	0.526	0.528
0.20000	0.651	0.653	0.655	0.656	0.658
0.25000	0.709	0.710	0.712	0.713	0.714
0.30000	0.765	0.766	0.767	0.768	0.769
0.35000	0.822	0.822	0.822	0.822	0.823
0.40000	0.879	0.878	0.878	0.877	0.877
0.45000	0.938	0.936	0.935	0.934	0.933
0.50000	1.000	0.997	0.995	0.993	0.991
0.55000	1.065	1.062	1.058	1.055	1.053
0.60000	1.137	1.132	1.127	1.123	1.119
0.65000	1.216	1.209	1.203	1.197	1.192
0.70000	1.305	1.296	1.288	1.281	1.274
0.75000	1.409	1.397	1.387	1.378	1.369
0.80000	1.535	1.520	1.506	1.494	1.483
0.85000	1.697	1.677	1.659	1.643	1.629
0.90000	1.928	1.899	1.874	1.852	1.832
0.95000	2.333	2.288	2.249	2.214	2.183
0.96000	2.468	2.417	2.373	2.334	2.299
0.97000	2.646	2.587	2.535	2.490	2.449
0.98000	2.904	2.833	2.770	2.715	2.666
0.99000	3.372	3.274	3.190	3.116	3.051
0.99500	3.874	3.747	3.637	3.541	3.456
0.99900	5.204	4.985	4.798	4.636	4.494
0.99990	7.606	7.186	6.832	6.530	6.270

Table generated using MINITAB

F-table for 16 Degrees of freedom in the denominator
numerator degrees of freedom

α	25	50	75	100	200
0.00010	0.143	0.157	0.160	0.162	0.167
0.00100	0.207	0.224	0.231	0.235	0.240
0.00500	0.276	0.296	0.304	0.308	0.314
0.01000	0.315	0.337	0.345	0.349	0.356
0.02000	0.364	0.386	0.394	0.399	0.405
0.03000	0.398	0.420	0.428	0.433	0.440
0.04000	0.425	0.447	0.456	0.460	0.467
0.05000	0.448	0.470	0.479	0.483	0.490
0.06000	0.469	0.491	0.499	0.503	0.510
0.07000	0.488	0.509	0.517	0.522	0.528
0.08000	0.505	0.526	0.534	0.538	0.545
0.09000	0.522	0.542	0.550	0.554	0.560
0.10000	0.537	0.557	0.565	0.569	0.575
0.20000	0.664	0.679	0.684	0.687	0.692
0.25000	0.719	0.730	0.734	0.737	0.740
0.30000	0.772	0.779	0.782	0.783	0.786
0.35000	0.824	0.827	0.828	0.829	0.830
0.40000	0.876	0.874	0.874	0.874	0.873
0.45000	0.929	0.922	0.920	0.919	0.917
0.50000	0.984	0.971	0.967	0.965	0.961
0.55000	1.043	1.022	1.016	1.012	1.007
0.60000	1.105	1.077	1.067	1.063	1.055
0.65000	1.174	1.136	1.123	1.117	1.107
0.70000	1.250	1.201	1.184	1.176	1.163
0.75000	1.338	1.274	1.253	1.242	1.226
0.80000	1.442	1.361	1.333	1.320	1.299
0.85000	1.574	1.468	1.432	1.415	1.388
0.90000	1.757	1.612	1.565	1.541	1.506
0.95000	2.069	1.850	1.780	1.745	1.694
0.96000	2.169	1.925	1.847	1.808	1.752
0.97000	2.300	2.020	1.932	1.889	1.824
0.98000	2.486	2.154	2.050	1.999	1.925
0.99000	2.813	2.381	2.249	2.185	2.091
0.99500	3.151	2.608	2.445	2.366	2.252
0.99900	3.993	3.141	2.896	2.779	2.612
0.99990	5.371	3.935	3.544	3.363	3.107

The table was generated using MINITAB

F-table for 17 Degrees of freedom in the denominator
numerator degrees of freedom

α	1	2	3	4	5
0.00010	0	0.06	0.075	0.085	0.094
0.00100	0	0.09	0.115	0.130	0.142
0.00500	0	0.14	0.162	0.182	0.197
0.01000	0	0.16	0.193	0.214	0.230
0.02000	0	0.20	0.233	0.256	0.274
0.03000	0	0.23	0.264	0.288	0.305
0.04000	0	0.26	0.290	0.314	0.332
0.05000	0	0.28	0.313	0.337	0.355
0.06000	0	0.30	0.334	0.359	0.377
0.07000	0	0.32	0.354	0.379	0.397
0.08000	0	0.34	0.374	0.398	0.415
0.09000	0	0.36	0.392	0.416	0.433
0.10000	0	0.38	0.410	0.433	0.450
0.20000	1	0.56	0.580	0.594	0.605
0.25000	1	0.66	0.666	0.672	0.678
0.30000	1	0.77	0.757	0.753	0.753
0.35000	1	0.89	0.854	0.839	0.832
0.40000	1	1.03	0.961	0.931	0.915
0.45000	2	1.19	1.081	1.032	1.005
0.50000	2	1.38	1.218	1.145	1.104
0.55000	3	1.61	1.376	1.272	1.214
0.60000	4	1.90	1.564	1.420	1.340
0.65000	5	2.26	1.793	1.596	1.487
0.70000	7	2.75	2.080	1.811	1.664
0.75000	10	3.42	2.458	2.083	1.883
0.80000	15	4.42	2.983	2.448	2.170
0.85000	27	6.09	3.780	2.977	2.575
0.90000	61	9.43	5.193	3.858	3.223
0.95000	247	19.44	8.683	5.832	4.590
0.96000	386	24.44	10.194	6.625	5.114
0.97000	686	32.77	12.503	7.785	5.861
0.98000	1545	49.44	16.609	9.730	7.070
0.99000	6181	99.44	26.786	14.115	9.642
0.99500	24725	199.44	42.942	20.311	13.032
0.99900	618187	999.46	126.924	46.453	25.669
0.99990	61803300	9997.71	591.665	148.674	65.890

The table was generated using MINITAB

F-table for 17 Degrees of freedom in the denominator

numerator degrees of freedom

α	6	7	8	9	10
0.00010	0.103	0.110	0.114	0.119	0.122
0.00100	0.152	0.160	0.167	0.173	0.179
0.00500	0.209	0.219	0.227	0.235	0.241
0.01000	0.243	0.254	0.263	0.271	0.278
0.02000	0.288	0.299	0.309	0.317	0.324
0.03000	0.320	0.332	0.342	0.350	0.357
0.04000	0.347	0.359	0.369	0.377	0.384
0.05000	0.370	0.382	0.392	0.400	0.408
0.06000	0.391	0.403	0.413	0.422	0.429
0.07000	0.411	0.423	0.433	0.441	0.448
0.08000	0.430	0.441	0.451	0.459	0.466
0.09000	0.447	0.459	0.468	0.476	0.483
0.10000	0.464	0.475	0.485	0.493	0.499
0.20000	0.614	0.622	0.628	0.634	0.638
0.25000	0.684	0.689	0.694	0.698	0.701
0.30000	0.755	0.757	0.759	0.761	0.763
0.35000	0.828	0.827	0.826	0.825	0.825
0.40000	0.905	0.899	0.895	0.892	0.889
0.45000	0.988	0.976	0.968	0.962	0.957
0.50000	1.077	1.060	1.046	1.036	1.028
0.55000	1.177	1.151	1.132	1.117	1.106
0.60000	1.289	1.253	1.227	1.207	1.191
0.65000	1.418	1.370	1.335	1.308	1.287
0.70000	1.571	1.508	1.461	1.426	1.397
0.75000	1.759	1.675	1.613	1.566	1.529
0.80000	2.001	1.886	1.803	1.741	1.691
0.85000	2.335	2.174	2.060	1.974	1.907
0.90000	2.854	2.614	2.445	2.320	2.223
0.95000	3.908	3.479	3.186	2.973	2.812
0.96000	4.299	3.793	3.450	3.203	3.016
0.97000	4.847	4.227	3.811	3.514	3.291
0.98000	5.711	4.899	4.364	3.985	3.704
0.99000	7.482	6.240	5.442	4.890	4.486
0.99500	9.708	7.867	6.718	5.938	5.378
0.99900	17.353	13.140	10.672	9.079	7.977
0.99990	38.583	26.423	19.932	16.025	13.467

Table generated using MINITAB

F-table for 17 Degrees of freedom in the denominator

numerator degrees of freedom

α	11	12	13	14	15
0.00010	0.127	0.130	0.133	0.134	0.135
0.00100	0.183	0.187	0.191	0.194	0.197
0.00500	0.247	0.251	0.256	0.260	0.263
0.01000	0.284	0.289	0.294	0.298	0.302
0.02000	0.330	0.336	0.341	0.345	0.349
0.03000	0.364	0.369	0.374	0.378	0.382
0.04000	0.391	0.396	0.401	0.406	0.410
0.05000	0.414	0.420	0.424	0.429	0.433
0.06000	0.435	0.441	0.445	0.450	0.454
0.07000	0.454	0.460	0.464	0.469	0.473
0.08000	0.472	0.478	0.482	0.486	0.490
0.09000	0.489	0.494	0.499	0.503	0.507
0.10000	0.505	0.510	0.515	0.519	0.523
0.20000	0.642	0.646	0.649	0.652	0.655
0.25000	0.704	0.707	0.709	0.711	0.713
0.30000	0.764	0.766	0.767	0.769	0.770
0.35000	0.825	0.825	0.826	0.826	0.826
0.40000	0.888	0.886	0.885	0.884	0.883
0.45000	0.953	0.950	0.947	0.945	0.943
0.50000	1.022	1.016	1.012	1.008	1.005
0.55000	1.096	1.088	1.082	1.076	1.071
0.60000	1.178	1.167	1.158	1.150	1.143
0.65000	1.270	1.255	1.243	1.232	1.223
0.70000	1.374	1.355	1.339	1.326	1.314
0.75000	1.499	1.474	1.453	1.435	1.420
0.80000	1.652	1.619	1.591	1.568	1.548
0.85000	1.853	1.809	1.772	1.740	1.714
0.90000	2.146	2.083	2.031	1.987	1.950
0.95000	2.685	2.582	2.498	2.428	2.368
0.96000	2.870	2.752	2.656	2.576	2.508
0.97000	3.118	2.979	2.866	2.772	2.692
0.98000	3.488	3.316	3.176	3.060	2.962
0.99000	4.180	3.939	3.745	3.585	3.452
0.99500	4.958	4.632	4.371	4.159	3.982
0.99900	7.174	6.567	6.092	5.712	5.401
0.99990	11.684	10.382	9.395	8.625	8.009

The table was generated using MINITAB

F-table for 17 Degrees of freedom in the denominator
numerator degrees of freedom

α	16	17	18	19	20
0.00010	0.139	0.142	0.143	0.142	0.144
0.00100	0.200	0.203	0.205	0.207	0.209
0.00500	0.266	0.269	0.272	0.274	0.277
0.01000	0.305	0.308	0.311	0.313	0.316
0.02000	0.352	0.356	0.359	0.361	0.364
0.03000	0.386	0.389	0.392	0.395	0.398
0.04000	0.413	0.416	0.419	0.422	0.425
0.05000	0.436	0.440	0.443	0.445	0.448
0.06000	0.457	0.460	0.463	0.466	0.469
0.07000	0.476	0.479	0.482	0.485	0.487
0.08000	0.494	0.497	0.500	0.502	0.505
0.09000	0.510	0.513	0.516	0.519	0.521
0.10000	0.526	0.529	0.532	0.534	0.536
0.20000	0.657	0.659	0.661	0.663	0.665
0.25000	0.715	0.716	0.718	0.719	0.721
0.30000	0.771	0.772	0.773	0.773	0.774
0.35000	0.826	0.827	0.827	0.827	0.827
0.40000	0.883	0.882	0.882	0.881	0.881
0.45000	0.941	0.940	0.938	0.937	0.936
0.50000	1.002	1.000	0.997	0.995	0.994
0.55000	1.067	1.063	1.060	1.057	1.054
0.60000	1.138	1.132	1.128	1.124	1.120
0.65000	1.215	1.209	1.202	1.197	1.192
0.70000	1.304	1.294	1.286	1.279	1.273
0.75000	1.406	1.394	1.384	1.374	1.366
0.80000	1.530	1.515	1.501	1.489	1.478
0.85000	1.690	1.670	1.652	1.635	1.621
0.90000	1.917	1.888	1.863	1.841	1.821
0.95000	2.316	2.271	2.232	2.197	2.166
0.96000	2.449	2.398	2.354	2.314	2.279
0.97000	2.624	2.565	2.513	2.468	2.427
0.98000	2.879	2.807	2.744	2.689	2.640
0.99000	3.339	3.241	3.157	3.083	3.018
0.99500	3.833	3.706	3.596	3.500	3.416
0.99900	5.142	4.924	4.737	4.576	4.435
0.99990	7.505	7.088	6.736	6.436	6.177

Table generated using MINITAB

F-table for 17 Degrees of freedom in the denominator
numerator degrees of freedom

α	25	50	75	100	200
0.00010	0.152	0.166	0.170	0.177	0.180
0.00100	0.217	0.235	0.243	0.246	0.253
0.00500	0.286	0.307	0.316	0.320	0.327
0.01000	0.326	0.348	0.357	0.361	0.369
0.02000	0.374	0.397	0.406	0.411	0.418
0.03000	0.408	0.431	0.440	0.445	0.452
0.04000	0.435	0.458	0.467	0.472	0.479
0.05000	0.458	0.481	0.490	0.494	0.502
0.06000	0.478	0.501	0.510	0.515	0.522
0.07000	0.497	0.520	0.528	0.533	0.540
0.08000	0.514	0.536	0.545	0.549	0.556
0.09000	0.530	0.552	0.560	0.565	0.571
0.10000	0.546	0.567	0.575	0.579	0.586
0.20000	0.671	0.687	0.692	0.696	0.700
0.25000	0.726	0.737	0.742	0.744	0.748
0.30000	0.777	0.785	0.788	0.790	0.793
0.35000	0.829	0.832	0.833	0.834	0.836
0.40000	0.880	0.878	0.878	0.878	0.878
0.45000	0.932	0.925	0.923	0.922	0.921
0.50000	0.987	0.974	0.969	0.967	0.964
0.55000	1.044	1.024	1.017	1.014	1.008
0.60000	1.106	1.077	1.068	1.063	1.055
0.65000	1.173	1.135	1.122	1.115	1.105
0.70000	1.248	1.198	1.181	1.173	1.160
0.75000	1.334	1.270	1.248	1.238	1.221
0.80000	1.437	1.354	1.327	1.313	1.292
0.85000	1.566	1.458	1.423	1.405	1.378
0.90000	1.746	1.600	1.552	1.528	1.492
0.95000	2.051	1.831	1.760	1.725	1.674
0.96000	2.150	1.904	1.825	1.787	1.729
0.97000	2.278	1.997	1.908	1.864	1.800
0.98000	2.460	2.127	2.022	1.971	1.896
0.99000	2.780	2.348	2.215	2.151	2.056
0.99500	3.111	2.568	2.405	2.326	2.211
0.99900	3.935	3.086	2.841	2.725	2.558
0.99990	5.284	3.856	3.468	3.288	3.034

The table was generated using MINITAB

F-table for 18 Degrees of freedom in the denominator

numerator degrees of freedom

α	1	2	3	4	5
0.00010	0	0.06	0.079	0.090	0.100
0.00100	0	0.10	0.118	0.134	0.146
0.00500	0	0.14	0.166	0.186	0.201
0.01000	0	0.17	0.196	0.218	0.235
0.02000	0	0.20	0.237	0.261	0.278
0.03000	0	0.23	0.267	0.292	0.310
0.04000	0	0.26	0.293	0.318	0.337
0.05000	0	0.28	0.316	0.342	0.360
0.06000	0	0.30	0.338	0.363	0.381
0.07000	0	0.32	0.358	0.383	0.401
0.08000	0	0.34	0.377	0.402	0.420
0.09000	0	0.36	0.396	0.420	0.438
0.10000	0	0.38	0.414	0.437	0.455
0.20000	1	0.57	0.584	0.598	0.609
0.25000	1	0.67	0.669	0.676	0.682
0.30000	1	0.78	0.760	0.757	0.757
0.35000	1	0.90	0.857	0.842	0.835
0.40000	1	1.04	0.964	0.934	0.918
0.45000	2	1.20	1.084	1.035	1.008
0.50000	2	1.39	1.220	1.147	1.106
0.55000	3	1.62	1.379	1.275	1.216
0.60000	4	1.90	1.566	1.422	1.342
0.65000	5	2.27	1.795	1.597	1.488
0.70000	7	2.75	2.082	1.811	1.664
0.75000	10	3.42	2.459	2.083	1.883
0.80000	15	4.43	2.983	2.447	2.169
0.85000	27	6.10	3.779	2.975	2.572
0.90000	62	9.44	5.190	3.853	3.217
0.95000	247	19.44	8.674	5.821	4.578
0.96000	387	24.44	10.183	6.612	5.100
0.97000	688	32.78	12.489	7.769	5.844
0.98000	1548	49.44	16.588	9.708	7.047
0.99000	6192	99.45	26.751	14.079	9.609
0.99500	24767	199.45	42.881	20.258	12.984
0.99900	619161	999.43	126.737	46.323	25.568
0.99990	61904212	9997.76	590.747	148.243	65.616

The table was generated using MINITAB

F-table for 18 Degrees of freedom in the denominator

numerator degrees of freedom

α	6	7	8	9	10
0.00010	0.106	0.113	0.118	0.124	0.128
0.00100	0.157	0.166	0.173	0.179	0.185
0.00500	0.214	0.224	0.233	0.241	0.248
0.01000	0.249	0.260	0.269	0.278	0.285
0.02000	0.293	0.305	0.315	0.324	0.331
0.03000	0.325	0.337	0.348	0.356	0.364
0.04000	0.352	0.364	0.375	0.383	0.391
0.05000	0.375	0.388	0.398	0.407	0.414
0.06000	0.397	0.409	0.419	0.428	0.435
0.07000	0.416	0.428	0.438	0.447	0.454
0.08000	0.435	0.447	0.457	0.465	0.472
0.09000	0.452	0.464	0.474	0.482	0.489
0.10000	0.469	0.481	0.490	0.498	0.505
0.20000	0.618	0.626	0.633	0.639	0.644
0.25000	0.688	0.694	0.698	0.702	0.706
0.30000	0.759	0.761	0.763	0.765	0.767
0.35000	0.832	0.830	0.829	0.829	0.829
0.40000	0.909	0.902	0.898	0.895	0.893
0.45000	0.991	0.979	0.971	0.964	0.960
0.50000	1.080	1.062	1.049	1.039	1.031
0.55000	1.179	1.153	1.134	1.119	1.107
0.60000	1.290	1.255	1.228	1.208	1.192
0.65000	1.419	1.371	1.336	1.309	1.287
0.70000	1.571	1.508	1.461	1.425	1.397
0.75000	1.758	1.673	1.611	1.564	1.527
0.80000	1.999	1.884	1.800	1.737	1.688
0.85000	2.330	2.170	2.055	1.968	1.901
0.90000	2.848	2.607	2.438	2.312	2.215
0.95000	3.895	3.466	3.173	2.960	2.798
0.96000	4.284	3.778	3.435	3.187	3.000
0.97000	4.829	4.209	3.793	3.496	3.273
0.98000	5.689	4.877	4.342	3.963	3.682
0.99000	7.450	6.208	5.411	4.859	4.456
0.99500	9.664	7.825	6.677	5.899	5.340
0.99900	17.267	13.063	10.601	9.012	7.913
0.99990	38.380	26.258	19.790	15.897	13.349

Table generated using MINITAB

F-table for 18 Degrees of freedom in the denominator
numerator degrees of freedom

α	11	12	13	14	15
0.00010	0.133	0.134	0.137	0.139	0.145
0.00100	0.190	0.194	0.198	0.202	0.205
0.00500	0.254	0.259	0.263	0.267	0.271
0.01000	0.291	0.296	0.301	0.305	0.309
0.02000	0.337	0.343	0.348	0.353	0.357
0.03000	0.370	0.376	0.381	0.386	0.390
0.04000	0.397	0.403	0.408	0.413	0.417
0.05000	0.421	0.427	0.432	0.436	0.440
0.06000	0.442	0.447	0.452	0.457	0.461
0.07000	0.461	0.466	0.471	0.476	0.480
0.08000	0.479	0.484	0.489	0.493	0.497
0.09000	0.495	0.501	0.506	0.510	0.514
0.10000	0.511	0.517	0.521	0.526	0.529
0.20000	0.648	0.652	0.655	0.658	0.661
0.25000	0.709	0.712	0.714	0.716	0.718
0.30000	0.769	0.770	0.772	0.773	0.774
0.35000	0.829	0.829	0.830	0.830	0.830
0.40000	0.891	0.890	0.889	0.888	0.887
0.45000	0.956	0.952	0.950	0.947	0.946
0.50000	1.024	1.019	1.014	1.010	1.007
0.55000	1.098	1.090	1.083	1.078	1.073
0.60000	1.179	1.168	1.159	1.151	1.144
0.65000	1.270	1.255	1.243	1.232	1.223
0.70000	1.373	1.354	1.338	1.324	1.313
0.75000	1.496	1.471	1.450	1.432	1.417
0.80000	1.648	1.615	1.587	1.564	1.544
0.85000	1.847	1.803	1.766	1.734	1.707
0.90000	2.138	2.075	2.022	1.978	1.940
0.95000	2.670	2.568	2.484	2.413	2.353
0.96000	2.854	2.736	2.640	2.559	2.491
0.97000	3.099	2.961	2.847	2.753	2.673
0.98000	3.465	3.293	3.153	3.037	2.939
0.99000	4.150	3.909	3.715	3.556	3.422
0.99500	4.920	4.594	4.334	4.122	3.945
0.99900	7.113	6.507	6.034	5.655	5.345
0.99990	11.575	10.278	9.296	8.529	7.916

The table was generated using MINITAB

F-table for 18 Degrees of freedom in the denominator
numerator degrees of freedom

α	16	17	18	19	20
0.00010	0.145	0.149	0.150	0.150	0.153
0.00100	0.208	0.211	0.213	0.216	0.218
0.00500	0.274	0.278	0.280	0.283	0.285
0.01000	0.313	0.316	0.319	0.322	0.325
0.02000	0.360	0.364	0.367	0.370	0.372
0.03000	0.394	0.397	0.400	0.403	0.406
0.04000	0.421	0.424	0.427	0.430	0.433
0.05000	0.444	0.447	0.451	0.453	0.456
0.06000	0.465	0.468	0.471	0.474	0.476
0.07000	0.484	0.487	0.490	0.493	0.495
0.08000	0.501	0.504	0.507	0.510	0.512
0.09000	0.517	0.521	0.523	0.526	0.529
0.10000	0.533	0.536	0.539	0.542	0.544
0.20000	0.663	0.665	0.667	0.669	0.671
0.25000	0.720	0.722	0.723	0.725	0.726
0.30000	0.776	0.777	0.778	0.778	0.779
0.35000	0.831	0.831	0.831	0.831	0.832
0.40000	0.886	0.886	0.885	0.885	0.885
0.45000	0.944	0.942	0.941	0.940	0.939
0.50000	1.004	1.002	1.000	0.998	0.996
0.55000	1.069	1.065	1.061	1.058	1.056
0.60000	1.138	1.133	1.128	1.124	1.120
0.65000	1.215	1.208	1.202	1.196	1.191
0.70000	1.302	1.293	1.285	1.277	1.271
0.75000	1.403	1.391	1.381	1.371	1.363
0.80000	1.526	1.511	1.497	1.484	1.473
0.85000	1.684	1.663	1.645	1.629	1.614
0.90000	1.907	1.879	1.853	1.831	1.811
0.95000	2.301	2.256	2.217	2.182	2.151
0.96000	2.432	2.381	2.337	2.297	2.262
0.97000	2.605	2.546	2.494	2.448	2.407
0.98000	2.856	2.784	2.721	2.666	2.617
0.99000	3.309	3.212	3.127	3.054	2.988
0.99500	3.797	3.670	3.560	3.464	3.380
0.99900	5.087	4.869	4.683	4.522	4.381
0.99990	7.416	7.000	6.650	6.351	6.094

Table generated using MINITAB

F-table for 18 Degrees of freedom in the denominator

numerator degrees of freedom

α	25	50	75	100	200
0.00010	0.160	0.175	0.182	0.185	0.193
0.00100	0.226	0.246	0.254	0.258	0.265
0.00500	0.295	0.318	0.327	0.332	0.339
0.01000	0.335	0.359	0.368	0.373	0.381
0.02000	0.383	0.408	0.417	0.422	0.430
0.03000	0.417	0.441	0.451	0.456	0.464
0.04000	0.444	0.468	0.478	0.483	0.490
0.05000	0.467	0.491	0.500	0.505	0.513
0.06000	0.487	0.511	0.520	0.525	0.533
0.07000	0.505	0.529	0.538	0.543	0.550
0.08000	0.522	0.546	0.554	0.559	0.566
0.09000	0.538	0.561	0.570	0.574	0.582
0.10000	0.554	0.576	0.584	0.589	0.596
0.20000	0.678	0.694	0.700	0.703	0.708
0.25000	0.731	0.744	0.748	0.751	0.755
0.30000	0.783	0.791	0.794	0.796	0.799
0.35000	0.833	0.837	0.838	0.839	0.841
0.40000	0.884	0.882	0.882	0.882	0.882
0.45000	0.935	0.928	0.926	0.925	0.924
0.50000	0.989	0.976	0.971	0.969	0.966
0.55000	1.046	1.025	1.018	1.015	1.009
0.60000	1.106	1.077	1.067	1.063	1.055
0.65000	1.172	1.134	1.120	1.114	1.104
0.70000	1.246	1.196	1.179	1.170	1.157
0.75000	1.331	1.266	1.244	1.233	1.216
0.80000	1.432	1.348	1.320	1.306	1.285
0.85000	1.559	1.450	1.414	1.396	1.369
0.90000	1.735	1.588	1.540	1.515	1.479
0.95000	2.035	1.814	1.743	1.707	1.655
0.96000	2.132	1.885	1.806	1.767	1.709
0.97000	2.257	1.975	1.886	1.842	1.777
0.98000	2.437	2.102	1.997	1.946	1.871
0.99000	2.750	2.317	2.184	2.120	2.025
0.99500	3.075	2.532	2.369	2.290	2.175
0.99900	3.883	3.036	2.792	2.676	2.509
0.99990	5.205	3.786	3.400	3.220	2.967

The table was generated using MINITAB

F-table for 19 Degrees of freedom in the denominator
numerator degrees of freedom

α	1	2	3	4	5
0.00010	0	0.07	0.080	0.092	0.103
0.00100	0	0.10	0.121	0.138	0.150
0.00500	0	0.14	0.169	0.190	0.206
0.01000	0	0.17	0.200	0.222	0.239
0.02000	0	0.21	0.240	0.265	0.283
0.03000	0	0.24	0.271	0.296	0.315
0.04000	0	0.26	0.297	0.322	0.341
0.05000	0	0.28	0.320	0.345	0.364
0.06000	0	0.31	0.341	0.367	0.386
0.07000	0	0.33	0.361	0.387	0.406
0.08000	0	0.35	0.381	0.406	0.424
0.09000	0	0.36	0.399	0.424	0.442
0.10000	0	0.38	0.417	0.441	0.459
0.20000	1	0.57	0.587	0.601	0.613
0.25000	1	0.67	0.673	0.679	0.686
0.30000	1	0.78	0.763	0.760	0.761
0.35000	1	0.90	0.860	0.845	0.838
0.40000	1	1.04	0.967	0.937	0.921
0.45000	2	1.20	1.087	1.038	1.010
0.50000	2	1.39	1.223	1.150	1.108
0.55000	3	1.62	1.381	1.277	1.218
0.60000	4	1.91	1.568	1.424	1.343
0.65000	5	2.27	1.796	1.599	1.489
0.70000	7	2.75	2.083	1.812	1.664
0.75000	10	3.42	2.459	2.083	1.882
0.80000	15	4.43	2.983	2.446	2.167
0.85000	27	6.10	3.778	2.973	2.569
0.90000	62	9.44	5.187	3.848	3.211
0.95000	248	19.44	8.667	5.811	4.567
0.96000	387	24.44	10.173	6.600	5.087
0.97000	689	32.78	12.476	7.754	5.829
0.98000	1550	49.45	16.570	9.688	7.027
0.99000	6201	99.45	26.719	14.048	9.579
0.99500	24803	199.45	42.826	20.211	12.942
0.99900	620167	999.48	126.572	46.205	25.476
0.99990	61991464	9997.72	589.959	147.859	65.372

The table was generated using MINITAB

F-table for 19 Degrees of freedom in the denominator

numerator degrees of freedom

α	6	7	8	9	10
0.00010	0.109	0.117	0.123	0.129	0.134
0.00100	0.161	0.171	0.178	0.185	0.191
0.00500	0.219	0.230	0.239	0.247	0.254
0.01000	0.253	0.265	0.275	0.283	0.291
0.02000	0.298	0.310	0.321	0.329	0.337
0.03000	0.330	0.343	0.353	0.362	0.370
0.04000	0.357	0.369	0.380	0.389	0.397
0.05000	0.380	0.393	0.403	0.412	0.420
0.06000	0.401	0.414	0.424	0.433	0.441
0.07000	0.421	0.433	0.444	0.453	0.460
0.08000	0.439	0.452	0.462	0.470	0.478
0.09000	0.457	0.469	0.479	0.487	0.495
0.10000	0.474	0.485	0.495	0.504	0.511
0.20000	0.622	0.631	0.637	0.643	0.648
0.25000	0.692	0.698	0.702	0.707	0.710
0.30000	0.763	0.765	0.767	0.769	0.771
0.35000	0.835	0.834	0.833	0.833	0.833
0.40000	0.911	0.905	0.901	0.898	0.896
0.45000	0.993	0.982	0.973	0.967	0.962
0.50000	1.082	1.064	1.051	1.041	1.033
0.55000	1.181	1.154	1.135	1.121	1.109
0.60000	1.292	1.256	1.229	1.209	1.193
0.65000	1.420	1.371	1.336	1.309	1.287
0.70000	1.571	1.507	1.460	1.424	1.396
0.75000	1.757	1.672	1.610	1.562	1.525
0.80000	1.997	1.881	1.798	1.734	1.685
0.85000	2.327	2.166	2.050	1.964	1.896
0.90000	2.841	2.600	2.431	2.305	2.207
0.95000	3.884	3.455	3.161	2.947	2.785
0.96000	4.271	3.765	3.421	3.173	2.986
0.97000	4.813	4.193	3.777	3.479	3.256
0.98000	5.669	4.857	4.321	3.943	3.662
0.99000	7.422	6.180	5.384	4.832	4.429
0.99500	9.624	7.788	6.641	5.864	5.305
0.99900	17.189	12.994	10.537	8.952	7.855
0.99990	38.197	26.109	19.662	15.783	13.244

Table generated using MINITAB

F-table for 19 Degrees of freedom in the denominator
numerator degrees of freedom

α	11	12	13	14	15
0.00010	0.137	0.142	0.145	0.148	0.150
0.00100	0.196	0.201	0.205	0.209	0.212
0.00500	0.260	0.265	0.270	0.274	0.278
0.01000	0.297	0.303	0.308	0.313	0.317
0.02000	0.344	0.350	0.355	0.360	0.364
0.03000	0.377	0.383	0.388	0.393	0.397
0.04000	0.404	0.410	0.415	0.420	0.424
0.05000	0.427	0.433	0.438	0.443	0.447
0.06000	0.448	0.454	0.459	0.463	0.468
0.07000	0.467	0.473	0.478	0.482	0.486
0.08000	0.484	0.490	0.495	0.500	0.504
0.09000	0.501	0.507	0.512	0.516	0.520
0.10000	0.517	0.523	0.527	0.532	0.536
0.20000	0.653	0.657	0.660	0.663	0.666
0.25000	0.713	0.716	0.719	0.721	0.723
0.30000	0.773	0.775	0.776	0.777	0.779
0.35000	0.833	0.833	0.833	0.834	0.834
0.40000	0.894	0.893	0.892	0.891	0.890
0.45000	0.958	0.955	0.952	0.950	0.948
0.50000	1.026	1.021	1.016	1.012	1.009
0.55000	1.099	1.091	1.085	1.079	1.074
0.60000	1.180	1.169	1.159	1.152	1.145
0.65000	1.270	1.255	1.243	1.232	1.223
0.70000	1.373	1.353	1.337	1.323	1.311
0.75000	1.494	1.469	1.448	1.430	1.414
0.80000	1.645	1.612	1.584	1.560	1.540
0.85000	1.842	1.797	1.760	1.728	1.701
0.90000	2.130	2.067	2.014	1.970	1.932
0.95000	2.658	2.555	2.470	2.400	2.339
0.96000	2.839	2.721	2.625	2.544	2.476
0.97000	3.082	2.944	2.830	2.736	2.656
0.98000	3.445	3.273	3.132	3.016	2.919
0.99000	4.123	3.882	3.688	3.529	3.396
0.99500	4.886	4.560	4.300	4.088	3.912
0.99900	7.057	6.453	5.981	5.603	5.294
0.99990	11.476	10.185	9.207	8.443	7.833

The table was generated using MINITAB

F-table for 19 Degrees of freedom in the denominator
numerator degrees of freedom

α	16	17	18	19	20
0.00010	0.151	0.155	0.157	0.158	0.161
0.00100	0.215	0.218	0.221	0.223	0.226
0.00500	0.282	0.285	0.288	0.291	0.293
0.01000	0.320	0.324	0.327	0.330	0.332
0.02000	0.368	0.371	0.375	0.378	0.380
0.03000	0.401	0.405	0.408	0.411	0.414
0.04000	0.428	0.431	0.435	0.438	0.440
0.05000	0.451	0.455	0.458	0.461	0.463
0.06000	0.472	0.475	0.478	0.481	0.484
0.07000	0.490	0.494	0.497	0.500	0.502
0.08000	0.508	0.511	0.514	0.517	0.519
0.09000	0.524	0.527	0.530	0.533	0.536
0.10000	0.539	0.543	0.546	0.548	0.551
0.20000	0.668	0.671	0.673	0.675	0.677
0.25000	0.725	0.727	0.728	0.730	0.731
0.30000	0.780	0.781	0.782	0.783	0.784
0.35000	0.834	0.835	0.835	0.835	0.836
0.40000	0.890	0.889	0.889	0.888	0.888
0.45000	0.947	0.945	0.944	0.943	0.942
0.50000	1.006	1.004	1.001	1.000	0.998
0.55000	1.070	1.066	1.063	1.060	1.057
0.60000	1.139	1.133	1.129	1.125	1.121
0.65000	1.215	1.208	1.201	1.196	1.191
0.70000	1.301	1.292	1.283	1.276	1.269
0.75000	1.401	1.389	1.378	1.369	1.360
0.80000	1.522	1.506	1.493	1.480	1.469
0.85000	1.678	1.657	1.639	1.622	1.608
0.90000	1.899	1.870	1.844	1.822	1.802
0.95000	2.287	2.242	2.203	2.168	2.137
0.96000	2.417	2.366	2.321	2.281	2.246
0.97000	2.587	2.528	2.476	2.430	2.390
0.98000	2.835	2.763	2.700	2.645	2.596
0.99000	3.282	3.185	3.101	3.027	2.961
0.99500	3.764	3.637	3.527	3.431	3.347
0.99900	5.037	4.820	4.634	4.473	4.333
0.99990	7.334	6.921	6.572	6.275	6.019

Table generated using MINITAB

F-table for 19 Degrees of freedom in the denominator
numerator degrees of freedom

α	25	50	75	100	200
0.00010	0.168	0.187	0.194	0.197	0.204
0.00100	0.234	0.256	0.264	0.269	0.276
0.00500	0.304	0.328	0.338	0.343	0.351
0.01000	0.343	0.369	0.378	0.384	0.392
0.02000	0.391	0.417	0.427	0.433	0.441
0.03000	0.425	0.451	0.461	0.466	0.475
0.04000	0.452	0.477	0.487	0.493	0.501
0.05000	0.474	0.500	0.510	0.515	0.523
0.06000	0.495	0.520	0.529	0.534	0.543
0.07000	0.513	0.538	0.547	0.552	0.560
0.08000	0.530	0.554	0.563	0.568	0.576
0.09000	0.546	0.569	0.578	0.583	0.591
0.10000	0.561	0.584	0.593	0.597	0.605
0.20000	0.684	0.701	0.707	0.710	0.716
0.25000	0.737	0.750	0.755	0.757	0.761
0.30000	0.787	0.796	0.799	0.801	0.804
0.35000	0.837	0.841	0.843	0.844	0.846
0.40000	0.887	0.886	0.886	0.886	0.886
0.45000	0.938	0.931	0.929	0.928	0.927
0.50000	0.991	0.978	0.973	0.971	0.968
0.55000	1.047	1.026	1.019	1.016	1.010
0.60000	1.106	1.077	1.067	1.062	1.055
0.65000	1.171	1.133	1.119	1.113	1.103
0.70000	1.244	1.193	1.176	1.167	1.154
0.75000	1.328	1.262	1.240	1.229	1.212
0.80000	1.427	1.343	1.315	1.300	1.279
0.85000	1.552	1.443	1.406	1.388	1.360
0.90000	1.726	1.577	1.529	1.504	1.468
0.95000	2.020	1.798	1.726	1.691	1.638
0.96000	2.115	1.867	1.788	1.749	1.691
0.97000	2.239	1.956	1.866	1.822	1.757
0.98000	2.415	2.080	1.975	1.923	1.847
0.99000	2.723	2.290	2.157	2.092	1.997
0.99500	3.042	2.499	2.336	2.257	2.142
0.99900	3.836	2.991	2.747	2.631	2.464
0.99990	5.134	3.722	3.337	3.159	2.907

The table was generated using MINITAB

F-table for 20 Degrees of freedom in the denominator
numerator degrees of freedom

α	1	2	3	4	5	
0.00010	0	0.07	0.084	0.096	0.105	
0.00100	0	0.10	0.123	0.141	0.154	
0.00500	0	0.14	0.172	0.193	0.210	
0.01000	0	0.17	0.202	0.226	0.243	
0.02000	0	0.21	0.243	0.268	0.287	
0.03000	0	0.24	0.274	0.299	0.319	
0.04000	0	0.26	0.300	0.326	0.345	
0.05000	0	0.29	0.323	0.349	0.368	
0.06000	0	0.31	0.344	0.370	0.390	
0.07000	0	0.33	0.364	0.390	0.409	
0.08000	0	0.35	0.384	0.409	0.428	
0.09000	0	0.37	0.402	0.427	0.446	
0.10000	0	0.39	0.420	0.445	0.463	
0.20000	1	0.57	0.590	0.604	0.616	
0.25000	1	0.67	0.675	0.683	0.689	
0.30000	1	0.78	0.766	0.763	0.764	
0.35000	1	0.90	0.863	0.848	0.841	
0.40000	1	1.04	0.970	0.940	0.923	
0.45000	2	1.20	1.089	1.040	1.012	
0.50000	2	1.39	1.225	1.152	1.110	
0.55000	3	1.62	1.383	1.278	1.220	
0.60000	4	1.91	1.570	1.425	1.344	
0.65000	5	2.27	1.798	1.600	1.490	
0.70000	7	2.75	2.084	1.812	1.664	
0.75000	10	3.43	2.460	2.083	1.881	
0.80000	15	4.43	2.983	2.445	2.166	
0.85000	27	6.10	3.778	2.970	2.566	
0.90000	62	9.44	5.185	3.844	3.206	
0.95000	248	19.45	8.660	5.802	4.558	
0.96000	388	24.45	10.165	6.589	5.076	
0.97000	690	32.78	12.464	7.740	5.815	
0.98000	1552	49.45	16.553	9.669	7.009	
0.99000	6209	99.44	26.691	14.020	9.552	
0.99500	24838	199.45	42.778	20.168	12.903	
0.99900	620884	999.44	126.420	46.100	25.395	1
0.99990	62066704	9997.50	589.214	147.513	65.153	3

The table was generated using MINITAB

F-table for 20 Degrees of freedom in the denominator
numerator degrees of freedom

α	6	7	8	9	10
0.00010	0.115	0.121	0.127	0.134	0.140
0.00100	0.166	0.175	0.183	0.190	0.197
0.00500	0.223	0.234	0.244	0.252	0.259
0.01000	0.258	0.270	0.280	0.289	0.296
0.02000	0.302	0.315	0.326	0.335	0.343
0.03000	0.334	0.347	0.358	0.367	0.376
0.04000	0.361	0.374	0.385	0.394	0.402
0.05000	0.384	0.397	0.408	0.417	0.425
0.06000	0.405	0.418	0.429	0.438	0.446
0.07000	0.425	0.438	0.448	0.458	0.465
0.08000	0.443	0.456	0.467	0.475	0.483
0.09000	0.461	0.473	0.484	0.492	0.500
0.10000	0.478	0.490	0.500	0.508	0.516
0.20000	0.626	0.634	0.641	0.647	0.653
0.25000	0.696	0.701	0.706	0.710	0.714
0.30000	0.766	0.768	0.770	0.773	0.775
0.35000	0.838	0.837	0.836	0.836	0.836
0.40000	0.914	0.908	0.904	0.901	0.899
0.45000	0.995	0.984	0.976	0.969	0.964
0.50000	1.084	1.066	1.053	1.042	1.034
0.55000	1.182	1.156	1.137	1.122	1.110
0.60000	1.293	1.257	1.230	1.210	1.194
0.65000	1.420	1.372	1.336	1.309	1.287
0.70000	1.571	1.507	1.460	1.424	1.395
0.75000	1.756	1.671	1.608	1.561	1.523
0.80000	1.995	1.879	1.795	1.732	1.682
0.85000	2.323	2.162	2.046	1.959	1.892
0.90000	2.836	2.594	2.424	2.298	2.200
0.95000	3.874	3.444	3.150	2.936	2.774
0.96000	4.259	3.752	3.408	3.160	2.973
0.97000	4.799	4.178	3.762	3.464	3.241
0.98000	5.650	4.839	4.303	3.924	3.643
0.99000	7.395	6.155	5.359	4.808	4.405
0.99500	9.588	7.754	6.608	5.831	5.274
0.99900	7.120	12.931	10.479	8.897	7.803
0.99990	8.033	25.976	19.546	15.679	13.149

Table generated using MINITAB

F-table for 20 Degrees of freedom in the denominator
numerator degrees of freedom

α	11	12	13	14	15
0.00010	0.143	0.145	0.150	0.153	0.155
0.00100	0.202	0.207	0.211	0.215	0.219
0.00500	0.266	0.271	0.277	0.281	0.285
0.01000	0.303	0.309	0.314	0.319	0.323
0.02000	0.350	0.356	0.361	0.366	0.370
0.03000	0.383	0.389	0.394	0.399	0.404
0.04000	0.409	0.416	0.421	0.426	0.430
0.05000	0.432	0.439	0.444	0.449	0.453
0.06000	0.453	0.459	0.465	0.469	0.474
0.07000	0.472	0.478	0.483	0.488	0.493
0.08000	0.490	0.496	0.501	0.506	0.510
0.09000	0.506	0.512	0.517	0.522	0.526
0.10000	0.522	0.528	0.533	0.537	0.542
0.20000	0.657	0.661	0.665	0.668	0.671
0.25000	0.717	0.720	0.723	0.725	0.728
0.30000	0.777	0.778	0.780	0.781	0.783
0.35000	0.836	0.836	0.837	0.837	0.837
0.40000	0.897	0.896	0.895	0.894	0.893
0.45000	0.961	0.957	0.955	0.952	0.951
0.50000	1.028	1.023	1.018	1.014	1.011
0.55000	1.101	1.093	1.086	1.080	1.075
0.60000	1.180	1.169	1.160	1.152	1.145
0.65000	1.269	1.255	1.242	1.232	1.222
0.70000	1.372	1.352	1.336	1.322	1.310
0.75000	1.493	1.467	1.446	1.428	1.412
0.80000	1.642	1.608	1.580	1.557	1.536
0.85000	1.837	1.792	1.755	1.723	1.696
0.90000	2.123	2.059	2.007	1.962	1.924
0.95000	2.646	2.543	2.458	2.387	2.327
0.96000	2.826	2.708	2.611	2.530	2.462
0.97000	3.067	2.928	2.815	2.720	2.640
0.98000	3.426	3.254	3.114	2.998	2.900
0.99000	4.099	3.858	3.664	3.505	3.371
0.99500	4.855	4.529	4.270	4.058	3.882
0.99900	7.007	6.404	5.933	5.556	5.248
0.99990	11.387	10.101	9.126	8.366	7.758

The table was generated using MINITAB

F-table for 20 Degrees of freedom in the denominator
numerator degrees of freedom

α	16	17	18	19	20
0.00010	0.161	0.162	0.164	0.165	0.168
0.00100	0.222	0.225	0.228	0.230	0.232
0.00500	0.289	0.292	0.295	0.298	0.301
0.01000	0.327	0.331	0.334	0.337	0.340
0.02000	0.375	0.378	0.382	0.385	0.388
0.03000	0.408	0.411	0.415	0.418	0.421
0.04000	0.434	0.438	0.441	0.445	0.447
0.05000	0.457	0.461	0.464	0.467	0.470
0.06000	0.478	0.481	0.485	0.488	0.491
0.07000	0.496	0.500	0.503	0.506	0.509
0.08000	0.514	0.517	0.520	0.523	0.526
0.09000	0.530	0.533	0.536	0.539	0.542
0.10000	0.545	0.549	0.552	0.554	0.557
0.20000	0.673	0.676	0.678	0.680	0.682
0.25000	0.729	0.731	0.733	0.734	0.736
0.30000	0.784	0.785	0.786	0.787	0.788
0.35000	0.838	0.838	0.839	0.839	0.839
0.40000	0.893	0.892	0.892	0.891	0.891
0.45000	0.949	0.948	0.946	0.945	0.944
0.50000	1.008	1.006	1.003	1.001	1.000
0.55000	1.071	1.067	1.064	1.061	1.058
0.60000	1.139	1.134	1.129	1.125	1.121
0.65000	1.214	1.207	1.201	1.195	1.190
0.70000	1.300	1.290	1.282	1.275	1.268
0.75000	1.399	1.386	1.376	1.366	1.358
0.80000	1.518	1.503	1.489	1.476	1.465
0.85000	1.672	1.651	1.633	1.617	1.602
0.90000	1.891	1.862	1.836	1.814	1.793
0.95000	2.275	2.230	2.190	2.155	2.124
0.96000	2.403	2.352	2.307	2.267	2.232
0.97000	2.571	2.512	2.460	2.414	2.373
0.98000	2.816	2.744	2.681	2.626	2.577
0.99000	3.258	3.161	3.077	3.003	2.937
0.99500	3.734	3.607	3.497	3.402	3.317
0.99900	4.991	4.775	4.589	4.429	4.289
0.99990	7.261	6.849	6.502	6.206	5.951

Table generated using MINITAB

F-table for 20 Degrees of freedom in the denominator

numerator degrees of freedom

α	25	50	75	100	200
0.00010	0.175	0.194	0.200	0.208	0.214
0.00100	0.242	0.265	0.274	0.279	0.287
0.00500	0.312	0.337	0.348	0.353	0.362
0.01000	0.351	0.378	0.388	0.394	0.403
0.02000	0.399	0.426	0.437	0.443	0.452
0.03000	0.432	0.460	0.470	0.476	0.485
0.04000	0.459	0.486	0.496	0.502	0.511
0.05000	0.482	0.508	0.519	0.524	0.533
0.06000	0.502	0.528	0.538	0.543	0.552
0.07000	0.520	0.546	0.556	0.561	0.569
0.08000	0.537	0.562	0.572	0.577	0.585
0.09000	0.553	0.577	0.587	0.592	0.600
0.10000	0.567	0.591	0.601	0.606	0.613
0.20000	0.689	0.707	0.713	0.717	0.722
0.25000	0.742	0.755	0.760	0.763	0.767
0.30000	0.792	0.801	0.804	0.806	0.809
0.35000	0.841	0.845	0.847	0.848	0.850
0.40000	0.890	0.889	0.889	0.889	0.890
0.45000	0.941	0.934	0.932	0.931	0.929
0.50000	0.993	0.979	0.975	0.973	0.970
0.55000	1.048	1.027	1.020	1.016	1.011
0.60000	1.107	1.077	1.067	1.062	1.055
0.65000	1.171	1.132	1.118	1.111	1.101
0.70000	1.243	1.191	1.174	1.165	1.152
0.75000	1.325	1.259	1.236	1.225	1.208
0.80000	1.423	1.338	1.309	1.295	1.273
0.85000	1.546	1.436	1.399	1.380	1.352
0.90000	1.717	1.568	1.518	1.494	1.457
0.95000	2.007	1.784	1.712	1.676	1.623
0.96000	2.101	1.852	1.772	1.732	1.674
0.97000	2.222	1.938	1.848	1.804	1.738
0.98000	2.396	2.059	1.954	1.902	1.826
0.99000	2.699	2.265	2.131	2.066	1.971
0.99500	3.013	2.470	2.306	2.227	2.111
0.99900	3.794	2.950	2.706	2.590	2.424
0.99990	5.070	3.664	3.281	3.103	2.852

The table was generated using MINITAB

F-table for 25 Degrees of freedom in the denominator

numerator degrees of freedom

α	1	2	3	4	5
0.00010	0	0.07	0.093	0.109	0.120
0.00100	0	0.11	0.134	0.154	0.169
0.00500	0	0.15	0.183	0.207	0.225
0.01000	0	0.18	0.214	0.239	0.259
0.02000	0	0.22	0.255	0.282	0.302
0.03000	0	0.25	0.285	0.313	0.334
0.04000	0	0.27	0.311	0.339	0.361
0.05000	0	0.30	0.334	0.362	0.384
0.06000	0	0.32	0.356	0.384	0.405
0.07000	0	0.34	0.376	0.404	0.424
0.08000	0	0.36	0.395	0.422	0.443
0.09000	0	0.38	0.414	0.440	0.461
0.10000	0	0.40	0.432	0.458	0.478
0.20000	1	0.58	0.601	0.617	0.629
0.25000	1	0.68	0.686	0.694	0.702
0.30000	1	0.79	0.776	0.774	0.775
0.35000	1	0.91	0.873	0.858	0.852
0.40000	1	1.05	0.979	0.949	0.933
0.45000	2	1.21	1.098	1.049	1.021
0.50000	2	1.40	1.234	1.160	1.118
0.55000	3	1.63	1.391	1.285	1.226
0.60000	4	1.92	1.577	1.431	1.349
0.65000	5	2.28	1.804	1.604	1.493
0.70000	7	2.76	2.089	1.815	1.665
0.75000	10	3.44	2.463	2.083	1.879
0.80000	15	4.44	2.984	2.442	2.160
0.85000	27	6.11	3.775	2.962	2.555
0.90000	62	9.45	5.175	3.828	3.187
0.95000	249	19.46	8.634	5.769	4.521
0.96000	390	24.46	10.132	6.548	5.032
0.97000	693	32.79	12.420	7.689	5.761
0.98000	1560	49.46	16.490	9.600	6.939
0.99000	6240	99.46	26.578	13.911	9.449
0.99500	24960	199.46	42.592	20.003	12.755
0.99900	624013	999.49	125.835	45.698	25.080
0.99990	62383640	9997.52	586.475	146.182	64.307

The table was generated using MINITAB

F-table for 25 Degrees of freedom in the denominator

numerator degrees of freedom

α	6	7	8	9	10
0.00010	0.130	0.141	0.148	0.154	0.159
0.00100	0.183	0.194	0.203	0.212	0.219
0.00500	0.240	0.253	0.264	0.274	0.282
0.01000	0.275	0.289	0.300	0.310	0.319
0.02000	0.319	0.334	0.346	0.356	0.365
0.03000	0.351	0.366	0.378	0.388	0.397
0.04000	0.378	0.392	0.404	0.415	0.424
0.05000	0.401	0.415	0.427	0.438	0.447
0.06000	0.422	0.436	0.448	0.458	0.467
0.07000	0.441	0.455	0.467	0.477	0.486
0.08000	0.460	0.473	0.485	0.495	0.503
0.09000	0.477	0.490	0.502	0.511	0.520
0.10000	0.494	0.507	0.518	0.527	0.536
0.20000	0.640	0.649	0.657	0.664	0.669
0.25000	0.709	0.715	0.720	0.725	0.729
0.30000	0.778	0.781	0.784	0.786	0.788
0.35000	0.849	0.848	0.848	0.848	0.848
0.40000	0.924	0.918	0.914	0.911	0.909
0.45000	1.004	0.993	0.984	0.978	0.973
0.50000	1.091	1.073	1.060	1.050	1.042
0.55000	1.188	1.162	1.142	1.127	1.115
0.60000	1.297	1.260	1.233	1.213	1.196
0.65000	1.422	1.373	1.337	1.309	1.287
0.70000	1.571	1.506	1.458	1.421	1.392
0.75000	1.753	1.666	1.603	1.554	1.516
0.80000	1.987	1.870	1.785	1.721	1.670
0.85000	2.311	2.147	2.031	1.943	1.874
0.90000	2.814	2.571	2.400	2.272	2.173
0.95000	3.834	3.403	3.108	2.893	2.729
0.96000	4.213	3.705	3.360	3.111	2.922
0.97000	4.744	4.122	3.705	3.406	3.182
0.98000	5.581	4.769	4.233	3.854	3.572
0.99000	7.296	6.058	5.263	4.713	4.311
0.99500	9.451	7.622	6.481	5.708	5.152
0.99900	16.852	12.692	10.258	8.688	7.604
0.99990	37.405	25.462	19.102	15.281	12.783

Table generated using MINITAB

F-table for 25 Degrees of freedom in the denominator

numerator degrees of freedom

α	11	12	13	14	15
0.00010	0.1656	0.170	0.173	0.178	0.183
0.00100	0.2260	0.231	0.237	0.242	0.246
0.00500	0.2901	0.296	0.302	0.308	0.312
0.01000	0.3272	0.334	0.340	0.345	0.350
0.02000	0.3733	0.380	0.386	0.392	0.397
0.03000	0.4058	0.412	0.419	0.424	0.430
0.04000	0.4322	0.439	0.445	0.451	0.456
0.05000	0.4550	0.461	0.468	0.473	0.478
0.06000	0.4753	0.482	0.488	0.493	0.498
0.07000	0.4940	0.500	0.506	0.512	0.516
0.08000	0.5113	0.517	0.523	0.529	0.533
0.09000	0.5276	0.534	0.539	0.545	0.549
0.10000	0.5431	0.549	0.555	0.560	0.564
0.20000	0.6748	0.679	0.683	0.686	0.690
0.25000	0.7336	0.736	0.739	0.742	0.745
0.30000	0.7911	0.793	0.795	0.796	0.798
0.35000	0.8490	0.849	0.850	0.850	0.851
0.40000	0.9082	0.907	0.906	0.905	0.904
0.45000	0.9700	0.966	0.964	0.962	0.960
0.50000	1.0354	1.030	1.025	1.021	1.018
0.55000	1.1059	1.097	1.091	1.085	1.080
0.60000	1.1832	1.171	1.162	1.154	1.147
0.65000	1.2696	1.254	1.241	1.231	1.221
0.70000	1.3686	1.348	1.332	1.318	1.305
0.75000	1.4857	1.459	1.438	1.419	1.403
0.80000	1.6299	1.596	1.567	1.543	1.522
0.85000	1.8192	1.773	1.735	1.703	1.675
0.90000	2.0953	2.031	1.977	1.932	1.893
0.95000	2.6014	2.497	2.412	2.340	2.279
0.96000	2.7752	2.656	2.559	2.477	2.408
0.97000	3.0083	2.868	2.754	2.659	2.578
0.98000	3.3554	3.182	3.042	2.925	2.827
0.99000	4.0051	3.764	3.570	3.411	3.278
0.99500	4.7356	4.411	4.152	3.941	3.766
0.99900	6.8148	6.217	5.750	5.376	5.071
0.99990	11.0448	9.776	8.816	8.066	7.467

The table was generated using MINITAB

F-table for 25 Degrees of freedom in the denominator
numerator degrees of freedom

α	16	17	18	19	20
0.00010	0.187	0.190	0.192	0.193	0.199
0.00100	0.250	0.254	0.257	0.260	0.263
0.00500	0.317	0.321	0.325	0.328	0.331
0.01000	0.355	0.359	0.363	0.367	0.370
0.02000	0.402	0.406	0.410	0.413	0.417
0.03000	0.434	0.438	0.442	0.446	0.449
0.04000	0.460	0.465	0.469	0.472	0.475
0.05000	0.483	0.487	0.491	0.494	0.498
0.06000	0.503	0.507	0.511	0.514	0.517
0.07000	0.521	0.525	0.529	0.532	0.535
0.08000	0.538	0.542	0.545	0.549	0.552
0.09000	0.553	0.557	0.561	0.564	0.567
0.10000	0.568	0.572	0.576	0.579	0.582
0.20000	0.693	0.695	0.698	0.700	0.702
0.25000	0.747	0.749	0.751	0.752	0.754
0.30000	0.799	0.801	0.802	0.803	0.804
0.35000	0.851	0.852	0.852	0.853	0.853
0.40000	0.904	0.904	0.903	0.903	0.903
0.45000	0.958	0.957	0.956	0.954	0.954
0.50000	1.015	1.012	1.010	1.008	1.006
0.55000	1.075	1.071	1.068	1.065	1.062
0.60000	1.141	1.135	1.130	1.126	1.122
0.65000	1.213	1.206	1.199	1.193	1.188
0.70000	1.294	1.285	1.276	1.269	1.262
0.75000	1.389	1.377	1.366	1.356	1.347
0.80000	1.504	1.488	1.473	1.461	1.449
0.85000	1.651	1.630	1.611	1.594	1.579
0.90000	1.860	1.830	1.804	1.781	1.761
0.95000	2.227	2.181	2.141	2.105	2.073
0.96000	2.349	2.297	2.252	2.212	2.176
0.97000	2.509	2.449	2.397	2.351	2.310
0.98000	2.743	2.670	2.607	2.551	2.502
0.99000	3.164	3.067	2.983	2.908	2.843
0.99500	3.618	3.491	3.382	3.286	3.202
0.99900	4.816	4.601	4.418	4.259	4.120
0.99990	6.978	6.573	6.231	5.940	5.689

Table generated using MINITAB

F-table for 25 Degrees of freedom in the denominator

numerator degrees of freedom

α	25	50	75	100	200
0.00010	0.20755	0.23094	0.24051	0.24896	0.25806
0.00100	0.27522	0.30501	0.31705	0.32353	0.33420
0.00500	0.34505	0.37708	0.39003	0.39691	0.40825
0.01000	0.38402	0.41670	0.42977	0.43695	0.44835
0.02000	0.43104	0.46385	0.47696	0.48401	0.49545
0.03000	0.46351	0.49599	0.50891	0.51591	0.52709
0.04000	0.48939	0.52139	0.53408	0.54094	0.55192
0.05000	0.51140	0.54287	0.55530	0.56200	0.57275
0.06000	0.53086	0.56170	0.57388	0.58041	0.59091
0.07000	0.54847	0.57866	0.59055	0.59696	0.60720
0.08000	0.56470	0.59420	0.60583	0.61206	0.62206
0.09000	0.57987	0.60866	0.61997	0.62606	0.63580
0.10000	0.59414	0.62223	0.63325	0.63918	0.64866
0.20000	0.71117	0.73129	0.73920	0.74344	0.75022
0.25000	0.76115	0.77682	0.78303	0.78638	0.79174
0.30000	0.80891	0.81978	0.82418	0.82659	0.83045
0.35000	0.85577	0.86141	0.86389	0.86526	0.86754
0.40000	0.90268	0.90262	0.90299	0.90329	0.90386
0.45000	0.95051	0.94411	0.94221	0.94134	0.94007
0.50000	1.00000	0.98660	0.98220	0.98000	0.97673
0.55000	1.05207	1.03076	1.02357	1.01992	1.01443
0.60000	1.10781	1.07746	1.06709	1.06183	1.05385
0.65000	1.16854	1.12769	1.11368	1.10659	1.09577
0.70000	1.23623	1.18290	1.16465	1.15537	1.14126
0.75000	1.31380	1.24524	1.22182	1.20998	1.19195
0.80000	1.40614	1.31815	1.28830	1.27323	1.25031
0.85000	1.52236	1.40812	1.36970	1.35031	1.32101
0.90000	1.68310	1.52942	1.47834	1.45276	1.41417
0.95000	1.95543	1.72734	1.65320	1.61634	1.56121
0.96000	2.04344	1.78944	1.70745	1.66680	1.60613
0.97000	2.15752	1.86868	1.77630	1.73063	1.66269
0.98000	2.32004	1.97929	1.87163	1.81867	1.74020
0.99000	2.60410	2.16667	2.03123	1.96516	1.86788
0.99500	2.89811	2.35332	2.18799	2.10795	1.99091
0.99900	3.62907	2.79021	2.54689	2.43109	2.26418
0.99990	4.82245	3.43771	3.06032	2.88485	2.63683

The table was generated using MINITAB

F-table for 50 Degrees of freedom in the denominator
numerator degrees of freedom

α	1	2	3	4	5
0.00010	0	0.09	0.115	0.138	0.1535
0.00100	0	0.13	0.158	0.183	0.2039
0.00500	0	0.17	0.207	0.236	0.2598
0.01000	0	0.20	0.238	0.269	0.2935
0.02000	0	0.24	0.279	0.311	0.3364
0.03000	0	0.27	0.309	0.342	0.3678
0.04000	0	0.29	0.335	0.368	0.3938
0.05000	0	0.31	0.358	0.391	0.4166
0.06000	0	0.34	0.380	0.412	0.4373
0.07000	0	0.36	0.400	0.432	0.4566
0.08000	0	0.38	0.419	0.450	0.4747
0.09000	0	0.40	0.437	0.468	0.4920
0.10000	0	0.41	0.455	0.485	0.5086
0.20000	1	0.60	0.623	0.642	0.6572
0.25000	1	0.70	0.708	0.718	0.7278
0.30000	1	0.81	0.797	0.797	0.7997
0.35000	1	0.93	0.893	0.880	0.8744
0.40000	1	1.07	0.999	0.969	0.9538
0.45000	2	1.23	1.117	1.067	1.0395
0.50000	2	1.42	1.251	1.176	1.1336
0.55000	3	1.65	1.406	1.299	1.2388
0.60000	4	1.94	1.591	1.442	1.3587
0.65000	5	2.30	1.816	1.612	1.4986
0.70000	7	2.78	2.098	1.819	1.6664
0.75000	10	3.46	2.469	2.082	1.8751
0.80000	15	4.46	2.984	2.434	2.1478
0.85000	28	6.13	3.768	2.945	2.5322
0.90000	63	9.47	5.155	3.795	3.1472
0.95000	252	19.48	8.581	5.699	4.4444
0.96000	394	24.48	10.064	6.464	4.9420
0.97000	700	32.81	12.331	7.584	5.6510
0.98000	1575	49.48	16.362	9.460	6.7972
0.99000	6303	99.48	26.354	13.690	9.2377
0.99500	25213	199.48	42.214	19.667	12.4533
0.99900	630309	999.53	124.662	44.884	24.4410
0.99990	63009264	9997.56	580.885	143.491	62.5951

The table was generated using MINITAB

F-table for 50 Degrees of freedom in the denominator

numerator degrees of freedom

α	6	7	8	9	10
0.00010	0.1698	0.1800	0.1931	0.2040	0.2129
0.00100	0.2218	0.2368	0.2501	0.2620	0.2724
0.00500	0.2794	0.2962	0.3106	0.3234	0.3347
0.01000	0.3138	0.3311	0.3461	0.3590	0.3707
0.02000	0.3574	0.3750	0.3901	0.4033	0.4150
0.03000	0.3888	0.4065	0.4216	0.4347	0.4463
0.04000	0.4147	0.4323	0.4472	0.4602	0.4717
0.05000	0.4374	0.4547	0.4695	0.4823	0.4935
0.06000	0.4578	0.4749	0.4895	0.5021	0.5131
0.07000	0.4768	0.4936	0.5079	0.5202	0.5310
0.08000	0.4945	0.5110	0.5250	0.5371	0.5477
0.09000	0.5114	0.5275	0.5412	0.5531	0.5634
0.10000	0.5276	0.5433	0.5567	0.5682	0.5783
0.20000	0.6700	0.6809	0.6902	0.6982	0.7053
0.25000	0.7366	0.7442	0.7509	0.7568	0.7621
0.30000	0.8035	0.8074	0.8112	0.8147	0.8179
0.35000	0.8725	0.8721	0.8724	0.8732	0.8741
0.40000	0.9450	0.9395	0.9359	0.9335	0.9318
0.45000	1.0226	1.0111	1.0029	0.9968	0.9921
0.50000	1.1068	1.0883	1.0747	1.0644	1.0562
0.55000	1.2001	1.1730	1.1530	1.1376	1.1254
0.60000	1.3052	1.2678	1.2400	1.2186	1.2015
0.65000	1.4264	1.3760	1.3387	1.3098	1.2869
0.70000	1.5698	1.5028	1.4534	1.4154	1.3851
0.75000	1.7457	1.6566	1.5914	1.5414	1.5017
0.80000	1.9717	1.8519	1.7648	1.6985	1.6461
0.85000	2.2837	2.1174	1.9978	1.9075	1.8368
0.90000	2.7697	2.5226	2.3481	2.2180	2.1171
0.95000	3.7537	3.3189	3.0204	2.8029	2.6371
0.96000	4.1191	3.6077	3.2599	3.0082	2.8176
0.97000	4.6306	4.0067	3.5874	3.2867	3.0607
0.98000	5.4378	4.6252	4.0883	3.7083	3.4254
0.99000	7.0915	5.8577	5.0654	4.5168	4.1155
0.99500	9.1697	7.3544	6.2216	5.4540	4.9022
0.99900	16.3068	12.2022	9.8045	8.2598	7.1927
0.99990	36.1250	24.4142	18.1943	14.4655	12.0315

Table generated using MINITAB

F-table for 50 Degrees of freedom in the denominator
numerator degrees of freedom

α	11	12	13	14	15
0.00010	0.220	0.229	0.236	0.243	0.248
0.00100	0.281	0.290	0.298	0.305	0.311
0.00500	0.344	0.353	0.362	0.369	0.376
0.01000	0.380	0.390	0.398	0.406	0.413
0.02000	0.425	0.434	0.443	0.450	0.457
0.03000	0.456	0.465	0.474	0.481	0.488
0.04000	0.481	0.490	0.499	0.506	0.513
0.05000	0.503	0.512	0.520	0.527	0.534
0.06000	0.522	0.531	0.539	0.546	0.553
0.07000	0.540	0.549	0.556	0.563	0.570
0.08000	0.557	0.565	0.572	0.579	0.585
0.09000	0.572	0.580	0.588	0.594	0.600
0.10000	0.587	0.595	0.602	0.608	0.614
0.20000	0.711	0.717	0.722	0.726	0.730
0.25000	0.766	0.771	0.774	0.778	0.781
0.30000	0.820	0.823	0.826	0.828	0.830
0.35000	0.875	0.876	0.877	0.878	0.879
0.40000	0.930	0.929	0.929	0.928	0.928
0.45000	0.988	0.985	0.982	0.980	0.979
0.50000	1.049	1.044	1.039	1.035	1.032
0.55000	1.115	1.107	1.099	1.093	1.088
0.60000	1.187	1.175	1.165	1.157	1.149
0.65000	1.268	1.252	1.239	1.227	1.217
0.70000	1.360	1.339	1.322	1.307	1.294
0.75000	1.469	1.442	1.419	1.400	1.383
0.80000	1.603	1.568	1.538	1.513	1.491
0.85000	1.779	1.732	1.693	1.659	1.630
0.90000	2.036	1.970	1.915	1.868	1.828
0.95000	2.506	2.401	2.313	2.240	2.177
0.96000	2.668	2.547	2.448	2.365	2.294
0.97000	2.884	2.743	2.627	2.530	2.448
0.98000	3.206	3.032	2.890	2.773	2.673
0.99000	3.809	3.569	3.375	3.215	3.081
0.99500	4.487	4.165	3.907	3.697	3.522
0.99900	6.416	5.829	5.370	5.002	4.701
0.99990	10.340	9.107	8.174	7.447	6.866

The table was generated using MINITAB

F-table for 50 Degrees of freedom in the denominator
numerator degrees of freedom

α	16	17	18	19	20
0.00010	0.252	0.260	0.262	0.268	0.273
0.00100	0.318	0.323	0.329	0.333	0.338
0.00500	0.383	0.389	0.394	0.399	0.404
0.01000	0.419	0.425	0.431	0.436	0.441
0.02000	0.464	0.470	0.475	0.480	0.485
0.03000	0.494	0.500	0.506	0.511	0.515
0.04000	0.519	0.525	0.530	0.535	0.539
0.05000	0.540	0.546	0.551	0.556	0.560
0.06000	0.559	0.564	0.569	0.574	0.578
0.07000	0.575	0.581	0.586	0.590	0.595
0.08000	0.591	0.596	0.601	0.606	0.610
0.09000	0.606	0.611	0.616	0.620	0.624
0.10000	0.620	0.625	0.629	0.633	0.637
0.20000	0.734	0.738	0.741	0.744	0.747
0.25000	0.784	0.787	0.789	0.791	0.794
0.30000	0.832	0.834	0.836	0.837	0.839
0.35000	0.880	0.880	0.881	0.882	0.883
0.40000	0.928	0.927	0.927	0.927	0.927
0.45000	0.977	0.976	0.975	0.974	0.973
0.50000	1.029	1.026	1.024	1.022	1.020
0.55000	1.084	1.080	1.076	1.073	1.070
0.60000	1.143	1.137	1.132	1.128	1.123
0.65000	1.208	1.201	1.194	1.188	1.182
0.70000	1.282	1.272	1.263	1.255	1.248
0.75000	1.368	1.355	1.343	1.333	1.323
0.80000	1.472	1.455	1.440	1.426	1.414
0.85000	1.605	1.582	1.562	1.545	1.529
0.90000	1.793	1.762	1.735	1.711	1.689
0.95000	2.124	2.076	2.035	1.998	1.965
0.96000	2.233	2.180	2.134	2.092	2.055
0.97000	2.378	2.317	2.263	2.216	2.173
0.98000	2.588	2.514	2.450	2.393	2.342
0.99000	2.967	2.869	2.784	2.709	2.642
0.99500	3.374	3.248	3.138	3.043	2.958
0.99900	4.451	4.239	4.058	3.901	3.765
0.99990	6.391	5.998	5.667	5.384	5.141

Table generated using MINITAB

F-table for 50 Degrees of freedom in the denominator
numerator degrees of freedom

α	25	50	75	100	200
0.00010	0.290	0.340	0.359	0.370	0.394
0.00100	0.358	0.409	0.432	0.445	0.467
0.00500	0.425	0.476	0.499	0.512	0.534
0.01000	0.461	0.513	0.535	0.548	0.569
0.02000	0.505	0.555	0.577	0.589	0.609
0.03000	0.535	0.584	0.604	0.616	0.636
0.04000	0.558	0.606	0.626	0.637	0.656
0.05000	0.578	0.625	0.644	0.655	0.674
0.06000	0.596	0.641	0.660	0.671	0.688
0.07000	0.612	0.656	0.674	0.685	0.702
0.08000	0.627	0.669	0.687	0.697	0.714
0.09000	0.640	0.682	0.699	0.709	0.725
0.10000	0.653	0.693	0.710	0.720	0.735
0.20000	0.758	0.787	0.798	0.805	0.816
0.25000	0.803	0.825	0.834	0.840	0.848
0.30000	0.845	0.861	0.868	0.872	0.878
0.35000	0.886	0.896	0.900	0.903	0.907
0.40000	0.928	0.930	0.932	0.933	0.934
0.45000	0.970	0.964	0.963	0.963	0.962
0.50000	1.013	1.000	0.995	0.993	0.990
0.55000	1.059	1.036	1.028	1.024	1.018
0.60000	1.107	1.074	1.062	1.056	1.047
0.65000	1.160	1.115	1.099	1.091	1.078
0.70000	1.219	1.160	1.139	1.129	1.112
0.75000	1.287	1.211	1.184	1.170	1.149
0.80000	1.367	1.270	1.236	1.219	1.191
0.85000	1.468	1.343	1.299	1.277	1.243
0.90000	1.607	1.440	1.383	1.354	1.310
0.95000	1.842	1.599	1.518	1.477	1.414
0.96000	1.917	1.649	1.559	1.514	1.446
0.97000	2.016	1.712	1.612	1.562	1.486
0.98000	2.155	1.800	1.684	1.627	1.540
0.99000	2.399	1.948	1.806	1.735	1.629
0.99500	2.652	2.096	1.924	1.840	1.714
0.99900	3.278	2.441	2.194	2.075	1.902
0.99990	4.299	2.949	2.578	2.403	2.155

The table was generated using MINITAB

F-table for 75 Degrees of freedom in the denominator

numerator degrees of freedom

α	1	2	3	4	5
0.00010	0	0.10	0.125	0.147	0.167
0.00100	0	0.13	0.166	0.194	0.216
0.00500	0	0.18	0.216	0.247	0.272
0.01000	0	0.20	0.247	0.279	0.305
0.02000	0	0.24	0.287	0.321	0.348
0.03000	0	0.27	0.318	0.352	0.379
0.04000	0	0.30	0.344	0.378	0.405
0.05000	0	0.32	0.367	0.401	0.428
0.06000	0	0.34	0.388	0.422	0.448
0.07000	0	0.36	0.408	0.441	0.467
0.08000	0	0.38	0.427	0.460	0.485
0.09000	0	0.40	0.445	0.478	0.502
0.10000	0	0.42	0.463	0.495	0.519
0.20000	1	0.61	0.631	0.650	0.666
0.25000	1	0.71	0.715	0.726	0.736
0.30000	1	0.82	0.804	0.804	0.807
0.35000	1	0.94	0.900	0.887	0.881
0.40000	1	1.08	1.005	0.976	0.960
0.45000	2	1.24	1.123	1.073	1.045
0.50000	2	1.43	1.256	1.181	1.138
0.55000	3	1.66	1.412	1.304	1.243
0.60000	4	1.94	1.596	1.446	1.361
0.65000	5	2.31	1.820	1.614	1.500
0.70000	7	2.79	2.101	1.820	1.666
0.75000	10	3.46	2.471	2.081	1.873
0.80000	15	4.47	2.984	2.432	2.143
0.85000	28	6.14	3.765	2.939	2.524
0.90000	63	9.48	5.148	3.784	3.133
0.95000	253	19.48	8.563	5.676	4.418
0.96000	395	24.48	10.041	6.436	4.910
0.97000	702	32.82	12.301	7.548	5.613
0.98000	1581	49.49	16.318	9.412	6.748
0.99000	6324	99.48	26.279	13.615	9.166
0.99500	25294	199.48	42.085	19.554	12.351
0.99900	632448	999.45	124.270	44.608	24.223
0.99990	63215868	9997.57	579.015	142.576	62.014

Table generated using MINITAB

F-table for 75 Degrees of freedom in the denominator
numerator degrees of freedom

α	6	7	8	9	10
0.00010	0.182	0.198	0.210	0.221	0.233
0.00100	0.236	0.252	0.267	0.280	0.292
0.00500	0.293	0.311	0.327	0.341	0.354
0.01000	0.327	0.346	0.362	0.376	0.389
0.02000	0.370	0.389	0.406	0.420	0.433
0.03000	0.402	0.420	0.437	0.451	0.463
0.04000	0.427	0.446	0.462	0.476	0.488
0.05000	0.450	0.468	0.484	0.498	0.510
0.06000	0.470	0.488	0.504	0.517	0.529
0.07000	0.489	0.506	0.522	0.535	0.547
0.08000	0.506	0.524	0.539	0.552	0.563
0.09000	0.523	0.540	0.555	0.567	0.578
0.10000	0.539	0.556	0.570	0.582	0.593
0.20000	0.680	0.691	0.701	0.710	0.717
0.25000	0.746	0.754	0.761	0.767	0.773
0.30000	0.812	0.816	0.820	0.824	0.828
0.35000	0.880	0.880	0.880	0.881	0.882
0.40000	0.952	0.946	0.943	0.940	0.939
0.45000	1.028	1.017	1.009	1.003	0.998
0.50000	1.111	1.093	1.079	1.069	1.060
0.55000	1.204	1.176	1.156	1.140	1.128
0.60000	1.307	1.270	1.241	1.220	1.202
0.65000	1.427	1.376	1.338	1.309	1.286
0.70000	1.569	1.501	1.451	1.413	1.382
0.75000	1.742	1.653	1.587	1.536	1.496
0.80000	1.966	1.845	1.757	1.690	1.637
0.85000	2.274	2.106	1.986	1.894	1.823
0.90000	2.754	2.505	2.330	2.198	2.097
0.95000	3.725	3.289	2.990	2.771	2.604
0.96000	4.086	3.574	3.225	2.972	2.781
0.97000	4.591	3.966	3.546	3.245	3.018
0.98000	5.388	4.575	4.038	3.657	3.374
0.99000	7.021	5.789	4.997	4.449	4.048
0.99500	9.073	7.262	6.132	5.366	4.815
0.99900	16.121	12.035	9.649	8.113	7.051
0.99990	35.690	24.058	17.884	14.186	11.774

Table generated using MINITAB

F-table for 75 Degrees of freedom in the denominator
numerator degrees of freedom

α	11	12	13	14	15
0.00010	0	0.10	0.125	0.147	0.167
0.00100	0	0.13	0.166	0.194	0.216
0.00500	0	0.18	0.216	0.247	0.272
0.01000	0	0.20	0.247	0.279	0.305
0.02000	0	0.24	0.287	0.321	0.348
0.03000	0	0.27	0.318	0.352	0.379
0.04000	0	0.30	0.344	0.378	0.405
0.05000	0	0.32	0.367	0.401	0.428
0.06000	0	0.34	0.388	0.422	0.448
0.07000	0	0.36	0.408	0.441	0.467
0.08000	0	0.38	0.427	0.460	0.485
0.09000	0	0.40	0.445	0.478	0.502
0.10000	0	0.42	0.463	0.495	0.519
0.20000	1	0.61	0.631	0.650	0.666
0.25000	1	0.71	0.715	0.726	0.736
0.30000	1	0.82	0.804	0.804	0.807
0.35000	1	0.94	0.900	0.887	0.881
0.40000	1	1.08	1.005	0.976	0.960
0.45000	2	1.24	1.123	1.073	1.045
0.50000	2	1.43	1.256	1.181	1.138
0.55000	3	1.66	1.412	1.304	1.243
0.60000	4	1.94	1.596	1.446	1.361
0.65000	5	2.31	1.820	1.614	1.500
0.70000	7	2.79	2.101	1.820	1.666
0.75000	10	3.46	2.471	2.081	1.873
0.80000	15	4.47	2.984	2.432	2.143
0.85000	28	6.14	3.765	2.939	2.524
0.90000	63	9.48	5.148	3.784	3.133
0.95000	253	19.48	8.563	5.676	4.418
0.96000	395	24.48	10.041	6.436	4.910
0.97000	702	32.82	12.301	7.548	5.613
0.98000	1581	49.49	16.318	9.412	6.748
0.99000	6324	99.48	26.279	13.615	9.166
0.99500	25294	199.48	42.085	19.554	12.351
0.99900	632448	999.45	124.270	44.608	24.223
0.99990	63215868	9997.57	579.015	142.576	62.014

Table generated using MINITAB

F-table for 75 Degrees of freedom in the denominator

numerator degrees of freedom

α	16	17	18	19	20
0.00010	0.182	0.198	0.210	0.221	0.233
0.00100	0.236	0.252	0.267	0.280	0.292
0.00500	0.293	0.311	0.327	0.341	0.354
0.01000	0.327	0.346	0.362	0.376	0.389
0.02000	0.370	0.389	0.406	0.420	0.433
0.03000	0.402	0.420	0.437	0.451	0.463
0.04000	0.427	0.446	0.462	0.476	0.488
0.05000	0.450	0.468	0.484	0.498	0.510
0.06000	0.470	0.488	0.504	0.517	0.529
0.07000	0.489	0.506	0.522	0.535	0.547
0.08000	0.506	0.524	0.539	0.552	0.563
0.09000	0.523	0.540	0.555	0.567	0.578
0.10000	0.539	0.556	0.570	0.582	0.593
0.20000	0.680	0.691	0.701	0.710	0.717
0.25000	0.746	0.754	0.761	0.767	0.773
0.30000	0.812	0.816	0.820	0.824	0.828
0.35000	0.880	0.880	0.880	0.881	0.882
0.40000	0.952	0.946	0.943	0.940	0.939
0.45000	1.028	1.017	1.009	1.003	0.998
0.50000	1.111	1.093	1.079	1.069	1.060
0.55000	1.204	1.176	1.156	1.140	1.128
0.60000	1.307	1.270	1.241	1.220	1.202
0.65000	1.427	1.376	1.338	1.309	1.286
0.70000	1.569	1.501	1.451	1.413	1.382
0.75000	1.742	1.653	1.587	1.536	1.496
0.80000	1.966	1.845	1.757	1.690	1.637
0.85000	2.274	2.106	1.986	1.894	1.823
0.90000	2.754	2.505	2.330	2.198	2.097
0.95000	3.725	3.289	2.990	2.771	2.604
0.96000	4.086	3.574	3.225	2.972	2.781
0.97000	4.591	3.966	3.546	3.245	3.018
0.98000	5.388	4.575	4.038	3.657	3.374
0.99000	7.021	5.789	4.997	4.449	4.048
0.99500	9.073	7.262	6.132	5.366	4.815
0.99900	16.121	12.035	9.649	8.113	7.051
0.99950	20.528	14.866	11.653	9.628	8.254
0.99990	35.690	24.058	17.884	14.186	11.774

Table generated using MINITAB

F-table for 75 Degrees of freedom in the denominator
numerator degrees of freedom

α	25	50	75	100	200
0.00010	0.328	0.386	0.414	0.435	0.464
0.00100	0.392	0.455	0.485	0.501	0.532
0.00500	0.457	0.519	0.548	0.564	0.593
0.01000	0.492	0.553	0.581	0.597	0.625
0.02000	0.534	0.593	0.619	0.635	0.661
0.03000	0.563	0.620	0.645	0.660	0.685
0.04000	0.585	0.641	0.665	0.679	0.703
0.05000	0.604	0.658	0.682	0.695	0.719
0.06000	0.621	0.673	0.696	0.709	0.732
0.07000	0.637	0.687	0.709	0.722	0.744
0.08000	0.651	0.700	0.721	0.733	0.754
0.09000	0.664	0.711	0.732	0.744	0.764
0.10000	0.676	0.722	0.742	0.754	0.773
0.20000	0.776	0.808	0.822	0.830	0.844
0.25000	0.818	0.844	0.855	0.861	0.872
0.30000	0.858	0.877	0.885	0.890	0.898
0.35000	0.897	0.909	0.914	0.917	0.923
0.40000	0.937	0.940	0.942	0.944	0.947
0.45000	0.976	0.972	0.971	0.970	0.970
0.50000	1.018	1.004	1.000	0.997	0.994
0.55000	1.061	1.037	1.029	1.025	1.018
0.60000	1.107	1.072	1.060	1.054	1.043
0.65000	1.157	1.110	1.093	1.084	1.070
0.70000	1.213	1.151	1.129	1.117	1.098
0.75000	1.277	1.197	1.169	1.154	1.130
0.80000	1.352	1.251	1.215	1.196	1.166
0.85000	1.447	1.317	1.271	1.247	1.209
0.90000	1.579	1.406	1.346	1.315	1.266
0.95000	1.800	1.550	1.465	1.421	1.354
0.96000	1.872	1.595	1.502	1.454	1.381
0.97000	1.964	1.653	1.548	1.495	1.414
0.98000	2.096	1.732	1.613	1.552	1.459
0.99000	2.326	1.867	1.719	1.646	1.534
0.99500	2.564	2.001	1.824	1.736	1.604
0.99900	3.154	2.313	2.062	1.940	1.760
0.99990	4.115	2.772	2.399	2.222	1.968

Table generated using MINITAB

F-table for 100 Degrees of freedom in the denominator
numerator degrees of freedom

α	1	2	3	4	5
0.00010	0	0.10	0.128	0.153	0.174
0.00100	0	0.14	0.171	0.200	0.223
0.00500	0	0.18	0.220	0.252	0.278
0.01000	0	0.21	0.251	0.285	0.311
0.02000	0	0.25	0.292	0.327	0.354
0.03000	0	0.28	0.322	0.357	0.385
0.04000	0	0.30	0.348	0.383	0.411
0.05000	0	0.32	0.371	0.406	0.433
0.06000	0	0.35	0.392	0.427	0.454
0.07000	0	0.37	0.412	0.446	0.473
0.08000	0	0.39	0.431	0.465	0.491
0.09000	0	0.41	0.450	0.482	0.508
0.10000	0	0.42	0.467	0.500	0.524
0.20000	1	0.61	0.635	0.655	0.671
0.25000	1	0.71	0.719	0.730	0.741
0.30000	1	0.82	0.808	0.808	0.812
0.35000	1	0.94	0.903	0.890	0.885
0.40000	1	1.08	1.008	0.979	0.964
0.45000	2	1.24	1.126	1.076	1.048
0.50000	2	1.43	1.259	1.184	1.141
0.55000	3	1.66	1.414	1.306	1.245
0.60000	4	1.95	1.598	1.448	1.363
0.65000	5	2.31	1.822	1.616	1.501
0.70000	7	2.79	2.103	1.821	1.666
0.75000	10	3.47	2.471	2.081	1.872
0.80000	15	4.47	2.984	2.430	2.141
0.85000	28	6.14	3.764	2.936	2.520
0.90000	63	9.48	5.144	3.778	3.126
0.95000	253	19.49	8.554	5.664	4.405
0.96000	396	24.49	10.030	6.421	4.895
0.97000	703	32.82	12.286	7.530	5.594
0.98000	1583	49.49	16.297	9.388	6.724
0.99000	6334	99.49	26.240	13.577	9.129
0.99500	25339	199.49	42.022	19.497	12.299
0.99900	633505	999.46	124.069	44.469	24.114
0.99990	63326804	9997.58	578.079	142.125	61.721

Table generated using MINITAB

F-table for 100 Degrees of freedom in the denominator
numerator degrees of freedom

α	6	7	8	9	10
0.00010	0.191	0.205	0.221	0.234	0.245
0.00100	0.243	0.261	0.276	0.290	0.303
0.00500	0.300	0.319	0.336	0.351	0.364
0.01000	0.334	0.354	0.371	0.386	0.399
0.02000	0.377	0.397	0.414	0.429	0.442
0.03000	0.408	0.428	0.445	0.460	0.473
0.04000	0.434	0.453	0.470	0.484	0.497
0.05000	0.456	0.475	0.492	0.506	0.519
0.06000	0.476	0.495	0.511	0.525	0.538
0.07000	0.495	0.513	0.529	0.543	0.555
0.08000	0.512	0.530	0.546	0.559	0.571
0.09000	0.529	0.547	0.562	0.575	0.586
0.10000	0.545	0.562	0.577	0.590	0.601
0.20000	0.685	0.697	0.707	0.716	0.724
0.25000	0.750	0.759	0.766	0.773	0.779
0.30000	0.816	0.821	0.825	0.829	0.833
0.35000	0.884	0.884	0.885	0.886	0.887
0.40000	0.955	0.950	0.946	0.944	0.943
0.45000	1.031	1.020	1.012	1.006	1.001
0.50000	1.114	1.095	1.082	1.071	1.063
0.55000	1.205	1.178	1.158	1.142	1.130
0.60000	1.309	1.271	1.242	1.221	1.203
0.65000	1.428	1.376	1.338	1.309	1.286
0.70000	1.568	1.500	1.450	1.411	1.380
0.75000	1.741	1.651	1.584	1.533	1.493
0.80000	1.963	1.841	1.753	1.686	1.632
0.85000	2.269	2.101	1.980	1.888	1.816
0.90000	2.746	2.497	2.320	2.189	2.086
0.95000	3.711	3.274	2.974	2.755	2.588
0.96000	4.070	3.557	3.207	2.954	2.762
0.97000	4.572	3.946	3.526	3.224	2.996
0.98000	5.364	4.550	4.013	3.632	3.348
0.99000	6.986	5.754	4.963	4.414	4.013
0.99500	9.025	7.216	6.087	5.322	4.772
0.99900	16.028	11.951	9.571	8.038	6.980
0.99990	35.472	23.878	17.729	14.046	11.644

Table generated using MINITAB

F-table for 100 Degrees of freedom in the denominator
numerator degrees of freedom

α	11	12	13	14	15
0.00010	0.255	0.264	0.275	0.281	0.291
0.00100	0.315	0.325	0.334	0.343	0.351
0.00500	0.376	0.387	0.397	0.406	0.414
0.01000	0.411	0.422	0.432	0.441	0.449
0.02000	0.454	0.465	0.475	0.484	0.492
0.03000	0.484	0.495	0.505	0.513	0.521
0.04000	0.509	0.519	0.529	0.537	0.545
0.05000	0.530	0.540	0.549	0.558	0.565
0.06000	0.549	0.558	0.567	0.576	0.583
0.07000	0.566	0.575	0.584	0.592	0.599
0.08000	0.582	0.591	0.600	0.607	0.615
0.09000	0.597	0.606	0.614	0.622	0.629
0.10000	0.611	0.620	0.628	0.635	0.642
0.20000	0.731	0.737	0.743	0.748	0.753
0.25000	0.784	0.789	0.793	0.797	0.801
0.30000	0.836	0.839	0.842	0.845	0.847
0.35000	0.888	0.890	0.891	0.892	0.893
0.40000	0.942	0.941	0.941	0.940	0.940
0.45000	0.997	0.994	0.992	0.990	0.988
0.50000	1.056	1.051	1.046	1.042	1.039
0.55000	1.119	1.111	1.104	1.098	1.092
0.60000	1.189	1.177	1.167	1.158	1.150
0.65000	1.266	1.250	1.237	1.225	1.214
0.70000	1.355	1.334	1.316	1.300	1.287
0.75000	1.460	1.432	1.409	1.389	1.371
0.80000	1.589	1.553	1.522	1.496	1.474
0.85000	1.758	1.710	1.670	1.635	1.605
0.90000	2.004	1.937	1.881	1.833	1.792
0.95000	2.456	2.349	2.261	2.186	2.123
0.96000	2.611	2.490	2.389	2.305	2.233
0.97000	2.819	2.677	2.560	2.462	2.379
0.98000	3.128	2.953	2.811	2.692	2.592
0.99000	3.707	3.466	3.272	3.111	2.977
0.99500	4.358	4.036	3.779	3.569	3.394
0.99900	6.210	5.627	5.171	4.806	4.507
0.99990	9.976	8.761	7.842	7.125	6.552

Table generated using MINITAB

F-table for 100 Degrees of freedom in the denominator
numerator degrees of freedom

α	16	17	18	19	20
0.00010	0.295	0.303	0.310	0.316	0.322
0.00100	0.359	0.366	0.373	0.379	0.385
0.00500	0.422	0.429	0.436	0.442	0.449
0.01000	0.457	0.464	0.471	0.477	0.483
0.02000	0.500	0.507	0.513	0.519	0.525
0.03000	0.529	0.536	0.542	0.548	0.554
0.04000	0.552	0.559	0.565	0.571	0.577
0.05000	0.572	0.579	0.585	0.591	0.596
0.06000	0.590	0.596	0.602	0.608	0.613
0.07000	0.606	0.612	0.618	0.624	0.629
0.08000	0.621	0.627	0.633	0.638	0.643
0.09000	0.635	0.641	0.646	0.651	0.656
0.10000	0.648	0.654	0.659	0.664	0.669
0.20000	0.757	0.761	0.765	0.768	0.771
0.25000	0.804	0.807	0.810	0.813	0.815
0.30000	0.850	0.852	0.854	0.856	0.857
0.35000	0.895	0.896	0.897	0.898	0.899
0.40000	0.940	0.940	0.940	0.940	0.940
0.45000	0.987	0.986	0.985	0.984	0.983
0.50000	1.036	1.033	1.031	1.029	1.027
0.55000	1.088	1.083	1.080	1.076	1.073
0.60000	1.144	1.138	1.132	1.128	1.123
0.65000	1.205	1.197	1.190	1.184	1.178
0.70000	1.275	1.265	1.255	1.247	1.239
0.75000	1.356	1.342	1.330	1.319	1.309
0.80000	1.454	1.436	1.420	1.406	1.394
0.85000	1.579	1.556	1.535	1.517	1.500
0.90000	1.757	1.725	1.697	1.672	1.650
0.95000	2.068	2.020	1.978	1.940	1.906
0.96000	2.171	2.117	2.070	2.028	1.990
0.97000	2.308	2.245	2.191	2.142	2.099
0.98000	2.505	2.431	2.365	2.308	2.256
0.99000	2.862	2.763	2.677	2.602	2.535
0.99500	3.245	3.119	3.009	2.913	2.828
0.99900	4.259	4.048	3.868	3.712	3.576
0.99990	6.085	5.697	5.371	5.092	4.852

Table generated using MINITAB

F-table for 100 Degrees of freedom in the denominator

numerator degrees of freedom

α	25	50	75	100	200
0.00010	0.348	0.414	0.450	0.471	0.509
0.00100	0.411	0.481	0.515	0.535	0.571
0.00500	0.474	0.543	0.575	0.594	0.629
0.01000	0.508	0.576	0.607	0.625	0.658
0.02000	0.549	0.614	0.644	0.661	0.692
0.03000	0.577	0.640	0.668	0.685	0.714
0.04000	0.599	0.660	0.687	0.703	0.731
0.05000	0.618	0.676	0.703	0.718	0.745
0.06000	0.635	0.691	0.716	0.731	0.757
0.07000	0.650	0.704	0.729	0.743	0.768
0.08000	0.663	0.716	0.740	0.754	0.778
0.09000	0.676	0.727	0.750	0.763	0.787
0.10000	0.688	0.738	0.760	0.773	0.795
0.20000	0.785	0.820	0.835	0.844	0.860
0.25000	0.826	0.854	0.866	0.873	0.885
0.30000	0.865	0.885	0.894	0.900	0.909
0.35000	0.903	0.916	0.922	0.925	0.931
0.40000	0.941	0.946	0.948	0.950	0.953
0.45000	0.980	0.976	0.975	0.975	0.975
0.50000	1.020	1.006	1.002	1.000	0.996
0.55000	1.062	1.038	1.029	1.025	1.018
0.60000	1.107	1.071	1.058	1.052	1.041
0.65000	1.155	1.107	1.089	1.080	1.065
0.70000	1.209	1.146	1.123	1.110	1.091
0.75000	1.271	1.190	1.160	1.144	1.119
0.80000	1.345	1.241	1.203	1.183	1.152
0.85000	1.437	1.304	1.256	1.231	1.191
0.90000	1.564	1.388	1.326	1.293	1.241
0.95000	1.779	1.524	1.437	1.391	1.320
0.96000	1.848	1.567	1.471	1.421	1.344
0.97000	1.938	1.621	1.514	1.459	1.374
0.98000	2.066	1.697	1.574	1.511	1.414
0.99000	2.288	1.824	1.673	1.597	1.481
0.99500	2.519	1.951	1.770	1.680	1.544
0.99900	3.090	2.245	1.991	1.867	1.682
0.99990	4.021	2.679	2.304	2.126	1.867

Table generated using MINITAB

F-table for 200 Degrees of freedom in the denominator

numerator degrees of freedom

α	2	3	4	5
0.00010	0.10	0.136	0.160	0.184
0.00100	0.14	0.178	0.208	0.233
0.00500	0.18	0.227	0.261	0.288
0.01000	0.21	0.258	0.293	0.321
0.02000	0.25	0.298	0.335	0.363
0.03000	0.28	0.329	0.365	0.394
0.04000	0.31	0.354	0.391	0.420
0.05000	0.33	0.377	0.414	0.442
0.06000	0.35	0.399	0.435	0.463
0.07000	0.37	0.419	0.454	0.481
0.08000	0.39	0.438	0.472	0.499
0.09000	0.41	0.456	0.490	0.516
0.10000	0.43	0.474	0.507	0.533
0.20000	0.62	0.640	0.661	0.678
0.25000	0.72	0.725	0.737	0.747
0.30000	0.83	0.813	0.814	0.818
0.35000	0.95	0.909	0.896	0.891
0.40000	1.09	1.013	0.984	0.969
0.45000	1.25	1.130	1.080	1.053
0.50000	1.44	1.264	1.188	1.145
0.55000	1.67	1.418	1.309	1.248
0.60000	1.95	1.601	1.450	1.365
0.65000	2.32	1.824	1.617	1.502
0.70000	2.80	2.105	1.822	1.666
0.75000	3.47	2.473	2.081	1.871
0.80000	4.48	2.984	2.428	2.137
0.85000	6.15	3.762	2.932	2.514
0.90000	9.49	5.139	3.769	3.115
0.95000	19.49	8.540	5.646	4.385
0.96000	24.49	10.012	6.400	4.871
0.97000	32.83	12.263	7.503	5.565
0.98000	49.49	16.264	9.352	6.687
0.99000	99.49	26.183	13.520	9.075
0.99500	199.50	41.925	19.410	12.221
0.99900	999.47	123.772	44.260	23.950
0.99990	9997.59	576.646	141.433	61.281

Table generated using MINITAB

F-table for 200 Degrees of freedom in the denominator
numerator degrees of freedom

α	6	7	8	9	10
0.00010	0.202	0.221	0.236	0.250	0.262
0.00100	0.255	0.274	0.291	0.306	0.320
0.00500	0.311	0.332	0.350	0.366	0.380
0.01000	0.345	0.366	0.384	0.400	0.414
0.02000	0.388	0.409	0.427	0.443	0.457
0.03000	0.419	0.439	0.457	0.473	0.487
0.04000	0.444	0.464	0.482	0.497	0.511
0.05000	0.466	0.486	0.503	0.519	0.532
0.06000	0.486	0.506	0.523	0.537	0.551
0.07000	0.504	0.524	0.540	0.555	0.568
0.08000	0.522	0.541	0.557	0.571	0.584
0.09000	0.538	0.557	0.573	0.586	0.599
0.10000	0.554	0.572	0.587	0.601	0.613
0.20000	0.693	0.705	0.716	0.725	0.733
0.25000	0.757	0.766	0.774	0.781	0.787
0.30000	0.823	0.828	0.832	0.836	0.840
0.35000	0.890	0.890	0.891	0.892	0.894
0.40000	0.960	0.955	0.952	0.950	0.949
0.45000	1.036	1.024	1.016	1.010	1.006
0.50000	1.118	1.099	1.085	1.075	1.066
0.55000	1.208	1.181	1.160	1.145	1.132
0.60000	1.310	1.272	1.244	1.222	1.204
0.65000	1.428	1.377	1.338	1.309	1.285
0.70000	1.568	1.499	1.449	1.409	1.378
0.75000	1.739	1.648	1.581	1.529	1.488
0.80000	1.958	1.836	1.747	1.679	1.625
0.85000	2.262	2.092	1.971	1.878	1.806
0.90000	2.734	2.484	2.306	2.174	2.071
0.95000	3.690	3.252	2.951	2.731	2.563
0.96000	4.045	3.531	3.180	2.926	2.734
0.97000	4.542	3.916	3.494	3.192	2.964
0.98000	5.326	4.513	3.974	3.593	3.309
0.99000	6.933	5.702	4.911	4.363	3.961
0.99500	8.952	7.146	6.019	5.255	4.705
0.99900	15.887	11.824	9.453	7.926	6.872
0.99990	35.142	23.607	17.493	13.833	11.447

Table generated using MINITAB

F-table for 200 Degrees of freedom in the denominator

numerator degrees of freedom

α	11	12	13	14	15
0.00010	0.272	0.285	0.296	0.303	0.312
0.00100	0.332	0.344	0.354	0.365	0.374
0.00500	0.393	0.405	0.416	0.426	0.435
0.01000	0.427	0.439	0.450	0.460	0.469
0.02000	0.470	0.481	0.492	0.502	0.511
0.03000	0.499	0.511	0.521	0.531	0.539
0.04000	0.523	0.534	0.545	0.554	0.562
0.05000	0.544	0.555	0.565	0.574	0.582
0.06000	0.562	0.573	0.583	0.591	0.599
0.07000	0.579	0.589	0.599	0.607	0.615
0.08000	0.595	0.605	0.614	0.622	0.630
0.09000	0.609	0.619	0.628	0.636	0.644
0.10000	0.623	0.633	0.641	0.649	0.657
0.20000	0.741	0.748	0.754	0.759	0.764
0.25000	0.793	0.798	0.803	0.807	0.811
0.30000	0.844	0.847	0.851	0.854	0.856
0.35000	0.895	0.897	0.898	0.900	0.901
0.40000	0.948	0.947	0.947	0.946	0.946
0.45000	1.002	0.999	0.997	0.995	0.993
0.50000	1.060	1.054	1.049	1.046	1.042
0.55000	1.122	1.113	1.106	1.100	1.094
0.60000	1.190	1.177	1.167	1.158	1.151
0.65000	1.265	1.249	1.235	1.223	1.213
0.70000	1.352	1.331	1.313	1.297	1.283
0.75000	1.455	1.427	1.403	1.383	1.365
0.80000	1.581	1.545	1.514	1.487	1.464
0.85000	1.747	1.698	1.657	1.622	1.592
0.90000	1.988	1.920	1.864	1.815	1.774
0.95000	2.430	2.323	2.234	2.159	2.095
0.96000	2.582	2.460	2.359	2.274	2.202
0.97000	2.786	2.642	2.525	2.427	2.343
0.98000	3.088	2.913	2.769	2.650	2.549
0.99000	3.655	3.414	3.219	3.058	2.923
0.99500	4.292	3.970	3.713	3.503	3.327
0.99900	6.105	5.524	5.070	4.706	4.408
0.99990	9.791	8.584	7.672	6.960	6.391

Table generated using MINITAB

F-table for 200 Degrees of freedom in the denominator
numerator degrees of freedom

α	16	17	18	19	20
0.00010	0.321	0.329	0.336	0.343	0.349
0.00100	0.382	0.390	0.398	0.405	0.412
0.00500	0.444	0.452	0.459	0.466	0.473
0.01000	0.478	0.486	0.493	0.500	0.507
0.02000	0.519	0.527	0.534	0.541	0.547
0.03000	0.547	0.555	0.562	0.569	0.575
0.04000	0.570	0.578	0.584	0.591	0.597
0.05000	0.590	0.597	0.603	0.610	0.616
0.06000	0.607	0.614	0.620	0.626	0.632
0.07000	0.623	0.629	0.636	0.641	0.647
0.08000	0.637	0.644	0.650	0.655	0.661
0.09000	0.651	0.657	0.663	0.668	0.674
0.10000	0.663	0.669	0.675	0.681	0.686
0.20000	0.769	0.773	0.777	0.781	0.785
0.25000	0.815	0.818	0.821	0.824	0.827
0.30000	0.859	0.861	0.863	0.865	0.867
0.35000	0.902	0.904	0.905	0.906	0.907
0.40000	0.947	0.947	0.947	0.947	0.947
0.45000	0.992	0.991	0.990	0.989	0.988
0.50000	1.039	1.037	1.034	1.032	1.030
0.55000	1.089	1.085	1.081	1.078	1.075
0.60000	1.144	1.138	1.132	1.127	1.123
0.65000	1.204	1.195	1.188	1.181	1.175
0.70000	1.271	1.260	1.251	1.242	1.234
0.75000	1.349	1.336	1.323	1.312	1.302
0.80000	1.444	1.426	1.410	1.396	1.383
0.85000	1.566	1.542	1.521	1.502	1.485
0.90000	1.737	1.705	1.677	1.652	1.629
0.95000	2.039	1.990	1.947	1.909	1.875
0.96000	2.139	2.085	2.036	1.994	1.955
0.97000	2.271	2.208	2.153	2.104	2.060
0.98000	2.463	2.387	2.322	2.263	2.211
0.99000	2.808	2.709	2.622	2.546	2.479
0.99500	3.179	3.052	2.942	2.845	2.760
0.99900	4.160	3.950	3.770	3.614	3.478
0.99990	5.928	5.543	5.218	4.941	4.703

Table generated using MINITAB

F-table for 200 Degrees of freedom in the denominator

numerator degrees of freedom

α	25	50	75	100	200
0.00010	0.378	0.465	0.505	0.536	0.590
0.00100	0.441	0.525	0.567	0.594	0.644
0.00500	0.502	0.583	0.623	0.647	0.693
0.01000	0.535	0.613	0.651	0.675	0.718
0.02000	0.574	0.649	0.685	0.706	0.747
0.03000	0.601	0.672	0.707	0.727	0.765
0.04000	0.622	0.691	0.724	0.743	0.780
0.05000	0.640	0.706	0.738	0.757	0.792
0.06000	0.656	0.720	0.750	0.768	0.802
0.07000	0.670	0.732	0.761	0.779	0.811
0.08000	0.683	0.743	0.771	0.788	0.819
0.09000	0.695	0.753	0.780	0.797	0.826
0.10000	0.707	0.763	0.789	0.805	0.833
0.20000	0.799	0.838	0.857	0.867	0.887
0.25000	0.838	0.870	0.884	0.893	0.908
0.30000	0.876	0.899	0.909	0.916	0.928
0.35000	0.912	0.927	0.934	0.938	0.946
0.40000	0.948	0.954	0.958	0.960	0.964
0.45000	0.985	0.982	0.981	0.981	0.982
0.50000	1.023	1.010	1.005	1.003	1.001
0.55000	1.063	1.039	1.030	1.025	1.017
0.60000	1.106	1.069	1.055	1.048	1.036
0.65000	1.152	1.102	1.083	1.073	1.056
0.70000	1.204	1.138	1.113	1.099	1.077
0.75000	1.263	1.178	1.146	1.128	1.100
0.80000	1.332	1.224	1.184	1.162	1.126
0.85000	1.420	1.282	1.230	1.203	1.158
0.90000	1.541	1.359	1.292	1.257	1.199
0.95000	1.745	1.483	1.390	1.341	1.262
0.96000	1.811	1.522	1.420	1.367	1.281
0.97000	1.897	1.571	1.458	1.400	1.305
0.98000	2.018	1.640	1.511	1.444	1.338
0.99000	2.230	1.756	1.599	1.518	1.391
0.99500	2.449	1.871	1.684	1.589	1.441
0.99900	2.992	2.139	1.878	1.749	1.551
0.99990	3.876	2.534	2.153	1.970	1.698

Table generated using MINITAB

Student's t-distribution, degrees of freedom

α	1	2	3	4	5
0.00010	-3183.54	-70.704	-22.203	-13.033	-9.677
0.00100	-318.31	-22.327	-10.214	-7.173	-5.893
0.00500	-63.66	-9.924	-5.841	-4.604	-4.032
0.01000	-31.82	-6.964	-4.540	-3.747	-3.364
0.02000	-15.89	-4.848	-3.481	-2.998	-2.756
0.03000	-10.58	-3.896	-2.950	-2.600	-2.421
0.04000	-7.92	-3.319	-2.605	-2.332	-2.190
0.05000	-6.31	-2.920	-2.353	-2.131	-2.015
0.06000	-5.24	-2.620	-2.156	-1.971	-1.872
0.07000	-4.47	-2.383	-1.995	-1.837	-1.752
0.08000	-3.89	-2.189	-1.858	-1.723	-1.649
0.09000	-3.44	-2.026	-1.741	-1.622	-1.557
0.10000	-3.08	-1.885	-1.637	-1.533	-1.475
0.20000	-1.38	-1.060	-0.978	-0.941	-0.919
0.25000	-1.00	-0.816	-0.764	-0.740	-0.726
0.30000	-0.73	-0.617	-0.584	-0.568	-0.559
0.35000	-0.51	-0.444	-0.424	-0.414	-0.408
0.40000	-0.32	-0.288	-0.276	-0.270	-0.267
0.45000	-0.16	-0.142	-0.136	-0.133	-0.132
0.50000	0.00	0.000	0.000	0.000	0.000
0.55000	0.16	0.142	0.136	0.133	0.132
0.60000	0.32	0.288	0.276	0.270	0.267
0.65000	0.51	0.444	0.424	0.414	0.408
0.70000	0.73	0.617	0.584	0.568	0.559
0.75000	1.00	0.816	0.764	0.740	0.726
0.80000	1.38	1.060	0.978	0.941	0.919
0.85000	1.96	1.386	1.249	1.189	1.155
0.90000	3.08	1.885	1.637	1.533	1.475
0.95000	6.31	2.920	2.353	2.131	2.015
0.96000	7.92	3.319	2.605	2.332	2.190
0.97000	10.58	3.896	2.950	2.600	2.421
0.98000	15.89	4.848	3.481	2.998	2.756
0.99000	31.82	6.964	4.540	3.747	3.364
0.99500	63.66	9.924	5.841	4.604	4.032
0.99600	79.57	11.113	6.322	4.907	4.261
0.99700	106.10	12.851	6.994	5.321	4.570
0.99900	318.32	22.327	10.214	7.173	5.893
0.99990	3183.01	70.694	22.202	13.033	9.677

This table was generated using MINITAB

Student's t-distribution, degrees of freedom

α	6	7	8	9	10
0.00010	-8.024	-7.063	-6.442	-6.010	-5.693
0.00100	-5.207	-4.785	-4.500	-4.296	-4.143
0.00500	-3.707	-3.499	-3.355	-3.249	-3.169
0.01000	-3.142	-2.997	-2.896	-2.821	-2.763
0.02000	-2.612	-2.516	-2.448	-2.398	-2.359
0.03000	-2.313	-2.240	-2.189	-2.150	-2.120
0.04000	-2.104	-2.045	-2.004	-1.972	-1.948
0.05000	-1.943	-1.894	-1.859	-1.833	-1.812
0.06000	-1.811	-1.770	-1.740	-1.717	-1.699
0.07000	-1.700	-1.664	-1.638	-1.618	-1.603
0.08000	-1.603	-1.571	-1.548	-1.531	-1.517
0.09000	-1.517	-1.489	-1.469	-1.453	-1.441
0.10000	-1.439	-1.414	-1.396	-1.383	-1.372
0.20000	-0.905	-0.896	-0.888	-0.883	-0.879
0.25000	-0.717	-0.711	-0.706	-0.702	-0.699
0.30000	-0.553	-0.549	-0.545	-0.543	-0.541
0.35000	-0.404	-0.401	-0.399	-0.397	-0.396
0.40000	-0.264	-0.263	-0.261	-0.260	-0.260
0.45000	-0.131	-0.130	-0.129	-0.129	-0.128
0.50000	0.000	0.000	0.000	0.000	0.000
0.55000	0.131	0.130	0.129	0.129	0.128
0.60000	0.264	0.263	0.261	0.260	0.260
0.65000	0.404	0.401	0.399	0.397	0.396
0.70000	0.553	0.549	0.545	0.543	0.541
0.75000	0.717	0.711	0.706	0.702	0.699
0.80000	0.905	0.896	0.888	0.883	0.879
0.85000	1.134	1.119	1.108	1.099	1.093
0.90000	1.439	1.414	1.396	1.383	1.372
0.95000	1.943	1.894	1.859	1.833	1.812
0.96000	2.104	2.045	2.004	1.972	1.948
0.97000	2.313	2.240	2.189	2.150	2.120
0.98000	2.612	2.516	2.448	2.398	2.359
0.99000	3.142	2.997	2.896	2.821	2.763
0.99500	3.707	3.499	3.355	3.249	3.169
0.99600	3.898	3.666	3.506	3.389	3.301
0.99700	4.151	3.886	3.704	3.572	3.472
0.99900	5.207	4.785	4.500	4.296	4.143
0.99990	8.024	7.063	6.441	6.010	5.693

This table generated using MINITAB.

Student's t-distribution, degrees of freedom

α	11	12	13	14	15
0.00010	-5.452	-5.263	-5.110	-4.984	-4.879
0.00100	-4.024	-3.929	-3.851	-3.787	-3.732
0.00500	-3.105	-3.054	-3.012	-2.976	-2.946
0.01000	-2.718	-2.681	-2.650	-2.624	-2.602
0.02000	-2.328	-2.302	-2.281	-2.263	-2.248
0.03000	-2.096	-2.076	-2.060	-2.046	-2.034
0.04000	-1.928	-1.912	-1.898	-1.887	-1.877
0.05000	-1.795	-1.782	-1.770	-1.761	-1.753
0.06000	-1.685	-1.673	-1.664	-1.655	-1.648
0.07000	-1.590	-1.580	-1.571	-1.564	-1.558
0.08000	-1.506	-1.497	-1.490	-1.483	-1.478
0.09000	-1.431	-1.423	-1.417	-1.411	-1.406
0.10000	-1.363	-1.356	-1.350	-1.345	-1.340
0.20000	-0.875	-0.872	-0.870	-0.868	-0.866
0.25000	-0.697	-0.695	-0.693	-0.692	-0.691
0.30000	-0.539	-0.538	-0.537	-0.536	-0.535
0.35000	-0.395	-0.394	-0.393	-0.393	-0.392
0.40000	-0.259	-0.259	-0.258	-0.258	-0.257
0.45000	-0.128	-0.128	-0.128	-0.127	-0.127
0.50000	0.000	0.000	0.000	0.000	0.000
0.55000	0.128	0.128	0.128	0.127	0.127
0.60000	0.259	0.259	0.258	0.258	0.257
0.65000	0.395	0.394	0.393	0.393	0.392
0.70000	0.539	0.538	0.537	0.536	0.535
0.75000	0.697	0.695	0.693	0.692	0.691
0.80000	0.875	0.872	0.870	0.868	0.866
0.85000	1.087	1.083	1.079	1.076	1.073
0.90000	1.363	1.356	1.350	1.345	1.340
0.95000	1.795	1.782	1.770	1.761	1.753
0.96000	1.928	1.912	1.898	1.887	1.877
0.97000	2.096	2.076	2.060	2.046	2.034
0.98000	2.328	2.302	2.281	2.263	2.248
0.99000	2.718	2.681	2.650	2.624	2.602
0.99500	3.105	3.054	3.012	2.976	2.946
0.99600	3.231	3.174	3.128	3.089	3.056
0.99700	3.393	3.329	3.277	3.234	3.197
0.99900	4.024	3.929	3.852	3.787	3.732
0.99990	5.452	5.263	5.110	4.984	4.879

This table was generated using MINITAB

Student's t-distribution, degrees of freedom

α	16	17	18	19	20
0.00010	-4.790	-4.714	-4.648	-4.589	-4.538
0.00100	-3.686	-3.645	-3.610	-3.579	-3.551
0.00500	-2.920	-2.898	-2.878	-2.860	-2.845
0.01000	-2.583	-2.566	-2.552	-2.539	-2.527
0.02000	-2.235	-2.223	-2.213	-2.204	-2.196
0.03000	-2.023	-2.015	-2.007	-2.000	-1.993
0.04000	-1.869	-1.861	-1.855	-1.849	-1.844
0.05000	-1.745	-1.739	-1.734	-1.729	-1.724
0.06000	-1.642	-1.637	-1.632	-1.627	-1.624
0.07000	-1.552	-1.548	-1.543	-1.540	-1.536
0.08000	-1.473	-1.469	-1.465	-1.462	-1.459
0.09000	-1.402	-1.398	-1.394	-1.392	-1.389
0.10000	-1.336	-1.333	-1.330	-1.327	-1.325
0.20000	-0.864	-0.863	-0.862	-0.860	-0.859
0.25000	-0.690	-0.689	-0.688	-0.687	-0.686
0.30000	-0.535	-0.534	-0.533	-0.533	-0.532
0.35000	-0.392	-0.391	-0.391	-0.391	-0.390
0.40000	-0.257	-0.257	-0.257	-0.256	-0.256
0.45000	-0.127	-0.127	-0.127	-0.127	-0.127
0.50000	0.000	0.000	0.000	0.000	0.000
0.55000	0.127	0.127	0.127	0.127	0.127
0.60000	0.257	0.257	0.257	0.256	0.256
0.65000	0.392	0.391	0.391	0.391	0.390
0.70000	0.535	0.534	0.533	0.533	0.532
0.75000	0.690	0.689	0.688	0.687	0.686
0.80000	0.864	0.863	0.862	0.860	0.859
0.85000	1.071	1.069	1.067	1.065	1.064
0.90000	1.336	1.333	1.330	1.327	1.325
0.95000	1.745	1.739	1.734	1.729	1.724
0.96000	1.869	1.861	1.855	1.849	1.844
0.97000	2.023	2.015	2.007	2.000	1.993
0.98000	2.235	2.223	2.213	2.204	2.196
0.99000	2.583	2.566	2.552	2.539	2.527
0.99500	2.920	2.898	2.878	2.860	2.845
0.99600	3.027	3.003	2.981	2.962	2.945
0.99700	3.165	3.137	3.113	3.092	3.073
0.99900	3.686	3.645	3.610	3.579	3.551
0.99990	4.790	4.714	4.647	4.589	4.538

This table was generated using MINITAB

Student's t-distribution, degrees of freedom

α	21	22	23	24	25
0.00010	-4.492	-4.451	-4.415	-4.381	-4.351
0.00100	-3.527	-3.504	-3.484	-3.466	-3.450
0.00500	-2.831	-2.818	-2.807	-2.796	-2.787
0.01000	-2.517	-2.508	-2.499	-2.492	-2.485
0.02000	-2.189	-2.182	-2.176	-2.171	-2.166
0.03000	-1.988	-1.982	-1.978	-1.974	-1.970
0.04000	-1.839	-1.835	-1.831	-1.828	-1.824
0.05000	-1.720	-1.717	-1.713	-1.710	-1.708
0.06000	-1.620	-1.617	-1.614	-1.612	-1.609
0.07000	-1.533	-1.531	-1.528	-1.526	-1.524
0.08000	-1.456	-1.454	-1.452	-1.450	-1.448
0.09000	-1.386	-1.384	-1.382	-1.381	-1.379
0.10000	-1.323	-1.321	-1.319	-1.317	-1.316
0.20000	-0.859	-0.858	-0.857	-0.856	-0.856
0.25000	-0.686	-0.685	-0.685	-0.684	-0.684
0.30000	-0.532	-0.532	-0.531	-0.531	-0.531
0.35000	-0.390	-0.390	-0.390	-0.389	-0.389
0.40000	-0.256	-0.256	-0.256	-0.256	-0.256
0.45000	-0.127	-0.127	-0.127	-0.127	-0.126
0.50000	0.000	0.000	0.000	0.000	0.000
0.55000	0.127	0.127	0.127	0.127	0.126
0.60000	0.256	0.256	0.256	0.256	0.256
0.65000	0.390	0.390	0.390	0.389	0.389
0.70000	0.532	0.532	0.531	0.531	0.531
0.75000	0.686	0.685	0.685	0.684	0.684
0.80000	0.859	0.858	0.857	0.856	0.856
0.85000	1.062	1.061	1.060	1.059	1.058
0.90000	1.323	1.321	1.319	1.317	1.316
0.95000	1.720	1.717	1.713	1.710	1.708
0.96000	1.839	1.835	1.831	1.828	1.824
0.97000	1.988	1.982	1.978	1.974	1.970
0.98000	2.189	2.182	2.176	2.171	2.166
0.99000	2.517	2.508	2.499	2.492	2.485
0.99500	2.831	2.818	2.807	2.796	2.787
0.99600	2.930	2.916	2.903	2.892	2.882
0.99700	3.056	3.040	3.026	3.014	3.002
0.99900	3.527	3.505	3.484	3.466	3.450
0.99990	4.492	4.451	4.415	4.381	4.351

This table generated using MINITAB

Student's t-distribution, degrees of freedom

α	26	27	28	29	30
0.00010	-4.324	-4.298	-4.275	-4.253	-4.233
0.00100	-3.435	-3.421	-3.408	-3.396	-3.385
0.00500	-2.778	-2.770	-2.763	-2.756	-2.750
0.01000	-2.478	-2.472	-2.467	-2.462	-2.457
0.02000	-2.162	-2.157	-2.153	-2.150	-2.146
0.03000	-1.966	-1.963	-1.960	-1.957	-1.954
0.04000	-1.821	-1.819	-1.816	-1.814	-1.812
0.05000	-1.705	-1.703	-1.701	-1.699	-1.697
0.06000	-1.607	-1.605	-1.603	-1.601	-1.600
0.07000	-1.522	-1.520	-1.518	-1.517	-1.515
0.08000	-1.446	-1.444	-1.443	-1.442	-1.440
0.09000	-1.377	-1.376	-1.375	-1.373	-1.372
0.10000	-1.314	-1.313	-1.312	-1.311	-1.310
0.20000	-0.855	-0.855	-0.854	-0.854	-0.853
0.25000	-0.684	-0.683	-0.683	-0.683	-0.682
0.30000	-0.530	-0.530	-0.530	-0.530	-0.530
0.35000	-0.389	-0.389	-0.389	-0.389	-0.389
0.40000	-0.255	-0.255	-0.255	-0.255	-0.255
0.45000	-0.126	-0.126	-0.126	-0.126	-0.126
0.50000	0.000	0.000	0.000	0.000	0.000
0.55000	0.126	0.126	0.126	0.126	0.126
0.60000	0.255	0.255	0.255	0.255	0.255
0.65000	0.389	0.389	0.389	0.389	0.389
0.70000	0.530	0.530	0.530	0.530	0.530
0.75000	0.684	0.683	0.683	0.683	0.682
0.80000	0.855	0.855	0.854	0.854	0.853
0.85000	1.057	1.056	1.055	1.055	1.054
0.90000	1.314	1.313	1.312	1.311	1.310
0.95000	1.705	1.703	1.701	1.699	1.697
0.96000	1.821	1.819	1.816	1.814	1.812
0.97000	1.966	1.963	1.960	1.957	1.954
0.98000	2.162	2.157	2.153	2.150	2.146
0.99000	2.478	2.472	2.467	2.462	2.457
0.99500	2.778	2.770	2.763	2.756	2.750
0.99600	2.872	2.863	2.855	2.848	2.841
0.99700	2.992	2.982	2.973	2.964	2.957
0.99900	3.435	3.421	3.408	3.396	3.385
0.99990	4.323	4.298	4.275	4.253	4.233

This table generated using MINITAB.

Student's t distribution, degrees of freedom

α	50	100	200
0.00010	-4.014	-3.861	-3.789
0.00100	-3.261	-3.173	-3.131
0.00500	-2.677	-2.625	-2.600
0.01000	-2.403	-2.364	-2.345
0.02000	-2.108	-2.080	-2.067
0.03000	-1.924	-1.902	-1.891
0.04000	-1.787	-1.768	-1.759
0.05000	-1.675	-1.660	-1.652
0.06000	-1.581	-1.568	-1.561
0.07000	-1.499	-1.487	-1.481
0.08000	-1.426	-1.415	-1.410
0.09000	-1.359	-1.350	-1.345
0.10000	-1.298	-1.290	-1.285
0.20000	-0.848	-0.845	-0.843
0.25000	-0.679	-0.676	-0.675
0.30000	-0.527	-0.526	-0.525
0.35000	-0.387	-0.386	-0.385
0.40000	-0.254	-0.254	-0.253
0.45000	-0.126	-0.125	-0.125
0.50000	0.000	0.000	0.000
0.55000	0.126	0.125	0.125
0.60000	0.254	0.254	0.253
0.65000	0.387	0.386	0.385
0.70000	0.527	0.526	0.525
0.75000	0.679	0.676	0.675
0.80000	0.848	0.845	0.843
0.85000	1.047	1.041	1.039
0.90000	1.298	1.290	1.285
0.95000	1.675	1.660	1.652
0.96000	1.787	1.768	1.759
0.97000	1.924	1.902	1.891
0.98000	2.108	2.080	2.067
0.99000	2.403	2.364	2.345
0.99500	2.677	2.625	2.600
0.99600	2.762	2.706	2.678
0.99700	2.870	2.807	2.777
0.99900	3.261	3.173	3.131
0.99990	4.014	3.861	3.789

This table was generated using MINITAB.

The Student's t distribution tables were generated using MINITAB with the following example commands:

```
set c1
.0005
.001
.01
.05
.1
.5
.75
.9
.95
.99
.999
.9995
end
invcdf c1;
t 5.
```

These commands would generate the t values for the significance levels contained in column 1 of the worksheet.

Values for the Poisson Distribution

K	Mean of the Poisson distribution				
0	**0.500**	**1.000**	**1.500**	**2.000**	**2.500**
1	0.6065	0.3679	0.2231	0.1353	0.0821
2	0.9098	0.7358	0.5578	0.4060	0.2873
3	0.9856	0.9197	0.8088	0.6767	0.5438
4	0.9982	0.9810	0.9344	0.8571	0.7576
5	0.9998	0.9963	0.9814	0.9473	0.8912
6	1.0000	0.9994	0.9955	0.9834	0.9580
7		0.9999	0.9991	0.9955	0.9858
8		1.0000	0.9998	0.9989	0.9958
9			1.0000	0.9998	0.9989
10				1.0000	0.9997
11					0.9999
12					1.0000

K	Mean of the Poisson distribution				
0	**3.000**	**3.500**	**4.000**	**4.500**	**5.000**
1	0.0498	0.0302	0.0183	0.0111	0.0067
2	0.1991	0.1359	0.0916	0.0611	0.0404
3	0.4232	0.3208	0.2381	0.1736	0.1247
4	0.6472	0.5366	0.4335	0.3423	0.2650
5	0.8153	0.7254	0.6288	0.5321	0.4405
6	0.9161	0.8576	0.7851	0.7029	0.6160
7	0.9665	0.9347	0.8893	0.8311	0.7622
8	0.9881	0.9733	0.9489	0.9134	0.8666
9	0.9962	0.9901	0.9786	0.9597	0.9319
10	0.9989	0.9967	0.9919	0.9829	0.9682
11	0.9997	0.9990	0.9972	0.9933	0.9863
12	0.9999	0.9997	0.9991	0.9976	0.9945
13	1.0000	0.9999	0.9997	0.9992	0.9980
14		1.0000	0.9999	0.9997	0.9993
15			1.0000	0.9999	0.9998
16				1.0000	0.9999
17					1.0000

The table entries show the Probability (K $\leq$ Mean of the distribution)

Table generated using MINITAB.

K	Mean of the Poisson distribution				
0	**5.500**	**6.000**	**6.500**	**7.000**	**7.500**
1	0.0041	0.0025	0.0015	0.0009	0.0006
2	0.0266	0.0174	0.0113	0.0073	0.0047
3	0.0884	0.0620	0.0430	0.0296	0.0203
4	0.2017	0.1512	0.1118	0.0818	0.0591
5	0.3575	0.2851	0.2237	0.1730	0.1321
6	0.5289	0.4457	0.3690	0.3007	0.2414
7	0.6860	0.6063	0.5265	0.4497	0.3782
8	0.8095	0.7440	0.6728	0.5987	0.5246
9	0.8944	0.8472	0.7916	0.7291	0.6620
10	0.9462	0.9161	0.8774	0.8305	0.7764
11	0.9747	0.9574	0.9332	0.9015	0.8622
12	0.9890	0.9799	0.9661	0.9467	0.9208
13	0.9955	0.9912	0.9840	0.9730	0.9573
14	0.9983	0.9964	0.9929	0.9872	0.9784
15	0.9994	0.9986	0.9970	0.9943	0.9897
16	0.9998	0.9995	0.9988	0.9976	0.9954
17	0.9999	0.9998	0.9996	0.9990	0.9980
18	1.0000	0.9999	0.9998	0.9996	0.9992
19		1.0000	0.9999	0.9999	0.9997
20			1.0000	1.0000	0.9999
21					1.0000

The table entries show the Probability (K $\leq$ Mean of the distribution)

Table generated using MINITAB

K	Mean of the Poisson distribution				
0	**8.000**	**8.500**	**9.000**	**9.500**	**10.000**
1	0.0003	0.0002	0.0001	0.0001	0.0000
2	0.0030	0.0019	0.0012	0.0008	0.0005
3	0.0138	0.0093	0.0062	0.0042	0.0028
4	0.0424	0.0301	0.0212	0.0149	0.0103
5	0.0996	0.0744	0.0550	0.0403	0.0293
6	0.1912	0.1496	0.1157	0.0885	0.0671
7	0.3134	0.2562	0.2068	0.1649	0.1301
8	0.4530	0.3856	0.3239	0.2687	0.2202
9	0.5925	0.5231	0.4557	0.3918	0.3328
10	0.7166	0.6530	0.5874	0.5218	0.4579
11	0.8159	0.7634	0.7060	0.6453	0.5830
12	0.8881	0.8487	0.8030	0.7520	0.6968
13	0.9362	0.9091	0.8758	0.8364	0.7916
14	0.9658	0.9486	0.9261	0.8981	0.8645
15	0.9827	0.9726	0.9585	0.9400	0.9165
16	0.9918	0.9862	0.9780	0.9665	0.9513
17	0.9963	0.9934	0.9889	0.9823	0.9730
18	0.9984	0.9970	0.9947	0.9911	0.9857
19	0.9993	0.9987	0.9976	0.9957	0.9928
20	0.9997	0.9995	0.9989	0.9980	0.9965
21	0.9999	0.9998	0.9996	0.9991	0.9984
22	1.0000	0.9999	0.9998	0.9996	0.9993
23		1.0000	0.9999	0.9999	0.9997
24			1.0000	0.9999	0.9999
25				1.0000	1.0000

The table entries show the Probability (K ≤ Mean of the distribution)
Table generated using MINITAB

χ^2 table					
	Degrees of Freedom				
α	1	2	3	4	5
0.00010	0.0000	0.0002	0.0052	0.0284	0.0822
0.00050	0.0000	0.0010	0.0153	0.0639	0.1581
0.00075	0.0000	0.0015	0.0200	0.0785	0.1867
0.00100	0.0000	0.0020	0.0243	0.0908	0.2102
0.00500	0.0000	0.0100	0.0717	0.2070	0.4117
0.01000	0.0002	0.0201	0.1148	0.2971	0.5543
0.02000	0.0006	0.0404	0.1848	0.4294	0.7519
0.03000	0.0014	0.0609	0.2451	0.5351	0.9031
0.04000	0.0025	0.0816	0.3002	0.6271	1.0313
0.05000	0.0039	0.1026	0.3518	0.7107	1.1455
0.06000	0.0057	0.1238	0.4012	0.7884	1.2499
0.07000	0.0077	0.1451	0.4487	0.8616	1.3472
0.08000	0.0101	0.1668	0.4949	0.9315	1.4390
0.09000	0.0128	0.1886	0.5401	0.9987	1.5264
0.10000	0.0158	0.2107	0.5844	1.0636	1.6103
0.20000	0.0642	0.4463	1.0052	1.6488	2.3425
0.25000	0.1015	0.5754	1.2125	1.9226	2.6746
0.30000	0.1485	0.7133	1.4237	2.1947	2.9999
0.35000	0.2059	0.8616	1.6416	2.4701	3.3251
0.40000	0.2750	1.0217	1.8692	2.7528	3.6555
0.45000	0.3573	1.1957	2.1095	3.0469	3.9959
0.50000	0.4549	1.3863	2.3660	3.3567	4.3515
0.55000	0.5707	1.5970	2.6430	3.6871	4.7278
0.60000	0.7083	1.8326	2.9462	4.0446	5.1319
0.65000	0.8735	2.0996	3.2831	4.4377	5.5731
0.70000	1.0742	2.4079	3.6649	4.8784	6.0644
0.75000	1.3233	2.7726	4.1083	5.3853	6.6257
0.80000	1.6424	3.2189	4.6416	5.9886	7.2893
0.85000	2.0723	3.7942	5.3170	6.7449	8.1152
0.90000	2.7055	4.6052	6.2514	7.7794	9.2364
0.95000	3.8415	5.9915	7.8147	9.4877	11.0705
0.96000	4.2179	6.4378	8.3112	10.0255	11.6443
0.97000	4.7093	7.0131	8.9473	10.7119	12.3746
0.98000	5.4119	7.8240	9.8374	11.6678	13.3882
0.99000	6.6349	9.2103	11.3449	13.2767	15.0863
0.99500	7.8795	10.5966	12.8382	14.8603	16.7496
0.99600	8.2838	11.0429	13.3164	15.3656	17.2790
0.99700	8.8075	11.6183	13.9314	16.0143	17.9576
0.99800	9.5496	12.4292	14.7956	16.9238	18.9074
0.99900	10.8277	13.8155	16.2663	18.4669	20.5150
0.99950	12.1157	15.2017	17.7300	19.9973	22.1052
0.99990	15.1372	18.4204	21.1075	23.5125	25.7445

Table generated with MINITAB

	Degrees of Freedom				
α	6	7	8	9	10
0.00010	0.1724	0.3000	0.4636	0.6608	0.8889
0.00050	0.2994	0.4849	0.7104	0.9717	1.2650
0.00075	0.3447	0.5482	0.7926	1.0730	1.3854
0.00100	0.3811	0.5985	0.8571	1.1519	1.4787
0.00500	0.6757	0.9893	1.3444	1.7349	2.1559
0.01000	0.8721	1.2390	1.6465	2.0879	2.5582
0.02000	1.1344	1.5643	2.0325	2.5324	3.0591
0.03000	1.3296	1.8016	2.3101	2.8485	3.4121
0.04000	1.4924	1.9971	2.5366	3.1047	3.6965
0.05000	1.6354	2.1674	2.7326	3.3251	3.9403
0.06000	1.7649	2.3205	2.9080	3.5215	4.1567
0.07000	1.8846	2.4611	3.0683	3.7004	4.3534
0.08000	1.9967	2.5921	3.2172	3.8661	4.5350
0.09000	2.1029	2.7157	3.3570	4.0214	4.7049
0.10000	2.2041	2.8331	3.4895	4.1682	4.8652
0.20000	3.0701	3.8223	4.5936	5.3801	6.1791
0.25000	3.4546	4.2549	5.0706	5.8988	6.7372
0.30000	3.8276	4.6713	5.5274	6.3933	7.2672
0.35000	4.1973	5.0816	5.9753	6.8763	7.7832
0.40000	4.5702	5.4932	6.4226	7.3570	8.2955
0.45000	4.9519	5.9125	6.8766	7.8434	8.8124
0.50000	5.3481	6.3458	7.3441	8.3428	9.3418
0.55000	5.7652	6.8000	7.8325	8.8632	9.8922
0.60000	6.2108	7.2832	8.3505	9.4136	10.4732
0.65000	6.6948	7.8061	8.9094	10.0060	11.0971
0.70000	7.2311	8.3834	9.5245	10.6564	11.7807
0.75000	7.8408	9.0371	10.2189	11.3888	12.5489
0.80000	8.5581	9.8033	11.0301	12.2421	13.4420
0.85000	9.4461	10.7479	12.0271	13.2880	14.5339
0.90000	10.6446	12.0170	13.3616	14.6837	15.9872
0.95000	12.5916	14.0671	15.5073	16.9190	18.3070
0.96000	13.1978	14.7030	16.1708	17.6083	19.0207
0.97000	13.9676	15.5091	17.0105	18.4796	19.9219
0.98000	15.0332	16.6224	18.1682	19.6790	21.1608
0.99000	16.8119	18.4753	20.0902	21.6660	23.2093
0.99500	18.5476	20.2778	21.9550	23.5893	25.1882
0.99600	19.0988	20.8491	22.5452	24.1973	25.8130
0.99700	19.8046	21.5802	23.2997	24.9740	26.6108
0.99800	20.7912	22.6007	24.3521	26.0564	27.7217
0.99900	22.4578	24.3220	26.1246	27.8771	29.5884
0.99950	24.1027	26.0178	27.8681	29.6656	31.4198
0.99990	27.8561	29.8779	31.8279	33.7189	35.5643

Table generated using MINITAB

	Degrees of Freedom				
α	**11**	**12**	**13**	**14**	**15**
0.00010	1.1453	1.4275	1.7333	2.0608	2.4082
0.00050	1.5868	1.9344	2.3051	2.6967	3.1075
0.00075	1.7263	2.0927	2.4819	2.8918	3.3206
0.00100	1.8339	2.2142	2.6172	3.0407	3.4827
0.00500	2.6032	3.0738	3.5650	4.0747	4.6009
0.01000	3.0535	3.5706	4.1069	4.6604	5.2293
0.02000	3.6087	4.1783	4.7654	5.3682	5.9849
0.03000	3.9972	4.6009	5.2210	5.8556	6.5032
0.04000	4.3087	4.9385	5.5838	6.2426	6.9137
0.05000	4.5748	5.2260	5.8919	6.5706	7.2609
0.06000	4.8104	5.4800	6.1635	6.8593	7.5661
0.07000	5.0240	5.7098	6.4088	7.1197	7.8410
0.08000	5.2209	5.9212	6.6343	7.3587	8.0930
0.09000	5.4046	6.1183	6.8442	7.5809	8.3271
0.10000	5.5778	6.3038	7.0415	7.7895	8.5468
0.20000	6.9887	7.8073	8.6339	9.4673	10.3070
0.25000	7.5841	8.4384	9.2991	10.1653	11.0365
0.30000	8.1479	9.0343	9.9257	10.8215	11.7212
0.35000	8.6952	9.6115	10.5315	11.4548	12.3809
0.40000	9.2373	10.1820	11.1291	12.0785	13.0297
0.45000	9.7831	10.7553	11.7288	12.7034	13.6790
0.50000	10.3410	11.3403	12.3398	13.3393	14.3389
0.55000	10.9199	11.9463	12.9717	13.9961	15.0197
0.60000	11.5298	12.5838	13.6356	14.6853	15.7332
0.65000	12.1836	13.2661	14.3451	15.4209	16.4940
0.70000	12.8987	14.0111	15.1187	16.2221	17.3217
0.75000	13.7007	14.8454	15.9839	17.1169	18.2451
0.80000	14.6314	15.8120	16.9848	18.1508	19.3107
0.85000	15.7671	16.9893	18.2020	19.4062	20.6030
0.90000	17.2750	18.5493	19.8119	21.0641	22.3071
0.95000	19.6751	21.0261	22.3620	23.6848	24.9958
0.96000	20.4120	21.7851	23.1423	24.4855	25.8162
0.97000	21.3416	22.7418	24.1249	25.4931	26.8479
0.98000	22.6179	24.0540	25.4715	26.8728	28.2595
0.99000	24.7250	26.2170	27.6882	29.1413	30.5779
0.99500	26.7569	28.2996	29.8194	31.3194	32.8013
0.99600	27.3977	28.9558	30.4904	32.0047	33.5003
0.99700	28.2156	29.7929	31.3459	32.8781	34.3909
0.99800	29.3537	30.9571	32.5350	34.0915	35.6276
0.99900	31.2642	32.9097	34.5278	36.1235	37.6973
0.99950	33.1366	34.8215	36.4768	38.1097	39.7186
0.99990	37.3669	39.1355	40.8659	42.5811	44.2625

Table generated using MINITAB

	Degrees of Freedom				
α	**16**	**17**	**18**	**19**	**20**
0.00010	2.7739	3.1567	3.5552	3.9683	4.3952
0.00050	3.5358	3.9802	4.4394	4.9123	5.3981
0.00075	3.7665	4.2282	4.7044	5.1940	5.6962
0.00100	3.9416	4.4161	4.9048	5.4068	5.9210
0.00500	5.1422	5.6972	6.2648	6.8440	7.4338
0.01000	5.8122	6.4078	7.0149	7.6327	8.2604
0.02000	6.6142	7.2550	7.9062	8.5670	9.2367
0.03000	7.1625	7.8324	8.5120	9.2004	9.8971
0.04000	7.5958	8.2878	8.9889	9.6983	10.4154
0.05000	7.9616	8.6718	9.3905	10.1170	10.8508
0.06000	8.2827	9.0083	9.7421	10.4833	11.2314
0.07000	8.5717	9.3109	10.0579	10.8120	11.5727
0.08000	8.8363	9.5878	10.3467	11.1124	11.8843
0.09000	9.0820	9.8446	10.6143	11.3906	12.1728
0.10000	9.3122	10.0852	10.8649	11.6509	12.4426
0.20000	11.1521	12.0023	12.8570	13.7158	14.5784
0.25000	11.9122	12.7919	13.6753	14.5620	15.4518
0.30000	12.6244	13.5307	14.4399	15.3517	16.2659
0.35000	13.3096	14.2406	15.1738	16.1089	17.0458
0.40000	13.9827	14.9373	15.8932	16.8504	17.8088
0.45000	14.6555	15.6328	16.6108	17.5894	18.5687
0.50000	15.3385	16.3382	17.3379	18.3377	19.3374
0.55000	16.0425	17.0646	18.0860	19.1069	20.1272
0.60000	16.7795	17.8244	18.8679	19.9102	20.9514
0.65000	17.5646	18.6330	19.6993	20.7638	21.8265
0.70000	18.4179	19.5110	20.6014	21.6891	22.7745
0.75000	19.3689	20.4887	21.6049	22.7178	23.8277
0.80000	20.4651	21.6146	22.7595	23.9004	25.0375
0.85000	21.7931	22.9770	24.1555	25.3288	26.4976
0.90000	23.5418	24.7690	25.9894	27.2036	28.4120
0.95000	26.2963	27.5871	28.8693	30.1435	31.4104
0.96000	27.1357	28.4449	29.7451	31.0367	32.3206
0.97000	28.1908	29.5226	30.8447	32.1577	33.4624
0.98000	29.6332	30.9950	32.3461	33.6873	35.0196
0.99000	32.0001	33.4085	34.8053	36.1907	37.5662
0.99500	34.2674	35.7182	37.1564	38.5820	39.9968
0.99600	34.9798	36.4434	37.8942	39.3320	40.7588
0.99700	35.8872	37.3667	38.8334	40.2863	41.7282
0.99800	37.1467	38.6478	40.1360	41.6096	43.0720
0.99900	39.2535	40.7889	42.3122	43.8188	45.3147
0.99950	41.3102	42.8765	44.4331	45.9702	47.4980
0.99990	45.9353	47.5534	49.1865	50.7818	52.3842

Table was generated using MINITAB.

	Degrees of Freedom				
α	25	30	35	40	45
0.00010	6.7066	9.2581	11.9957	14.8831	17.8940
0.00050	7.9910	10.8044	13.7875	16.9062	20.1366
0.00075	8.3671	11.2527	14.3032	17.4854	20.7758
0.00100	8.6493	11.5880	14.6878	17.9164	21.2507
0.00500	10.5197	13.7867	17.1918	20.7065	24.3110
0.01000	11.5240	14.9535	18.5089	22.1643	25.9013
0.02000	12.6973	16.3062	20.0274	23.8376	27.7203
0.03000	13.4840	17.2076	21.0348	24.9437	28.9194
0.04000	14.0978	17.9083	21.8154	25.7989	29.8447
0.05000	14.6114	18.4927	22.4650	26.5093	30.6123
0.06000	15.0587	19.0004	23.0284	27.1245	31.2762
0.07000	15.4586	19.4534	23.5303	27.6720	31.8664
0.08000	15.8229	19.8654	23.9861	28.1686	32.4014
0.09000	16.1594	20.2452	24.4058	28.6255	32.8932
0.10000	16.4734	20.5992	24.7967	29.0505	33.3504
0.20000	18.9398	23.3641	27.8359	32.3450	36.8844
0.25000	19.9393	24.4776	29.0540	33.6603	38.2910
0.30000	20.8670	25.5078	30.1782	34.8720	39.5847
0.35000	21.7524	26.4881	31.2458	36.0207	40.8095
0.40000	22.6156	27.4416	32.2821	37.1340	41.9950
0.45000	23.4724	28.3858	33.3065	38.2329	43.1638
0.50000	24.3366	29.3360	34.3356	39.3354	44.3351
0.55000	25.2218	30.3073	35.3858	40.4589	45.5275
0.60000	26.1430	31.3159	36.4746	41.6222	46.7607
0.65000	27.1183	32.3815	37.6231	42.8477	48.0585
0.70000	28.1719	33.5302	38.8592	44.1649	49.4517
0.75000	29.3389	34.7997	40.2228	45.6161	50.9850
0.80000	30.6752	36.2502	41.7780	47.2686	52.7288
0.85000	32.2825	37.9903	43.6400	49.2439	54.8105
0.90000	34.3816	40.2560	46.0588	51.8051	57.5053
0.95000	37.6525	43.7730	49.8019	55.7586	61.6563
0.96000	38.6417	44.8336	50.9282	56.9460	62.9010
0.97000	39.8805	46.1599	52.3352	58.4281	64.4535
0.98000	41.5662	47.9618	54.2440	60.4365	66.5553
0.99000	44.3144	50.8922	57.3424	63.6914	69.9570
0.99500	46.9285	53.6720	60.2755	66.7673	73.1663
0.99600	47.7464	54.5403	61.1905	67.7259	74.1656
0.99700	48.7858	55.6429	62.3517	68.9418	75.4323
0.99800	50.2249	57.1676	63.9563	70.6208	77.1802
0.99900	52.6224	59.7033	66.6220	73.4080	80.0780
0.99950	54.9526	62.1621	69.2045	76.1061	82.8779
0.99990	60.1648	67.6341	74.9546	82.1172	89.0804

Table generated with MINITAB

	Degrees of Freedom			
α	50	55	60	65
0.00010	21.0093	24.214	27.497	30.848
0.00050	23.4610	26.866	30.340	33.877
0.00075	24.1572	27.617	31.144	34.730
0.00100	24.6739	28.173	31.738	35.362
0.00500	27.9907	31.735	35.535	39.383
0.01000	29.7067	33.570	37.485	41.444
0.02000	31.6639	35.659	39.699	43.779
0.03000	32.9509	37.030	41.150	45.307
0.04000	33.9426	38.085	42.266	46.480
0.05000	34.7643	38.958	43.188	47.450
0.06000	35.4743	39.712	43.984	48.286
0.07000	36.1050	40.381	44.690	49.027
0.08000	36.6762	40.987	45.328	49.697
0.09000	37.2010	41.543	45.915	50.312
0.10000	37.6886	42.060	46.459	50.883
0.20000	41.4492	46.036	50.641	55.262
0.25000	42.9421	47.610	52.294	56.990
0.30000	44.3133	49.055	53.809	58.573
0.35000	45.6100	50.420	55.239	60.066
0.40000	46.8638	51.739	56.620	61.506
0.45000	48.0986	53.037	57.978	62.921
0.50000	49.3349	54.335	59.335	64.335
0.55000	50.5923	55.654	60.713	65.769
0.60000	51.8916	57.016	62.135	67.249
0.65000	53.2576	58.447	63.628	68.801
0.70000	54.7228	59.980	65.227	70.462
0.75000	56.3336	61.665	66.982	72.285
0.80000	58.1638	63.577	68.972	74.350
0.85000	60.3460	65.855	71.341	76.807
0.90000	63.1671	68.796	74.397	79.973
0.95000	67.5048	73.312	79.082	84.820
0.96000	68.8038	74.662	80.482	86.268
0.97000	70.4229	76.345	82.226	88.069
0.98000	72.6132	78.619	84.581	90.500
0.99000	76.1537	82.292	88.381	94.420
0.99500	79.4896	85.749	91.955	98.102
0.99600	80.5274	86.824	93.065	99.245
0.99700	81.8424	88.185	94.472	100.691
0.99800	83.6558	90.062	96.411	102.683
0.99900	86.6592	93.169	99.621	105.974
0.99950	89.5571	96.166	102.721	109.135
0.99990	95.9531	102.788	109.626	116.029

Table generated using MINITAB

Factors for Control Charts

Observations in Sample (n)	Chart for Averages			Chart for Standard Deviations						Chart for Ranges						
	Factors for Control Limits			Factors for Central Line		Factors For Control Limits				Factors For Central Line			Factors For Control Limits			
	A	A_2	A_3	c_4	$1/c_4$	B_3	B_4	B_5	B_6	d_2	$1/d_2$	d_3	D_1	D_2	D_3	D_4
2	2.121	1.880	2.659	0.7979	1.2533	0	3.267	0	2.606	1.128	0.8862	0.853	0	3.686	0	3.267
3	1.732	1.023	1.954	0.8862	1.1284	0	2.568	0	2.276	1.693	0.5908	0.888	0	4.358	0	2.575
4	1.500	0.729	1.628	0.9213	1.0854	0	2.266	0	2.088	2.059	0.4857	0.880	0	4.698	0	2.282
5	1.342	0.577	1.427	0.9400	1.0638	0	2.089	0	1.964	2.326	0.4299	0.864	0	4.918	0	2.114
6	1.225	0.483	1.287	0.9515	1.0510	0.030	1.970	0.029	1.874	2.534	0.3946	0.848	0	5.079	0	2.004
7	1.134	0.419	1.182	0.9594	1.0424	0.118	1.882	0.113	1.806	2.704	0.3698	0.833	0.205	5.204	0.076	1.924
8	1.061	0.373	1.099	0.9650	1.0363	0.185	1.815	0.179	1.751	2.847	0.3512	0.820	0.388	5.307	0.136	1.864
9	1.000	0.337	1.032	0.9693	1.0317	0.239	1.761	0.232	1.707	2.970	0.3367	0.808	0.547	5.393	0.184	1.816
10	0.949	0.308	0.975	0.9727	1.0281	0.284	1.716	0.276	1.669	3.078	0.3249	0.797	0.686	5.469	0.223	1.777
11	0.905	0.285	0.927	0.9754	1.0253	0.321	1.679	0.313	1.637	3.173	0.3152	0.787	0.811	5.535	0.256	1.744
12	0.866	0.266	0.886	0.9776	1.0230	0.354	1.646	0.346	1.610	3.258	0.3069	0.778	0.923	5.594	0.283	1.717
13	0.832	0.249	0.850	0.9794	1.0210	0.382	1.618	0.374	1.585	3.336	0.2998	0.770	1.025	5.647	0.307	1.693
14	0.802	0.235	0.817	0.9810	1.0194	0.406	1.594	0.399	1.563	3.407	0.2935	0.763	1.118	5.696	0.328	1.672
15	0.775	0.223	0.789	0.9823	1.0180	0.428	1.572	0.421	1.544	3.472	0.2880	0.756	1.203	5.740	0.347	1.653
16	0.750	0.212	0.763	0.9835	1.0168	0.448	1.552	0.440	1.526	3.532	0.2831	0.750	1.282	5.782	0.363	1.637
17	0.728	0.203	0.739	0.9845	1.0157	0.466	1.534	0.458	1.511	3.588	0.2787	0.744	1.356	5.820	0.378	1.622
18	0.707	0.194	0.718	0.9854	1.0148	0.482	1.518	0.475	1.496	3.640	0.2747	0.739	1.424	5.856	0.391	1.609
19	0.688	0.187	0.698	0.9862	1.0140	0.497	1.503	0.490	1.483	3.689	0.2711	0.733	1.489	5.889	0.404	1.596
20	0.671	0.180	0.680	0.9869	1.0132	0.510	1.490	0.504	1.470	3.735	0.2677	0.729	1.549	5.921	0.415	1.585
21	0.655	0.173	0.663	0.9876	1.0126	0.523	1.477	0.516	1.459	3.778	0.2647	0.724	1.606	5.951	0.425	1.575
22	0.640	0.167	0.647	0.9882	1.0120	0.534	1.466	0.528	1.488	3.819	0.2618	0.720	1.660	5.979	0.435	1.565
23	0.626	0.162	0.633	0.9887	1.0114	0.545	1.455	0.539	1.438	3.858	0.2592	0.716	1.711	6.006	0.443	1.557
24	0.612	0.157	0.619	0.9892	1.0109	0.555	1.445	0.549	1.429	3.859	0.2567	0.712	1.759	6.032	0.452	1.548
25	0.600	0.135	0.606	0.9896	1.0105	0.565	1.435	0.559	1.420	3.931	0.2544	0.708	1.805	6.056	0.459	1.541

For values of n greater than 25 see the formulas contained in the ***Manual on Presentation of Data and Control Chart Analysis*** By ASTM.

Table of convolution integers for Savitzky-Golay Smoothing

Convolution Integers for quadratic smoothing

Points	25	23	21	19	17	15	13	11	9	7	5
-12	-253										
-11	-138	-42									
-10	-33	-21	-171								
-9	62	-2	-76	-136							
-8	147	15	9	-51	-21						
-7	222	30	84	24	-6	-78					
-6	287	43	149	89	7	-13	-11				
-5	322	54	204	144	18	42	0	-36			
-4	387	63	249	189	27	87	9	9	-21		
-3	422	70	284	224	34	122	16	44	14	-2	
-2	447	75	309	249	39	147	21	69	39	3	-3
-1	462	78	324	264	42	162	24	84	54	6	12
0	467	79	329	269	43	167	25	89	59	7	17
1	462	78	324	264	42	162	24	84	54	6	12
2	447	75	309	249	39	147	21	69	39	3	-3
3	422	70	284	224	34	122	16	44	14	-2	
4	387	63	249	189	27	87	9	9	-21		
5	322	54	204	144	18	42	0	-36			
6	287	43	149	89	7	-13	-11				
7	222	30	84	24	-6	-78					
8	147	15	9	-51	-21						
9	62	-2	-76	-136							
10	-33	-21	-171								
11	-138	-42									
12	-253										

Table reproduced by permission of the American Chemical Society, Washington D.C.

Table of convolution integers for Savitzky-Golay Smoothing

Convolution Integers for first derivative, quadratic model

Points	25	23	21	19	17	15	13	11	9	7	5
-12	-12										
-11	-11	-11									
-10	-10	-10	-10								
-9	-9	-9	-9	-9	-9						
-8	-8	-8	-8	-8	-8	-8					
-7	-7	-7	-7	-7	-7	-7	-7				
-6	-6	-6	-6	-6	-6	-6	-6	-6			
-5	-5	-5	-5	-5	-5	-5	-5	-5	-5		
-4	-4	-4	-4	-4	-4	-4	-4	-4	-4	-4	
-3	-3	-3	-3	-3	-3	-3	-3	-3	-3	-3	-3
-2	-2	-2	-2	-2	-2	-2	-2	-2	-2	-2	-2
-1	-1	-1	-1	-1	-1	-1	-1	-1	-1	-1	-1
0	0	0	0	0	0	0	0	0	0	0	0
1	1	1	1	1	1	1	1	1	1	1	1
2	2	2	2	2	2	2	2	2	2	2	2
3	3	3	3	3	3	3	3	3	3	3	3
4	4	4	4	4	4	4	4	4	4	4	
5	5	5	5	5	5	5	5	5	5		
6	6	6	6	6	6	6	6	6			
7	7	7	7	7	7	7	7				
8	8	8	8	8	8	8					
9	9	9	9	9	9						
10	10	10	10								
11	11	11									
12	12										

Table of convolution integers for Savitzky-Golay Smoothing

Convolution integers for second derivative, quadratic model

Points	25	23	21	19	17	15	13	11	9	7	5
-12	92										
-11	69	77									
-10	48	56	190								
-9	29	37	133	51							
-8	12	20	82	34	40						
-7	-3	5	37	19	25	91					
-6	-16	-8	-2	6	12	52	22				
-5	-27	-19	-35	-5	1	19	11	15			
-4	-36	-28	-62	-14	-8	-8	2	6	28		
-3	-43	-35	-83	-21	-15	-29	-5	-1	7	5	
-2	-48	-40	-98	-26	-20	-48	-10	-6	-8	0	2
-1	-51	-43	-107	-29	-23	-53	-13	-9	-17	-3	-1
0	-52	-44	-110	-30	-24	-56	-14	-10	-20	-4	-2
1	-51	-43	-107	-29	-23	-53	-13	-9	-17	-3	-1
2	-48	-40	-98	-26	-20	-48	-10	-6	-8	0	2
3	-43	-35	-83	-21	-15	-29	-5	-1	7	5	
4	-36	-28	-62	-14	-8	-8	2	6	28		
5	-27	-19	-35	-5	1	19	11	15			
6	-16	-8	-2	6	12	52	22				
7	-3	5	37	19	25	91					
8	12	20	82	34	40						
9	29	37	133	51							
10	48	56	190								
11	69	77									
12	92										

Kolmogorov-Smirnov Test Statistics for Distribution Testing

	Significance Level				
Number of Points	**0.01**	**0,05**	**0.10**	**0.15**	**0.20**
4	0.417	0.381	0.352	0.319	0.300
5	0.405	0.337	0.315	0.299	0.285
6	0.364	0.319	0.294	0.277	0.265
7	0.348	0.300	0.276	0.258	0.247
8	0.331	0.285	0.261	0.244	0.233
9	0.311	0.271	0.249	0.233	0.223
10	0.294	0.258	0.239	0.224	0.215
11	0.284	0.249	0.230	0.217	0.206
12	0.275	0.242	0.223	0.212	0.199
13	0.268	0.234	0.214	0.202	0.190
14	0.261	0.227	0.207	0.194	0.183
15	0.257	0.220	0.201	0.187	0.177
16	0.250	0.213	0.195	0.182	0.173
17	0.245	0.206	0.189	0.177	0.169
18	0.239	0.200	0.184	0.173	0.166
19	0.235	0.195	0.179	0.169	0.163
20	0.231	0.190	0.174	0.166	0.160
25	0.200	0.173	0.158	0.147	0.142
30	0.187	0.161	0.144	0.136	0.131
n>30	$\frac{1.628}{\sqrt{n}}$	$\frac{1.358}{\sqrt{n}}$	$\frac{1.224}{\sqrt{n}}$	$\frac{1.138}{\sqrt{n}}$	$\frac{1.073}{\sqrt{n}}$

INDEX

A

B

C

D

E

L

M

N

O

P

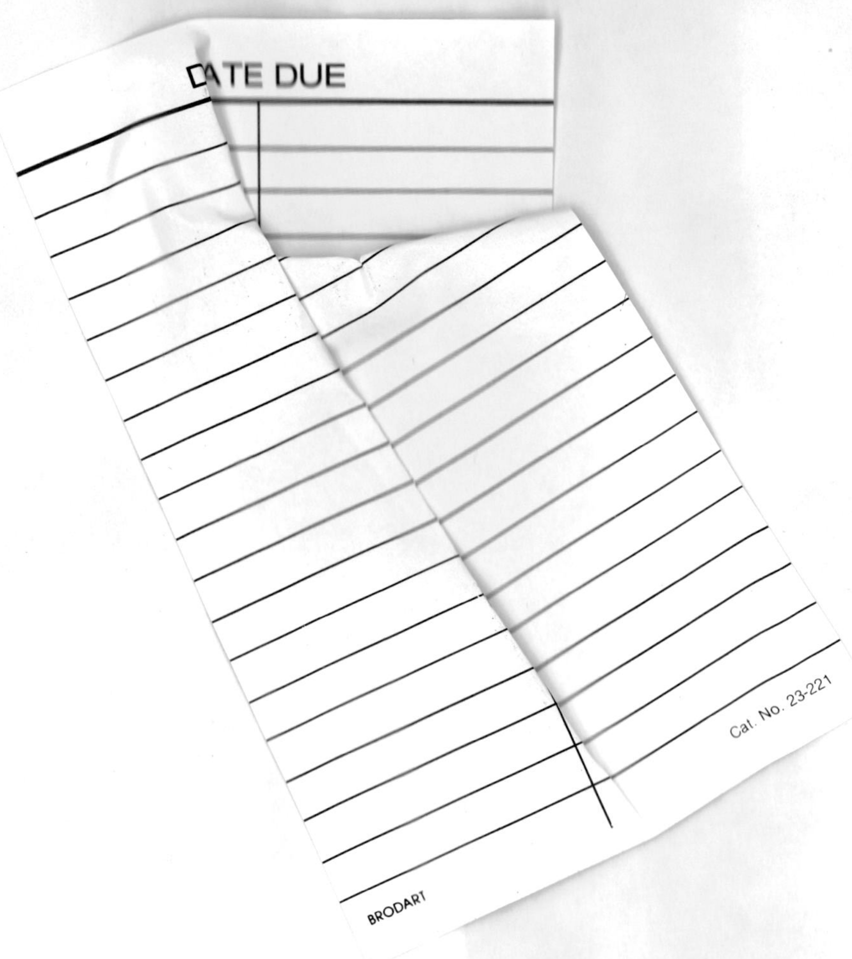

DATE DUE
Cat. No. 23-221
BRODART